Enforcing Ecocide

"Who serves and protects the planet? This eye-opening volume demonstrates the destructive impacts of security forces imposing an ecologically devastating and socially harmful economic order. Beware of green militarization and the policing of a green transition."

—Professor Philippe Le Billon, *Department of Geography, University of British Columbia*

"*Enforcing Ecocide* makes a powerful case for breaking with discipline in the face of entwined crises of militarization and ecological destruction. The authors weave academic and activist knowledge from around the globe to bring needed attention to the political ecology of policing."

—Dawn Marie Paley, *journalist and author of* Drug War Capitalism *and* Guerra Neoliberal

"*Enforcing Ecocide* is a highly original, timely, and provocative volume that provides new insights into the ways in which a plethora of policing actors are driving forces behind global socio-ecologically degradation. Covering new ground within the field of policing research, and security studies more broadly, the contributions to this volume offer fresh assessments of the resulting practical as well as normative consequences and challenges. Global in scope, critical in outlook, the book will be an indispensable and thought-provoking reading for students and scholars of policing, as well as for anyone seeking to contribute to making our planet a more socio-ecologically just place for humans and non-humans alike."

—Professor Markus-Michael Müller, *Department of Social Sciences and Business, Roskilde University*

"This book explores and explains the 'violence work' involved in maintaining and enforcing ecocide. Dunlap and Brock have brought together key thinkers to create a challenging set of essays that span a range of cases, critically interrogating the enforcement actions against those defending their lands, environments and rights. This impressive book is essential for understanding how and why state formation, coloniality and policing are integral to the continuing enforcement of ecological catastrophe, and how the effects are differentially felt and experienced by marginalised communities. This is an important critical intervention, which will shape political ecology for years to come."

—Professor Rosaleen Duffy, *Politics and International Relations, The University of Sheffield*

"*Enforcing Ecocide* critically explores the role of policing in the everyday violence which underpins industrial society. It provides a needed historical, political and more global account of the sources, drivers and resistance to ecocide in diverse settings from the Middle East and South Asia to North America and Europe. I recommend it to anyone interested in reclaiming a viable future."

—Professor Peter Newell, *Department of International Relations, University of Sussex*

"While there is a growing debate whether and to what extent ecological destruction is a driver of global insecurity, often involving policing responses to curtail violent socio-political consequences of human-made environmental degradation, the ways in which policing itself is a factor in exacerbating ecological destruction have been overlooked. *Enforcing Ecocide* provides original and important insights into this domain. The book showcases how a multitude of state and non-state policing actors enforce, and thereby exacerbate, environmental insecurities – for humans, non-humans and our planet as a whole. Combining empirically rich case studies from around the globe, with compelling theorization, the book blazes a new trail for understanding a so-far neglected dimension of planetary ecological crisis."

—Associate Professor Louise Wiuff Moe, *Department of Social Sciences and Business, Roskilde University*

"*Enforcing Ecocide: Power, Policing and Planetary Militarization* is an important, thought-provoking work. It offers critical analyses of the political ecology of policing through relevant case studies from across the world, showing the multifaceted facets and ways recurrent forms of ecological destruction and social injustice are designed, reinforced and legitimated by repressive political orders. Building on (anarchist) political-ecological approaches, linked to empirically grounded data, the book offers insightful, eye-opening analyses of how coercive institutional structures, accelerating corporate profit-making, and repressive policing are intimately interlinked in the production of planetary-scale crises and disasters. Anyone who is concerned with debates over ecocides and repressive policing should read this book."

—Professor Anja Nygren, *University of Helsinki*

Alexander Dunlap · Andrea Brock
Editors

Enforcing Ecocide

Power, Policing & Planetary Militarization

palgrave
macmillan

Editors
Alexander Dunlap
Centre for Development
and the Environment
University of Oslo
Oslo, Norway

Andrea Brock
Department of International
Relations, School of Global Studies
University of Sussex
Brighton, UK

ISBN 978-3-030-99645-1   ISBN 978-3-030-99646-8 (eBook)
https://doi.org/10.1007/978-3-030-99646-8

Cover illustration: ZUMA Press, Inc

This Palgrave Macmillan imprint is published by the registered company Springer Nature Switzerland AG
The registered company address is: Gewerbestrasse 11, 6330 Cham, Switzerland

This book is dedicated to Andrea's dad, Micheal "Hell" Carfagna, Joe Hammake, and Matt "Limo" Markus; and all people struggling to defend their habitats and to stop ecological destruction across the world.

Foreword: A Long Hot Summer (From Now On)

For the residents of Portland, Oregon, the summer of 2020 was colored by more than one hundred consecutive nights of rioting. Angry crowds broke windows, looted goods, set fires, and fought with police. The police—including local cops from several jurisdictions, sheriff's deputies, state police, and federal marshals—made nearly a thousand arrests and used force (by the Department of Justice's estimate) more than 6,000 times, sometimes indiscriminately attacking protestors, bystanders, and journalists, and filling entire neighborhoods with teargas.

The riots began in May, after the Minneapolis police were recorded callously suffocating an unarmed Black man named George Floyd, killing him. Protests took place across the United States and even abroad—some violent, mostly nonviolent; some revolutionary in aim, more reformist. The very scope of the unrest shows that the fires burning in the cities were not caused simply by a single precipitating incident, but by the accumulated trauma produced by the entire system of policing, and white supremacy more generally. History provided the fuel for rebellion; an individual death was just enough to ignite it.

In Portland, nightly unrest continued until September, when the smoke from nearby wildfires blanketed the city, turning the skies yellow and rendering the air unbreathable. While not precisely caused by climate change, these fires were certainly exacerbated by it: rising temperatures and drier weather, plus the overgrown underbrush from decades of fire suppression, turned the forests into a virtual tinderbox just waiting for the right spark. More than one million acres burned.

These two bookends, the event triggering the unrest and the conditions ending it, may not at first glance seem to be related. But the essays in this collection suggest that they very much are. The state uses violence—that of the police, military, private mercenaries, and vigilante auxiliaries—to seize land, control resources, guard extractive industries, and repress the environmental movement. State violence contributes to environmental destruction in more direct ways as well, polluting the air, water, and soil, and destroying the habitat of innumerable species—sometimes deliberately. (Just think of the phrase "scorched earth.") State violence also *depends* on the exploitation of the natural environment for resources such as fuel and the rare metals used in producing high-tech weaponry. Concurrently, environmental exploitation requires and thus produces state violence—to impose property relations, to remove the inhabitants of exploitable areas, to discipline local populations, to overcome resistance, and to protect the resulting system of inequality. What emerges from this volume, taken as a whole, is an understanding of the relationship between extractive industry and monopolies of violence, not merely as compatible or symbiotic pursuits, but as inter-derivable programs united by a shared logic. Call it colonization, call it primitive accumulation, or call it enclosure, but the results are much the same: death on a potentially planetary scale.

If that is right, then it is no wonder that the standard proposals to address our climate crisis, whether based on state intervention or marketplace reasoning, are at best transparently inadequate and often actively counter-productive. Contributors to this collection point to the experience of New Orleans during and after Hurricane Katrina as an "indicative example" of how states will respond as the effects of climate change become more acute: the abdication of any responsibility to protect the population and widespread violence in the defense of property relations, followed by land grabs and displacement at a scale tantamount to ethnic cleansing. But New Orleans also provides a glimpse of another possible future, prefigured by the mutual aid networks and, to adapt a phrase, the extraordinary sense of community that can arise in the midst of a disaster.[1] These dual and contrasting responses are typical of crises that produce institutional failure. In emergencies, most people react in ways

[1] See: Rebecca Solnit, *A Paradise Built in Hell: The Extraordinary Communities that Arise in Disaster* (New York: Penguin, 2010).

that are sensible, practical, solidaristic, and humane. They come together to help each other, protect each other, and share resources in ways that would be virtually unthinkable within the confines of capitalism. Those who don't—the people who resort to hoarding and murderous violence—tend in fact to be the very people trying desperately to preserve the *status quo ante*. Of course, that makes sense if we remember that hoarding and violence are the essence of capitalism and the state.

Enforcing Ecocide collects evidence from around the world to illustrate the fundamental connections between state violence, capitalist accumulation, and environmental destruction. It thus helps bring into focus the precise nature of the problem we are facing. It is now far too late to prevent catastrophic climate change; any realistic strategy must turn its energies to mitigation and adaptation. The state will attempt to adapt via increased militarism and authoritarianism, because the first imperative of state power is to preserve state power. In our present circumstances, that demand may be at odds with the survival of humanity.

But, also: survival is not enough. We want to live, but we also want lives worth living. We want a world worth living in. We want peace, freedom, equality, community, beauty, justice, joy. A society that took seriously these desires—these values—would also, I believe, be one with the best possible chance of adapting to climate change and providing a future suitable to human life.

The world we have known is ending. That process will be painful, no matter what comes next. But what follows will be to some large extent up to us, what we do now and in the years to come. The stakes could not be higher.

Portland, Oregon — Kristian Williams

Kristian Williams is the author of *Our Enemies in Blue: Police and Power in America* (South End Press, 2007; AK Press, 2015), co-editor of *Life During Wartime: Resisting Counterinsurgency* (AK Press, 2013), *Fire the Cops! Essays, Lectures, and Journalism* (Kersplebebed, 2014), and the forthcoming book *Gang Politics: Revolution, Repression, and Crime* (AK Press). He is presently at work on a history of policing in Portland, Oregon.

Acknowledgments

This book was written and put together under strenuous circumstances, from the Covid-19 pandemic resulting in lockdowns and deaths to climate breakdown and the flooding of our hometowns, including heartbreaks and deaths of loved ones.

We would like to acknowledge this situation, but also the lesser known struggles in defense of land, habitats, and dignity against megaprojects and socio-ecological degradation in the name of "development", "energy security", and overall profit-seeking from people and the land alike. This volume would not be possible without the people struggling and in resistance against the multifarious forces of repression across the world.

We want to acknowledge the support, feedback, and care from our friends Amber Huff, Nathan Stephens Griffin, Olfee Kitty, and cat partners Sabotage and Hecate—and everyone who has been there with and for us over the past two years. We are grateful for your love, support, and friendship.

Contents

Notes on Contributors

Tina Beigi is a Ph.D. candidate at McGill University, working in the boundary waters of political economy, ecological economics, and thermodynamics. She is part of the Leadership for the Ecozoic (L4E) global partnership, which aims to define and facilitate a transition to the Ecozoic, a future founded on mutually enhancing human–Earth relationships.

Andrea Brock is a political ecologist/economist at the University of Sussex, UK. She researches corporate power, state violence, and mining/infrastructure projects. She is particularly interested in the repression, policing, and co-optation of environmental defenders, and how corporate and state actors manage resistance and engineer consent. Her recent work develops an anarchist political ecology approach to fracking and coal mining, and she is involved in a number of anti-extractivist struggles.

Matthew J. Burke explores sociopolitical and ecological dimensions of energy transitions. Matthew is a Lecturer and Postdoctoral Associate at the University of Vermont, and a Postdoctoral Fellow with the Leadership for the Ecozoic project. Matthew completed a Ph.D. in Renewable Resources—Environment at McGill University, examining energy democracy in theory and practice.

Wendela de Vries studied Political Science at the University of Amsterdam. She is a long-time researcher and campaigner against arms trade and military industry at the independent peace organization Stop

Wapenhandel, of which she is co-founder. She is part of the Steering Group of the European Network Against Arms Trade and coordinates an international working group on arms, militarism, and climate justice.

Alexander Dunlap is a postdoctoral research fellow at the Centre for Development and the Environment, University of Oslo. His work has critically examined police-military transformations, market-based conservation, wind energy development, and extractive projects more generally in Latin America and Europe. He has published two books: Renewing Destruction: Wind Energy Development, Conflict and Resistance in an American Context (Rowman & Littlefield, 2019) and, the co-authored, The Violent Technologies of Extraction (Palgrave, 2020).

Anwesha Dutta is a Senior Researcher at the Chr. Michelsen Institute, Bergen, Norway.

Lucy Everett (she/her) is a recent graduate from McGill University with an Honours BA in Environment and Development. Born and raised on occupied Coast Salish territory before moving to Tio'tia:ke (Montreal) for school on occupied Kanien'keha:ka and Anishinaabe lands, she is a mixed white settler (British, Scottish, Mennonite) and Red River Metis through her paternal grandmother. She is passionate about a sustainable and just future for all.

Peter Gelderloos is a social movement participant and author of numerous books. This includes *How Nonviolence Protects the State* (2005, South End Press), *Anarchy Works* (2010, Ardent Press), *The Failure of Non-Violence* (2013, Left Bank Books), *Worshiping Power: An Anarchist View of Early State Formation* (2017, AK Press), and *The Solutions Are Already Here: Strategies of Ecological Revolution from Below* (2022, Pluto Press).

Jen Gobby is an activist-scholar and educator based on unceded Abenaki territory in rural Quebec. She is now an Affiliate Assistant Professor at Concordia University in the Department of Geography, Planning and Environment. She is the author of the book *More Powerful Together: Conversations with Climate Activists and Indigenous Land Defenders* and director of Research for the Front Lines, a new initiative that connects grassroots communities and organizers on the front lines of the fight for environmental and climate justice in Canada with researchers who have skills and time to offer.

Jan Goodey is a Senior Lecturer in data journalism and a UK-based environmental activist. He helps run an allotment co-op and orchard from the south coast. From the 1990s onward he has campaigned for communities and ancient woodlands threatened by road programs, airport expansion, and high-speed rail. His data work has supported divestment campaigns from fossil fuels through to meat, dairy, and pesticides industries.

Michael Hennessy Picard is Lecturer in International Environmental Law at the University of Edinburgh Law School.

Maziar Samiee researches the post-financial crisis transformations of corporate power and public debt in the EU at the University of Sussex. Before moving to the UK in 2010, Maziar was an activist in their home country of Iran, participating in various democratic and social justice struggles, such as with the Iranian feminist and environmental movements, including with the NGO Green Front.

Daniel Selwyn is a researcher and educator with the London Mining Network, which is an alliance of more than 30 organizations working to expose human rights abuses and environmental crimes committed by mining companies based in London. His research focuses on the intersections between resource extraction, the global arms trade, British imperialism, and climate (in)justice.

Trishant Simlai is a Ph.D. candidate in the Department of Geography and Selwyn College at the University of Cambridge, UK.

Nina L. Smolyar is a graduate fellow at the Gund Institute for the Environment and Leadership for the Ecozoic project. She researches decolonial economics, Indigenous sovereignty, and participatory action methods for collective liberation and systems transformation. Having worked in various environmental and justice organizations, she completed a self-designed and self-directed MA thesis on conflict transformation in intentional community, at Goddard College.

List of Figures

CHAPTER 1

Introduction: Enforcing Ecological Destruction

Alexander Dunlap and Andrea Brock

> Regardless of what color policemen [sic] are, the suits they wear, what they call themselves, they are all the same. They are the same for the simple reason that a policeman exists in society as a behavior control mechanism. The basic principles of what is done, how it is done, and why it is done are the same.
>
> —Lauren Goin, US Agency for International Development, Office of Public Safety Director, 1973–74

A. Dunlap (✉)
Centre for Development & the Environment, University of Oslo, Oslo, Norway
e-mail: alexander.dunlap@sum.uio.no

A. Brock
Department of International Relations, School of Global Studies, University of Sussex, Brighton, UK
e-mail: a.brock@sussex.ac.uk

A. Dunlap and A. Brock (eds.), *Enforcing Ecocide*,
https://doi.org/10.1007/978-3-030-99646-8_1

The current conjuncture of ecological crises goes far beyond climate change, which has come to dominate media headlines and intergovernmental negotiations. The planet is burning, flooding, and heating, habitats are being destroyed and degraded, and species are disappearing at unprecedented rates (Hickel, 2020). The effects of these crises are increasingly apparent in the Global North, where human and nonhuman lives and livelihoods are being lost through floods and record heatwaves (Oltermann, 2021; Bekiempis, 2021). Meanwhile, communities in the Global South, and particularly indigenous communities, have been witnessing for centuries the effects of anthropogenic climatic changes (Hanaček et al., 2022), human-induced ecological degradation (Peluso and Watts, 2001), and ecocide (Short, 2016; Crook and Short, 2021). Postponing adequate and coherent action is no longer an option.

Ecological degradation and ecocide are the outcome of colonial-capitalist development (Hickel, 2020). This entails a long history of exploitation, extraction, enclosure, and dispossession (Rodeny, 2009/1972; Garleano, 1997/1973) which continues through the guise of "green" investment, sustainable development, and low-carbon technologies today (Fairhead et al., 2012; Sovacool, 2021; Dunlap, 2021a). Acts of ecological degradation (such as deforestation, mining, or dredge fishing), loss of land, pollution, and emissions accumulate on a global scale, spawning climate catastrophe and mass extinction. It is this deliberate destruction and degradation of ecosystems, which has cumulative climatic effects, that we refer to when we speak about *ecocide*.

The term *ecocide* has emerged from discussions on the purposeful ecological destruction of the jungles in Laos, Cambodia, and Vietnam to eliminate the shelter of the North Vietnamese army (Short, 2016). The US government deployed carpet-bombing[1] and used massive amounts of napalm, Agent Orange,[2] and other chemical weapons to eliminate the jungle as a means of defeating the North Vietnamese (Short, 2016). To get an idea of the scale, Clayton R. Koppes (1985: 131) reminds us that from "1962 to 1971, the United States Air Force dropped 18.85 million gallons of herbicides – about two-thirds of it Agent Orange – on the forests and crops of South Vietnam". This covered an area of 3.6 million

[1] The mass-bombing of an entire area or city, akin to laying a "carpet" of unguided bombes over an entire area.

[2] This is a herbicide and defoliant chemical, which is equal parts of two herbicides, 2,4,5-T and 2,4-D.

acres, 8.6% of the country's total land surface (Koppes, 1985). Since the 1970s, campaigners have fought to include ecocide into international law (Higgins, 2012). Despite its inclusion into drafts of the Code of Crimes Against the Peace and Security of Mankind [sic] and being "considered as a missing method of genocide that could be written into the Genocide Convention" (Short, 2016: 49), in the end, ecocide is still not part of any international legal framework.[3]

Yet, the term *ecocide* continues to be as relevant as ever. Governments and companies continue to wage war on ecosystems and natural environments: From increasing mineral extraction for computational and low-carbon technologies, to drilling for oil in order to power automobiles and industrial production of goods, not to forget the clearing of forests for agriculture and urbanization. All the while, the burning of fossil fuels for energy consumption and the production of lethal chemicals for mining and industry continues. Industrial society remains on an ecocidal path. The spread of "infrastructural harm" caused by energy production and consumption, transport, communication, and (low-carbon) infrastructures has not slowed, but accelerated across the world (Dunlap, 2021). In fact, ecocidal concerns have long taken place on a planetary scale as scientists and international institutions warned humanity that mass extinction is on the rise. Summarizing these statistics, Jason Hickel (2020: 6–16) reminds us: "40% of the planet's soils are now seriously degraded"; earthworm biomass has dropped by 83%; 85% of global fish stocks are depleted or facing collapse; mammal populations have dropped by half; and dead zones from chemical run off, nitrogen, and phosphorous spread along the coastlines of industrialized region across the world. Moreover, deforestation is at an all-time high. Forest fires, record heatwaves (resulting in human and nonhuman death), floods, rising sea levels, and erratic weather are experienced at a greater and increasing frequency (Hickel, 2020). Concerns of "biological annihilation" via ecological destruction have replaced concerns of nuclear war, Hickel argues (2020: 9). The Russian inviasion of Ukraine in 2022 has triggered new fears of nuclear attack, while once again illustrating the ecocidal effects of warfare on human and nonhuman communities, food production, and ecosystems. Yet ecocide is not *new*—it has been part and parcel of colonization and

[3] See also Crook and colleagues (2018) to understand more about genocide-ecocide and criminology. Moreover, we recognize that including ecocide into international law, could also serve as another pretext for imperial powers to justify invading other countries.

repression of indigenous communities for centuries. The extermination of the bison on the plains of Turtle Island (North America) to break and undermine the Sioux, Kiowa, and Comanches resistance to settler frontier expansion (Isenberg, 2000) is just one famous example of many.

This raises the question of how ecological and climate crises are made possible and continue despite an abundance of scientific evidence and opposition. Of course, we can easily answer this question with vague signifiers, identifying colonialism, capitalism, techno-industrialism, modernity, and development as the culprits. More still, these reinforce and are maintained by patriarchy, classism, racism, and (hetero/homo) normativities that are the product of civilization and state formation (Bæden, 2014; Scott, 2017; Gelderloos, 2017). In themselves, however, these general answers do not explain the persistence of socio-ecological crises and extinction, let alone introduce practical strategies to resist and intervene to subvert ecological catastrophe.

Instead, this book sets out to illustrate how these crises are facilitated; revealing, through a range of case studies and contributions, the many facets and ways ecologically destructive activities are planned, executed, and—as the title suggests—enforced. The contributors point to the diversity of actors, mechanisms, power relations, and technologies involved in causing ecological destruction, with the aim to open ruptures/possibilities and support resistance. *Enforcing Ecocide* contends that armed and policing forces—or institutions of scientific violence—generate multiple forms of ecological harm, culminating in climate change, ecological degradation, and mass extinction.

Enforcing Ecological Catastrophe

With the term "enforcement", we refer to the ways, techniques, materials, networks, and actors involved in the ongoing ecocide on the ground. Enforcement responds to or preempts resistance and attacks against the dominant order, which threatens people's livelihoods, habitats, and sociocultural fabrics. According to Global Witness (2021) 227 lethal attacks against land defenders were recorded in 2020, amounting to an average of more than four people a week. This number ignores armed groups, criminal groups, and other killings executed by police or mercenary forces that are linked to people defending their territory against mining and other destructive projects (see Gelderloos, this volume).[4] The Environmental Justice Atlas documents 2,743 environmental conflicts (Scheidel et al., 2020), testament to the resistance by people and efforts from governments and companies at imposing ecologically destructive projects. Enforcement refers to the *violence work* that is necessary for enforcing ecologically destructive relationships, projects, and outcomes. *Violence work* involves the labor of maintaining order and repressing dissent (Huggins et al., 2002; Seigel, 2018). This book focuses on the laborious actions undertaken to target people who are defending their habitats, encounter riot police, and experience harassment, intimidation, armed coercion, surveillance, as well as the resulting stigmatization and criminalization attached to environmental self-defense. Indigenous peoples, while diverse and replete with disagreement and differing politics within communities, are the disproportionate targets of this repression (Temper et al., 2020; Menton and Le Billon, 2021). Environmentalists, anarchists, citizens, and others become targets of authorities when resisting megaprojects and processes of economic development. These policing forces, we could say, are the knights of racial capitalism and ecological degradation, the arbiters of private property, and the guardians of extraction.

This violence work, however, is also *scientific*—tested, measured, and calculated. *Scientific violence*, remembering Rodney's (2009: 260) discussion of (neo)colonial management, is the schooling and practice of coercive management to administer a population. This is a reminder that practices of colonial warfare are integral to dispensing brutal repression, disciplining populations, and facilitating the formation of colonial societies, which entails the merging of police and military powers (Neocleous, 2014; Schrader, 2019). The "main concern of the police is the constitution of order rather than defeating the enemy", Mark Neocleous (2014: 14) reminds us, asserting that "we need to grasp the exercise of police power in constant war against the 'enemies of order'". The internal and external "permanent war" of state formation organizes and spreads (through colonialism) a system of production and consumption that channels desires, dehumanizes, and divides people as "resources" to be

[4] In matters of recording the kill counts by repressive forces, take, for example, the official counting of people murdered by the police in the United States. Only recently, it was confirmed that more than 17,000 people murdered by the police were underreported in official USA National Vital Statistics System (NVSS) between 1980 and 2019 (Naghavi et al., 2021).

integrated and managed into the working order of the global industrial system (Dunlap, 2014: 72; Álvarez and Coolsaet, 2020). The policing of women, non-binary, and racialized bodies, genders, sexualities, and moralities is key to upholding this order; disciplining populations in both public life and their private intimacies and family lives (Elliott-Cooper, 2021). The working class—white or racialized—is also not exempt from police violence. This indicates the "social war"[5]—the reshaping, disciplining, and reconfiguring of peoples' bodies, social relationships, and habits—that is carried out by governments and companies to indoctrinate and affirm the order of accumulation.

We view a connection between violence work and the scientific development and schooling of policing techniques designed to instill submission and docility. Our focus on policing, specifically, emphasizes the institutionalized and professionalized praxis of disciplining, incentivizing, and managing people and nature, through direct and indirect means. With institutions of policing, moreover, we refer not only to "the police". Instead, this includes the military, various police forces, private contractors, criminal groups, mercenaries, and other armed or unarmed forces tasked with the job of securing forms of "insecurity" that threaten the reproduction of industrial and financial capitalisms across geographic space and into everyday life (Paley, 2014; Granovsky-Larsen and Paley, 2019). This includes acknowledging how infrastructures frequently fulfill a policing roll (Rodgers and O'Neill, 2012; Kallianos et al., forthcoming). They support policing operations that directly and indirectly facilitate and maintain "death conditions" (Kaur, 2021). This entails, in different contexts, depriving people of amenities, facilitating environmental toxification, and eliminating the means of subsistence (Pulido and De Lara, 2018). This is characteristic of counterinsurgency campaigns past and present (Verweijen and Dunlap, 2021). This book contends that policing is a crucial and oddly underemphasized mechanism integral to maintaining a socio-ecologically destructive order and enforces this socio-economic imperative via coercion, deception, and disciplinary action. Policing, in other words, is responsible for ecological calamity and climate change.

The good news, however, is that since ecocide is the product of human actions, embedded in human-made political and economic structures, it

[5] See Dunlap (2014, 2019a, b) and Dunlap and Correa-Arce (2021) for a greater elaboration on social warfare.

can be stopped and changed. The purpose of this edited volume is to reveal the enforced nature of ecocide and climate change, demonstrating the plight and struggle of people in resistance but also the negligence of environmental policy that ignores and accepts this repression against land defenders and the corresponding ecological cost.

Exploring Catastrophe

This trajectory of ecological catastrophe has not only been enforced, but it has been celebrated in the name of progress and modernity since the onset of colonization and industrialization. From the conquering of "wilderness" to the domination and domestication of "nature"—including indigenous peoples who were dehumanized and "naturalized" as "primitive savages" or sexualized as "exotic others" (Davidov, 2012)—enforcing ecological catastrophe has been infused by ideologies of white supremacy, patriarchy, and Western superiority. Human separation from and subjugation of nature—human and nonhuman—and subsequent constructions and reductions of nonhuman natures to "resources" by states and empires thus lie at the roots of socio-ecological catastrophe (Foucault, 2007/1978; Bookchin, 1982; Scott, 1998; Sullivan, 2017; Hickel, 2020). State formation, thinking of Patrick Wolfe (2006), is the structure of conquest, which entails enormous human, nonhuman, and climatic costs.

Capitalist, development, and, to a degree, even Marxist theory[6] has been rather explicit about industrial development, affirming this destructive and utopian modernist trajectory. National security planner and development theorist Walter Rostow (1960), attempting to counter Marx's historical materialism and Soviet communism, formalized this pathway and packaged it for international export as *The Stages of Economic Growth*. Ecological factors, or concern for the environment, were—and still are—not of central importance in comparison to state formation, industrialization, market development, competition, and, as Rostow's final stage suggests, "high mass consumption". But this modernization entails mass consumption and fixation on economic growth, as Hickel

[6] We must acknowledge a great deal of Marxian theory has departed from this to various degrees (see Holloway, 2010; Vergara-Camus, 2014), which includes concerted ecological interpretations of Marx (see Anreucci and Engel-DiMauro, 2019; Kovel, 2007; Franquesa, 2018; please read Morris, 2015 on Kovel, 2007).

(2020) reminds us. Modernist development came at an extraordinary socio-ecological cost to habitats and Indigenous cultures North and South of the globe—now extending to the entire biosphere, as the planet is heating, species are dying at unprecedented rates, and many people are suffering radical insecurity from climate catastrophe.

Yet, this celebrated trajectory of capitalist progress continues. As we are writing this introduction, the Conference of the Parties (COP) 26 roars on with representatives across the world still celebrating this political, economic, and technological trajectory formalized by Rostow. The only difference is that these processes of industrial development are now branded as "green", "environmentally friendly", "low-carbon", and ambiguously "sustainable" (see Brock, 2020a; Dunlap and Sullivan, 2020; Huff, 2021). While international climate negotiations and green capitalism are highly contested (see Böhm and Sullivan, 2021), we remain concerned about how this celebration of capitalist conservation, "green extractivism", and low-carbon infrastructures extends existing modernist and capitalist ideologies (Brock, 2020a; Dunlap, 2019a, 2021b). The beliefs in economic growth and (extractive) technological development as an ecological and climatic remedy is evident in the fixation on net-zero policy aims, market-based policies like carbon trading, technological fixes, and "nature-based solutions". This narrative perpetuates the existing development model, reinforcing destructive ecological pathways and an (extractive) capitalist vision, while attempting to silence, or worse absorb, the opposition of people fighting for social and ecological change.

While many of us remain enchanted by scientific progress or the aspirations and convenience of industrial lifestyles (Alexander, 2008), such lifestyles would not be possible without the threat and exercise of force, the production and flows of weapons, and coercion and political manipulation to facilitate the extractive and industrial projects behind the production of goods and amenities. The social construction of "resources" (Sullivan, 2017; Simpson, 2019), which reduces the lives of water, trees, mountains, wind, and wetlands to commodities, and their extraction—or killing—which initiates their integration into production processes, remain essential to state formation and capitalism (Dunlap and Jakobsen, 2020). Extraction, to be clear, refers to the removal—cutting, digging, slicing, and mining—of large quantities of mineral, hydrocarbons, biological, human, and kinetic energetic resources. *Extractivism* is the ideology within which extraction is embedded and tied to ideas of progress, state power and growth (Gudynas, 2009, 2021). *Extractivism* involves acquiring large volumes of raw material for export, profiting national or international business classes (or governments, in the case of neo-extractivism), while breaking down existing sectors as well as social and ecological relationships (Gudynas, 2009). *Extractivism* results in a high intensity of environmental degradation and, finally, corresponding deleterious labor opportunities and conditions (Gudynas, 2009; see also Lang and Mokrani, 2013/2011).[7] *Extractivism* remains central to understanding ecological and climate catastrophe, to which the enforcement of extractive activities is integral.

Resistance to centralized authority (Scott, 2009, 2017; Gelderloos, 2017), enclosures (Merchant, 1983; Federici, 2009, 2004), colonialism (Rodney, 2009; Galeano, 1997), slavery (Sakolsky and Koehnline, 1993; Linebaugh and Rediker, 2013; Trebitch, 2021), patriarchy (Federici, 2009, 2004), and the destruction of habitats, cultural practices, and ecosystems have always accompanied "development" and state formation (Dunlap and Jakobsen, 2020). Against this resistance, the military, police, and private contractors spy, intimidate, and pull the trigger against land defenders and communities. These actors, and the institutions behind them, enforce ecocidal projects against the opposition fighting plantations, coal, copper, and hydrocarbon extraction, as well as a variety of infrastructure projects (see Granovsky-Larsen and Paley, 2019; Brock, 2020b; Kroger, 2021; Shapiro and McNeish, 2021; Verweijen and Dunlap, 2021). This extends to low-carbon, or so-called "renewable energy" projects that have similarly negative socio-ecological impacts, generating sustained and fierce resistance (Siamanta, 2017, 2019; Franquesa, 2018; Dunlap, 2019a, 2021a). These institutions and modalities of development progressively enforce cultural erasure, livelihood vulnerability, the fragmentation of social relationships, and habitat destruction. The connection between socio-ecological degradation and policing is intimate, ubiquitous, and largely underacknowledged.

Policing has rippling interconnected and unequal effects across the world. The ecological and, following Nixon (2011), "slow violence" of policing forces is often rendered secondary in comparison to the brute force of "fast violence", the kneeling on people's necks, the torture of

[7] Recent debates have emerged over the expansive use of extractivism (see Dunlap, 2021b; Gudynas, 2021), specifically over the sectoral boundaries and analytical usefulness of the term, which Markus Kröger (2022: 47–49) partially resolves by offering a scale of different intensities of extractivism.

prisoners, and the killing of environmental defenders. Let the Ashville police department, in North Carolina, be a reminder of this on a small scale. In an effort to disable anti-police protests in the aftermath of the murder of George Floyd, Ashville police pre-empted protests by raiding and destroying the supplies of a medical tent set up by residents to help people beaten, teargassed, and wounded by "less-lethal" police munitions. In their effort to suppress the logistical support of protests—attempting to eliminate their means to protest— they used knives to rip holes into all the water bottles in the medical tent. The police thus poured an enormous amount of drinking water into the street, while producing a large pile of plastic in its wake.[8] This is a common practice at the US–Mexico border (Warren, 2019), where border patrol wreck hydration stations set up in the desert by activists, church groups, and NGOs. These instances are important reminders that these small-scale acts are common, normalized, and accumulate over time, alongside other repressive activities. The discourse of climate change tends to ignore and accept the ecological cost of these systematic acts of policing and border enforcement, but also the political violence this entails.

The psychological and social impacts of policing on people are well known (Huggins et al., 2002; Williams, 2007 [2004]; Bachmann et al., 2015; Camp and Heatherton, 2016; Seigel, 2018; Tahir, 2019). The impacts of policing on ecologies and nonhumans, on the other hand, are rarely discussed—if not entirely ignored. The political ecology of policing deserves greater attention and development. Ignoring the political ecology of policing, we presume, is the result of colonial legacies, modern ontologies, and existing cultures that treat our ecosystems, habitats, and nonhuman neighbors as dead material or "resources" to be plundered (Sullivan, 2017; Simpson, 2019). Celebrating a total liberation ethic (Pellow, 2014; Springer et al., 2021; Anonymous, 2019), this book rejects the ethos of disrespecting, degrading, and destroying planetary habitats. A total liberation ethic (Pellow, 2014), which anarchist political ecology articulates, attempts to unravel the institutionalization of abuse through research inherently critical of authoritarian organizations, claims, mechanisms, and coercive interventions (Brock, 2020b; Dunlap,

[8] See images and video here: https://www.scarymommy.com/cops-destroy-medic-tent-asheville/.

2020c; Springer et al., 2021; Brock and Stephens-Griffin, 2021). Kirkpatrick Sale (2000/1991: 122), referring to Murray Bookchin (1982), narrates cogently this concern of anarchist political ecology:

> [S]ocieties that dominate nature also dominate people. Where there is the idea that a massive dam should be built to control a river's flow, there is the idea that people should be enslaved to build it; where there is the belief that a giant metropole may serve itself by despoiling the surrounding countryside and devouring its raw materials, there are castes and hierarchies to ensure that this is accomplished.

This introduction, and other chapters, are written in the spirit of an anarchist political ecology approach (Springer et al., 2021; Brock, 2020b; Dunlap, 2020c). The purpose is to examine the role of policing forces in enforcing ecologically destructive developmental regimes, but also the indirect ecological costs of policing operations—whether military war making, private contractors protecting oil and gas extraction, police forces patrolling cities—across the world. The political ecology of policing remains an important research frontier, capable of further revealing the relationship between ecological degradation and political repression and its role in driving planetary crisis and catastrophe.

Anthropocentrism in Policing Literature

While there is some exceptional work examining police violence, social control, and its overlap with militarization (Williams, 2007 [2004]; Williams et al., 2013; Neocleous, 2014; Bachmann et al., 2015; Hönke and Müller, 2016; Schrader, 2019; Tahir, 2019), the environment remains missing from these analyses (see Brock and Stephens-Griffin, 2021, for a recent exception). Ecosystems and ecologies—the very foundations of life sustenance—have been invisibilized into the backgrounds of the wider political economy, the urban environments, the racism integral to policing, and, of course, the violent practices of policing themselves. Yet, struggles against prisons, policing, and pollution (Thompson, 2018), as environmental justice organizers and abolitionists are showing us on the ground (Braz and Gilmore, 2006), are inextricably linked and interconnected. The harms associated with policing, prisons, and pollution particularly affect people of color, as political ecologists Pulido and De Lara (2018) have shown. The ecological cost of policing and its impacts on nonhuman life is largely missing in the study of policing. While the military has received greater attention in this regard (Thomas, 1994; Smith, 2017; Belcher et al., 2020), the role of policing forces as driver of ecological and climate catastrophe is underexplored.

This might be due to the anthropocentrism that dominates the social sciences, including political science, international relations, geography, and anthropology. Political ecology, specifically the political ecology of counterinsurgency (Peluso and Vandergeest, 2011; Dunlap, 2019b; Brock, 2020b), might offer an exception or, more accurately, a gateway to further explore policing in greater depths. Examining the militarization of environments and warfare (Peluso and Watts, 2001; Belcher et al., 2020), political ecology has excelled at unraveling conservation militarization (Lundstrum, 2014; Duffy, 2016; Büscher and Fletcher, 2018; Fairhead, 2018; Verweijen and Marijnen, 2018) and the assemblage of repression organized by state and extraction companies to acquire mineral and hydrocarbon resources (Brock and Dunlap, 2018; Becerril, 2018; Verweijen and Dunlap, 2021; Brock, 2020b; Huff and Orengo, 2020; Granovsky-Larsen and Santos, 2021). With the exception of Oliver Belcher and colleagues (2020) and other works examining the military (Thomas, 1994; Smith, 2017), research has yet to extend to the multi-faceted importance of *scientific violence* in the production of climate and ecological catastrophe. The violence, operational disposition, and manipulation of police forces distract from their ecological impacts. Despite occasional outrage, many continue to accept the commonly held notion that the police ensure public order and safety. Even when uprisings "break the spell" or the illusion of the police, the ecological costs are forgotten or, at least, understandably ignored for immediate confrontation with police forces during uprisings. This violence is grounded in histories of racist practices of urban development (Gibson, 2007), which reinforce environmental racism (Braz and Gilmore, 2006; Pulido and De Lara, 2018). From mines to dams and energy infrastructures, racial and technological capitalism accepts the sacrifice of rural and Indigenous populations in the name of modernity and development (Scott and Smith, 2016; Granovsky-Larsen and Paley, 2019; Dunlap, 2019b; Verweijen and Dunlap, 2021). This entails creating inhospitable environments—directly and indirectly—that encourage "social death" (Short, 2016; Dunlap, 2018; Kaur, 2021), which structure environments in order to create depression, anxiety, self-hatred, drug abuse, and suicidal tendencies (Duran and Duran, 1995). This book seeks to open a doorway to this ecological gap, illuminating the role policing institutions play in the production of violence, misery, and ecological and climate catastrophe.

The next section will explore further the relationship between colonialism, counterinsurgency, and policing. This section emphasizes the scientific and intentional nature of repression, as we explore how violence, counterinsurgency, and policing were exported and developed through colonialism. Scientific violence remains a coercive art, with trauma-inducing effects not only for people but also the ecosystems supplying the means to create violent technologies and to fuel their operations. We now turn toward the intimate relationships of warfare and policing, before exploring and introducing the contributions to this book.

Colonialism, the State, and Counterinsurgency-Policing

The relationships between state formation and colonialism is rather important, and oddly missing—or, more accurately, decentered—from decolonial theory in the academy.[9] Colonial power has roots in, and continuity with, ancient civilizations (Scott, 2017; Gelderloos, 2017). To become a colonial power and culture, one had to first colonize from "within"—implementing colonial violence and order at "home" (Churchill, 2002; Gelderloos, 2017; Dunlap and Jakobsen, 2020). Alongside the military—responsible for quelling anti-authoritarian uprisings and land defense—the police remain among the most potent "internal colonial" mechanisms and institutions. They make state formation possible and enforce a particular political economic order, which enables and spreads colonial exploration and business ventures. The constructions of racial difference and "the enemy" (Foucault, 2003; Arendt, 1962; Rodney, 2009), key to the invention and spread of racism, and the general "Othering" and indifference to neighbors in other watersheds, bioregions, or countries—in the Global North or South—has allowed for divide-and-conquer strategies by those in power. Racism justified

[9] We can only speculate this has to do with the very nature of the academy as a statist institution, as well as the refashioning of authoritarian politics through identities or academic catchphrases, where ambiguity and the naturalization of authoritarian structures (state, university, or otherwise) allow the regrading of authoritarianisms, "left" or "right" wing.

colonization, plundering, and extraction of these territories by imperial powers, in the name of "enlightenment" and "development".

Numerous genealogies of the police (Foucault, 2002, 2007; Williams, 2007 [2004]; Neocleous, 2014) examine their origins, but a focus on policing as tactics, maneuvers, and strategies reveals the important role of actors and institutions that existed long before the formalization of policing powers in police forces or even militaries. These include Catholic priests who were integral to the domestication of peasants and the normalization of social hierarchy during Roman imperial rule (Perlman, 2010 [1983]). Policing became instrumental to enforcing industrialization, extraction, and "progress" through the promotion and enforcement of authority, a particular work ethic, the criminalization and stigmatization of idleness, and enterprise (Anthony, 2001/1977; Daggett, 2019; Brock and Stephens-Griffith, 2021). The police, workhouse, prisons, and penal colonies were central mechanisms for state consolidation and industrialization (Foucault, 1995, 2007; Redfield, 2005) which worked to reinforce processes of internal and extra-territorial colonization. The threat of force from the military alongside the deployment of police in internal colonial contexts (e.g. state formation/consolidation) performs a similar function to (the threat of) prison, which complements conditioning people for factory proletarianization. The internal colonization that created Europe and the extra-territorial colonialism that made Europe wealthy are marked by different temporalities and intensities of violence, the latter being outright genocidal (Moses, 2008; Moses and Stone, 2013). Patrick Wolfe (2006) documents the colonial genocidal process in three phases: starting with initial confrontation or invasion of territories; followed by incarceration programs that involve displacement or resettlement; and finally, assimilation initiatives that aim to integrate indigenous populations into colonial structures. Campaigns of extermination give way to assimilation programs, which later intensify and transform into processes of (statist) self-management. This is where colonial political and economic structures, values, and organizations become normalized, creating self-identification and thereby making the colonial and/or capitalist political economy self-reinforcing and managing (Dunlap, 2018). This produces colonial cultures, power, and, by extension, imperial possibilities.

These processes naturally entail a process of ecological degradation, if not—as mentioned above with the bison example—concerted efforts at *ecocide* to defeat and pacify people wedded to the land (see Boot,

2013). It is for this reason that Crook and Short (2014, 2021; Short, 2016) coin the term "genocide-ecocide nexus", to indicate how genocide and ecocide are two sides of the same intricate program of land control and pacification. Ecocide has been integral to the efforts to control and coerce populations, human and nonhuman; to enforce settlement, industrial agriculture and domestication; to displace; and to "discipline" people into wage labor (Brock and Stephens-Griffith, 2021). The connection between state formation, ecocide, colonialism, and policing should in no way be underestimated or ignored.

The "techniques employed up to 1940", remarked Foucault (1980/1972: 18), "relied primarily on the policy of imperialism (the army/the colonies), whereas those employed since then are closer to a fascist model (police, internal surveillance, confinement)". The violence of colonial "small" or counterinsurgency wars "boomeranged" back to Europe (Arendt, 1962, 1951: 155; Dunlap, 2014), where police and military powers would merge, taking on and later normalizing "fascist" qualities of political policing within Europe. Aimé Césair (2001/1955: 36) reminds us "that before they [Europeans] were its victims [of political terror], they were accomplices [during colonialism]; that they tolerated that Nazism before it was inflicted on them". The existence of colonial/fascist violence allowed and tolerated by authorities and people in the colonies "came home" and, following the theory of the Genocide Machine (Davis and Zannis, 1973), would evolve by economizing and advancing technological capacities to articulate with precision—or "smartness"—the proliferation of a productive statist model toward total extractivism and warfare (Dunlap and Jakobsen, 2020). The green economy emerges as an expression of ecological "economizing", articulating increasingly complicated and "smart" processes of ecological extractivism through commodification of nature, financialization, digital monitoring (e.g. "smart" censors), and kinetic energy extraction via low-carbon infrastructures. The green economy is a renewal of extractivism and ecological conquest.

Colonialism, the application of scientific violence, state formation, and the commodification and marketization of nature as resources affirmed processes that were and are detrimental for collective socio-ecological well-being. The colonial boomerangs of scientific violence, Jana Hönke and Markus-Michael Müller (2016: 3) remind us, no longer just move north to south, but are "multi-directional travelling of practices across the globe as well as the active agency and participation of seemingly 'marginal' actors in producing and co-constituting what is conventionally thought as 'Western' policing practice, knowledge and institutions". Said differently, the colonial and scientific violence runs rampant within and between countries across the world to maintain the extractive capitalist political economy. This, as Hönke and Müller indicate, has required active participation by people indigenous to particular territories, which illustrates the importance of colonial intermediaries or cadres to normalize colonial values, spatial layouts, and socio-ecological practices. When habitats—homes, neighbors, friends—are relegated and degraded into resources, it initiates a process that makes extractivism acceptable. It is not long before people begin to betray themselves, each other, and their habitats, submitting to the allure of technology and authority of colonial systems.

Colonial warfare—past and present—remains the dominant style of warfare. Criticizing how narrowly international relations scholars categorize "war", Mark Neocleous (2014) illuminates an important historical sleight of hand. Designating "colonial warfare as 'unconventional' or a 'small wars' affair", he explains, "dismiss[es] what has in fact been by far the most common form of warfare in the modern world" (Neocleous, 2014: 5). In reality, colonial warfare is *total*, remaining an exposition in the art of political occupation, cultural reconfiguration, and extractivism. Colonial warfare, and the governance toolbox of counterinsurgency warfare that developed from it, offers the methodology by which political structure and order is imposed. Conquest or invasion, Wolfe's (2006: 388) words continue to echo, "is a structure not an event".

The creation of industrial infrastructures—no matter how useful and enchanting—comes at serious socio-ecological costs that are the result of mining, manufacturing, and refining raw materials; operational impacts; and decommissioning. Continuing Wolfe's (2006) understanding of the colonial genocide process, Dunlap and Correa-Arce (2021: 7) argue that "invasion is infrastructural": modernist infrastructures form material and energy intensive ecosystems that (slowly) absorb and condition their inhabitants. Modern infrastructures thereby perform a "psychogeographical colonization" (ibid.), embodying and reinforcing particular ontologies, epistemologies, and cultural values. The imposition of colonial values—ecological domination, authoritarian control, technological and economic fetishization—requires infrastructural arrangements, but also flexible infrastructures such as the police and military to impose them. Infrastructures, extraction, and policing forces form a catastrophic trilemma, which makes invasion permanent.

Alongside elite education, community development (Dunlap and Fairhead, 2014), and global university restructuring (Berman, 1983), overseas police assistance is another crucial mechanism of enforcement. The US, England, and France have long been involved in police reform and professionalization across the world during colonial, but also (neo)colonial times, through the guise of development (Huggins, 1998; Neocleous, 2014; McCoy, 2005; Schrader, 2019). Numerous police academies, alongside military academies, serve the purpose of "training the trainers" (see Schrader, 2019: 171). To illustrate this: by 1972, the US Agency for Intentional Development's (USAID) Office for Public Safety (OPS) had trained 1.5 million police officers across the world, according to their own numbers (Schrader, 2019: 181). They trained not only police departments but also extreme rightwing groups in techniques of counterinsurgency including bomb making and interrogation techniques such as genital electrocution.[10] The OPS describes these activities as "functional specialization of personnel, use of modern technology, neutrality in law enforcement, responsible use of discretion, and a measure of autonomous self-regulation" (Schrader, 2019: 11).[11] The OPS had advisers across the world, operating under the banner of development assistance, which, as Schrader (2019) documents, created space—in addition to special (counterinsurgency) warfare schools—to facilitate training to execute "Dirty War" activities in police departments. While the OPS was terminated in 1974 by the US Congress, in the 2000s, USAID (2011) began reviving police assistance under the banner of "Assistance to Civilian Law Enforcement in Developing Countries" (see Hochmüller and Müller, 2017). The spread of scientific violence and political terror would percolate and later boomerang back to police departments in the US—in what Kristian Williams (2007 [2004]: 218) famously formulated as "community policing + militarization = counterinsurgency".

Practices of population management and ordering—known as population-centric counterinsurgency—are instrumental to engineering and sustaining the colonial order of capitalist political economy. The total

[10] See Costa-Gavras' then controversial film *État de Siège* (1972, *State of Siege*) to see a cinematic depiction.

[11] This relates to colonial warrior General Brigadier Kitson's conceptual use of the judicial system: The "law should be used as just another weapon in the government's arsenal, and in this case it becomes little more than a propaganda cover for the disposal of unwanted members of the public" (1971: 69).

warfare approach of counterinsurgency, as Schrader (2019: 14) argues, is "a misnomer, because the insurgency to be countered was one that had not yet occurred". This required governments, not without struggle, to disingenuously refashion legal definitions of terrorism to include non-violent political action (Lubbers, 2012; Del Gandio and Nocella, 2014; Brock and Dunlap, 2018). The governmental self-serving definitions of terrorism designed to criminalize militant protest also complements the "watering down" of conceptions of nonviolence, which creates internal divisions and further dilutes political movements (Gelderloos, 2013, 2020). The lessons of colonial warfare have been institutionalized and are constantly enforced by every means to maintain the capitalist political economy and the governing framework of states, which are instrumental to propelling ecological and climate crises.

While this review is brief, this history and continued operations illustrate the high body counts of policing, including from conventional and dirty wars, but also the everyday maintenance of structural violence, infrastructural regimes, racism, and countless other discriminatory practices impossible to accurately calculate. *Enforcing Ecocide*, however, seeks to highlight that this brutality, manipulation, and violence simultaneously result in generational and climatic environmental impacts related to the killing of trees and animals, river pollution, wildfires, habitat loss, mass die-offs, and extinctions. The political terrorism emanating from the institutions of scientific violence and armed actors is matched with ecological terrorism paving the way toward ecocide and climate breakdown. We now start mapping their ecological costs.

The Book Contributors

This volume presents a range of case studies from across the world, documenting and discussing the various socio-ecological impacts of policing and militarization. It is organized into four parts. The first part, "Hydrocarbon Militarization", discusses three cases studies related to war, contamination, and struggles against the development of hydrocarbon infrastructure. The second part, "Enforcing Extraction", explores the policing and extraction of mineral and hydrocarbon resources necessary for policing operations, technologies, and equipment. The third part, "Policing Ecosystems" delves into how military, police, and private security forces control resources, fight oppositional moments, and degrade ecosystems. Lastly, the fourth part, "Looking forward", puts forward some ideas about decarbonization by demilitarization. These parts and contributors offer engaging works that reveal the destructive and ecologically taxing realities of militarizing and policing lands and their inhabitants.

The second chapter, "A Postcolonial History of Accumulation by Contamination in the Gulf" by ***Michael Hennessy Picard and Tina Beigi*** examines the energy–security nexus through the lens of primitive accumulation in the Gulf States of the Middle East. Picard and Beigi explore the fraught and conflictive history of oil exploration and development in Iraq. The chapter argues that the corollary to the enforcement of capital accumulation is the *contamination* of human ecology, cannibalizing the daily life of many people who have to live in war rubbles and oil spills. The concept of accumulation by contamination provides a critical framework to demonstrate the way imperialism deploys sophisticated techniques of contamination to expropriate people and colonize their resources. This chapter reveals how political control, repression, and successive rounds of oil wars are responsible for capital accumulation and contamination. In other words, as Max Liboiron (2021) recently argued, *Pollution is Colonialism*. Militarization for oil control and extraction, Picard and Beigi show, has transformed Iraq and its neighboring states into lands inundated with industrial and chemical wastes from oil refinery burnings, massive oil spills, lethal and radioactive munitions. "This contamination was the product of the dynamic relationship between firebombing from above and oil drilling from below", they explain. This chapter is a testament to the socio-ecological harms of war on people and the planet, reminding us of the necessity to, as the British punk rock band *Crass*[12] exclaims, *Fight War, Not Wars.*

Moving across the border, Chapter 3 explores the history of environmental struggle in Iran. "Beyond Rentier State and Climate Conflict: Clashing Environmental Imaginaries and Ecological Oppression in Iran" by ***Maziar Samiee*** discusses the struggle of environmental movements in Iran, but also unpacks the wider "rentier state" dynamics and faults within the "climate conflict" discourse. Within these narratives, Samiee explains, "the Iranian state seems trapped in a vicious cycle of rent-seeking, ecological damage, and political dissent that eventually undermines its political and economic structures". Yet, he argues, rentier state and climate conflict

[12] *Crass*, "Fight War, Not Wars" (1978), https://www.youtube.com/watch?v=EV4a786hZzE.

framings are, in actuality, advancing projects of extraction and ecological degradation. Samiee locates the state as central to the production of ecological degradation and climate change. The problem of the state, however, extends to how it constructs mythology and imagination. This chapter stresses the need to focus the state at the center of analysis, exploring not only the violent processes of coercion executed by governments, but also the imposition of a particular environmental imaginary onto people and ecologies. It investigates the "transformations of hegemonic ecologies of states in the postcolonial era" in order to better understand the social control mechanisms and imaginaries that support and facilitate ecological degradation and climate change.

Jumping across the Atlantic to Turtle Island in North America, Chapter 4, "Policing Indigenous Land Defense and Climate Activism: Learnings from the Frontlines of Pipeline Resistance in Canada" by ***Jen Gobby and Lucy Everett*** examines the process of hydrocarbon pipeline development in settler colonial Canada. Against the promoted self-image of Canada as environmentally friendly and fighting against climate change, Gobby and Everett offer a completely different image by exploring the history and current reality of fossil fuel pipeline development across Indigenous territories in Canada. Acknowledging the fierce and systematic resistance of peoples against pipeline development, this chapter focuses on the immense amounts of resources and strategies dedicated to surveilling, policing, coercing, and exhausting land defenders to force through hydrocarbon pipelines. Through the "conflict transformation framework" that is based on three forms of power—*discursive power*, *institutional power*, and *relational power*—Gobby and Everett examine the coercive strategies of the Canadian government and the hydrocarbon companies. They employ a fourth category of analysis, *material power*, to conceptualize the power wielded by financial interests and the power of physical force. Through this approach, Gobby and Everett show how the policing of anti-pipeline movements is actively enforcing the violation of Indigenous rights, entrenching Canada's fossil fuel dependency, and solidifying a path of climate catastrophe. This chapter documents in detail the continued process of land grabbing and settler colonial expansion of the Canadian state, and how these are entangled with *ecocide*.

The second part, *Enforcing Extraction*, begins with a chapter that goes straight to the heart of modern warfare: Mining. Chapter 5, "Global Britain and London's Mega-mining Corporations: Colonial Ecocide, Extractive Zones, and Frontiers of Martial Mining" by ***Daniel Selwyn,*** delves into the realities of the mineral extraction necessary for military vehicles, aircraft, and equipment. *Martial mining* problematizes the "intimate and interdependent relationship between the arms trade, industrial resource extraction, and widespread ecological degradation integral to the operations and technologies of racial capitalism on local to global scales". Through a special focus on London-based mining companies, Selwyn explores the violent colonial histories of transnational mining corporations, embedding them in ongoing mass extinction events such as Indigenous genocides and chattel slavery. This allows him to explore and situate contemporary mining conflicts in the South African platinum belt in Marikana and the Grasberg mine in West Papua and the social and ecological relations of corporate imperialist plunder and neo-colonial state governance around them. Demonstrating the inherently racist ecological destruction necessary to produce weapons of war, Selwyn reminds us that without war and "counterinsurgency strategies, the appropriation of land, extraction of natural resources, and repression of resistance would be impossible to maintain".

Continuing on the theme of extractivism for repressive operations, Chapter 6, "The Self-Reinforcing Cycle of Ecological Degradation and Repression: Revealing the Ecological Cost of Policing and Militarization" by ***Alexander Dunlap***, examines the immense amount of resources required for *political violence* and *social control*. This chapter explores and starts to map the extractive cost of the military and police. While limited in scope, the chapter still reveals how the military and police, associated with facilitating imperialism and preserving class structures and white supremacy, are also inherently an ecological and extractive problem. This means that "all the harassment, fines, arrests, beatings and murders, underlining the George Floyd rebellions across the United States, which resonated across the world, are a fundamentally ecological issue". Consequently, he argues, anti-police outrage and rioting can also be viewed as against the extractivism and ecological destruction that support police power and brutality. The chapter examines military and police budgets, along with the material requirements and ecological impacts of the military and police, to argue that militarization and policing are underestimated and central contributors to ecological harms and climate catastrophe.

In Chapter 7, "Oil, Arms and Emissions: The Role of the Military in a Changing Climate", ***Wendela de Vries*** critically examines the relationship between (Western) militaries, their *power projection* strategies, and ecological harm. Legitimated by "superior" Western values, she argues, Western militaries protect the extraction economy and defend access to natural resources, rare earth minerals, and sea lines of communication at all (ecological) costs. Meanwhile, Western arms industries profit from arms sales that serve to violently repress anti-extractive protests in the global South, as well as border militarization to "manage" climate migrants. Arms control treaties continue to allow for overruling human rights and ecological health for corporate profit. Yet, militaries have started to position themselves as part of the solution to climate catastrophe, essentially propagating, she argues, a "twenty-first century adaptation of the White Man's Burden and... involvement of western militaries in climate conflicts in vulnerable areas". Meanwhile, their ecological "bootprint" remains. Deconstructing arguments for "greening" the military (for instance through energy-efficiency measures), de Vries advocates for non-military ways of addressing conflict, including disarmament treaties and abolishing power projection strategies. To address climate catastrophe, she concludes, we need de-militarization, not green militarization.

The third part, *Policing Ecosystems*, investigates the on-the-ground policing in and around—and at the detriment of—human and nonhuman habitats. The section starts with ***Anwesha Dutta and Trishant Simlai***'s chapter, "If the Army Cuts Trees, Why Can't We? Resource Extraction, Hunting and the Impacts of Militaries on Biodiversity Conservation". Here, Dutta and Simlai unravel the complexities associated with military presence in North and Northeast India, in highly biodiverse areas that are also simultaneously inhabited by indigenous populations and have emerged as sites for counterinsurgency operations. Communities in these areas, they show, are entangled in violence and environmental injustices related to the imposition of conservation enclosures that are militarized and generate local insecurities. "[M]ilitary entanglements with both the biophysical environment as well as communities inhabiting those spaces is not always fraught with spectacular forms of violence or overt destruction of nature but is expressed through quotidian ways of resource extraction (in connivance with local syndicates or contractors), occupying territory leading to evictions and displacement or blockades of essential animal corridors". The result is habitat destruction by armed forces, human and nonhuman displacement, and the participation of security forces in legal and illegal resource extraction, from constructing infrastructure to hunting animals. At the same time, armies' environmental operations such as flood and forest fire management are often glorified, "pushing the negative consequences of their ecological presence to the backburner". Despite the lack of transparency and nondisclosure of information that is inherent to military operations, the chapter thus illustrates the multiple ways in which the military enforces ecocidal processes in biodiverse areas inhabited by Indigenous peoples in India.

Chapter 9, "Policing the High Speed 2 (HS2) Train Line: Repression and Collusion Along Europe's Biggest Infrastructure Project" by ***Andrea Brock and Jan Goodey*** turns to ecological policing in the UK. The authors examine the role of police and private security forces, corporate-state collaboration, and policing technologies used against local communities, ecologists, and activists to enforce the high-speed railway project between London and the North of the country. While promoted as environmentally and socially beneficial by the government, Brock and Goodey argue that the project constitutes a fundamentally extractivist green capitalist megaproject. It relies on coercive policing, grounded in open-source intelligence gathering by private firms, a-priori criminalization and deterrence through corporate injunctions, and the silencing of dissent and manipulation of the political narrative through nondisclosure agreements and pressure on landowners and activists. The case illustrates the importance of repressive forces deployed in the service of protecting industrial interests. The chapter demonstrates how policing selectively enforces some regulations, while ignoring others, particularly legislation that defends local ecologies and wildlife. "Subsuming of protesters' rights and ecological health to the rights of developers", Brock and Goodey argue, "is thus not about *abuse* of police/security powers. Protecting extractive and infrastructure projects – upholding private property rights and industrial interests – is the *aim* of policing". Repression, thus, is essential to causing ecological harm. Public–private security partnerships, the deterrence and repression of resistance, the avoidance of negative media coverage through Non-Disclosure Agreements (NDAs), and securing a positive "green" narrative, however weak and contested, in turn, are integral to facilitating the development of the project.

Zooming out, and providing an overview of environmental policing, the next chapter, "Ecological Terror and Pacification: Counterinsurgency for the Climate Crisis", explores counterinsurgency and the repressive techniques used against land defenders across the world. Here, ***Peter Gelderloos*** offers a comprehensive overview of the repression and violence that enforces the present path of ecological and climate catastrophe. This chapter contends that the first responses from governments and leading economic interests to the ecological crisis are meant to bolster measures of social control and to prepare for increased resistance. The chapter documents a long history of state repression and control against people fighting to defend their land and territory, which extends to people saving animals from testing labs and attacking destructive industries. Gelderloos highlights the histories of Indigenous, anarchist, and autonomist struggle frequently forgotten by the majority in the academy, implicitly showing the necessity of anti-authoritarian resistance to ecological and climate catastrophe. Gelderloos offers an important and searing critique of the non-governmental organization (NGO) sector and their role in policing, and reveals structural issues with accounting for harm to land defenders. "The very NGOs who evince a concern for human rights are crucial to this counterinsurgency operation, celebrating the activists who use tactics deemed legitimate by the government", explains Gelderloos. "By not celebrating every bit as vociferously the lives and the resistance of those who make the hard choice to resist with illegal and effective methods, these NGOs are signaling who is legitimate and who is subhuman, they are creating the division between good actors and bad actors that the militaries, death squads, and counterinsurgency experts rely on, and they are doing so in a way that most people who think they care about the environment or human rights will find credible". This chapter, and criticism, offers important insights and issues in need of self-reflection not just in the NGO sector, but within academia and beyond.

Lastly, the book concludes with a short Part IV, *Looking Forward*, and a contribution by ***Matthew Burke and Nina L. Smolyar***. In their chapter, "Demilitarize for a Just Transition", Burke and Smolyar systematically examine and rebut claims about the role of the military in a just green transition. Indeed, they show how the US military contributes to ecological catastrophe and prevents the realization of an ecological society. *Militarism*, they affirm, undermines the ecological basis of well-being and diverts funds from other human and ecological needs, and only aggravates climate catastrophe. Militarism shapes geopolitics and undermines global cooperation, preventing meaningful climate action, while requiring resources. In the authors' words: "Military dominance requires extraction of everything, and more of it". Having shown that militaries are major obstacles to just and green transitions, Burke and Smolyar call for their systematic repurposing to contribute to a just transition; the reclaiming of military assets and repositioning them under civilian control for non-military purposes. This involves spending reductions and reallocation, economic conversion, and the civilian repurposing of the military, among others. This is not an easy task; they concede, "[t]he vested interests, from fossil fuel corporations, military contractors, Pentagon elites, and DoD [Department of Defense] officials, to federal, state, and local politicians will do all in their substantial power to resist cuts and downsizing. Yet no real transition to a peaceful and ecological future is possible without demilitarization". This chapter thus concludes the book with a vision of "decarbonization through demilitarization", with implications for just and ecological transitions at all levels.

Conclusion

Policing (and militarization) is/are key drivers of ecological and climate catastrophe. Drawing on a variety of (anarchist) political ecology approaches, ecological marxism, and other critical theories, this book contends that the objectification, commodification, and degradation of ecosystems are closely entangled with the coercive and abusive hierarchies instrumental to states, colonial ventures, corporate profit-making, and the necessity of war and policing. Policing and militarization are essential to state formation, expanding and evolving violent practices through colonial ventures, where policing—and the art of scientific violence supporting it—only intensifies, evolves, and proliferates across the globe. At the same time, state formation, coloniality, and policing are integral to ecocide. The earth is a closed ecosystem, which means that unless the violence and trauma are reconciled, they will continue to mutate, circulate, and transform across the planet. This violence expresses itself through—and is essential to—the accumulation of ecological degradation, resulting in climatic changes and mass extinction. Rooted in human-nature binaries and utopian/dystopian[13] visions of civilization and progress, the wonders of modernist development have come to create a dire ecological and climatic situation. This trajectory, we must remember, can be changed.

As this volume shows, policing and militarization deserve greater critical attention. Belcher and colleagues (2020: 76) have argued that "that social movements concerned with climate change must be every bit as

[13] Depending on the eye of the beholder. See Caroline Merchant (1983) for a discussion of utopias.

vociferous in contesting US military interventionism". While we agree, we would further extend this concern to include the institutions of *scientific violence* and *violence work* more generally—including police forces, paramilitaries, security firms, border patrols, intelligence agencies, and prisons. The enchantment and addiction of modern industrial life should not stop our critical vigilance to understanding the political ecology of policing, which explores and challenges the socio-ecological injustice demanded by bureaucratic, capitalist, and ecologically destructive political orders. It appears that the institutional turn toward "greening" the military, police, and prisons, organizing various regimes of "sustainable violence" (Dunlap, 2017), is an emerging research frontier necessitating greater knowledge development. Ongoing and accelerating ecological and climate catastrophes reveal that it will take more than "decarbonizing the military and police" to stop ecological harm, avert climate catastrophe, and take steps toward meaningful ecological repair. Green weapons, biodegradable bullets, and solar-powered aircrafts are not the direction toward socio-ecological sustainability, but a repackaging of the existing violent and ecologically degrading reality that brought us up to this catastrophic point. This ill-fated trajectory, we worry, necessitates insurrectionary ecological and abolitionist struggle to end the war against the earth and its inhabitants.

Bibliography

Alexander BK. (2008) *The Globalization of Addiction: A Study in Poverty of the Spirit.* New York: Oxford University Press.

Álvarez L and Coolsaet B. (2020) Decolonizing environmental justice studies: A Latin American perspective. *Capitalism Nature Socialism* 31: 50–69.

Andreucci D and Engel-Di Mauro S. (2019) Capitalism, socialism and the challenge of degrowth: Introduction to the symposium. *Capitalism Nature Socialism* 30: 176–188.

Anonymous. (2019) *Total Liberation.* London: Active Distriubtion. Availiable at: https://theanarchistlibrary.org/library/total-liberation-anonymous-english.

Anthony PD. (2001 [1977]) *The Ideology of Work.* Oxon: Travistock Press/Routledge.

Arendt H. (1962 [1951]) *The Origins of Totalitarianism.* New York: The World Publishing Company.

Bachmann J, Bell C and Holmqvist C. (2015) *War, Police and Assemblages of Intervention.* Oxon: Routledge.

Bædan. (2014) Bædan 2: A Queer. *Journal of Heresy* 2: 1–198.

Becerril MSW. (2018) Mining conflicts in Peru: Civil resistance and corporate counterinsurgency. *Journal of Resistance Studies* 4: 94–127.

Bekiempis, V. (2021) Record-breaking US Pacific north-west heatwave killed almost 200 people. Accessed July 19, 2021. https://www.theguardian.com/us-news/2021/jul/08/pacific-northwest-heatwave-deaths.

Belcher O, Bigger P, Neimark B, et al. (2020) Hidden carbon costs of the "everywhere war": Logistics, geopolitical ecology, and the carbon boot-print of the US military. *Transactions of the Institute of British Geographers* 45: 65–80.

Berman E. (1983) *The Ideology of Philanthropy: The Influence of the Carnegie, Ford, and Rockefeller Foundations on American Foreign Policy*. Albany: State University of New York Press.

Bigger P and Neimark BD. (2017) Weaponizing nature: The geopolitical ecology of the US Navy's biofuel program. *Political Geography* 60: 13–22.

Böhm S and Sullivan S. (2021) *Negotiating Climate Change in Crisis*. London: Open Book Publishers.

Bolger M, Marin D, Tofighi-Niaki A, et al. (2021) *'Green mining' is a myth: The case for cutting EU resource consumption*. Available at: https://friendsoftheearth.eu/wp-content/uploads/2021/09/Methodology-considerations-Annex-to-green-mining-is-a-myth.pdf.

Bookchin M. (1982) *The Ecology of Freedom: The Emergence and Dissolution of Hierarchy*. Palo Alto, CA: Cheshire Books.

Boot M. (2013) *Invisible Armies: An Epic History of Guerrilla Warfare from Ancient Times to the Present*. New York: Liveright Publishing Corporation.

Braz, R. and Gilmore, C. (2006) Joining Forces: Prisons andEnvironmental Justice in Recent California Organizing. *RadicalHistory Review* 96: 95–111.

Brock A. (2020a) Securing accumulation by restoration—Exploring spectacular corporate conservation, coal mining and biodiversity compensation in the German Rhineland. *Environment and Planning E: Nature and Space*: 2514848620924597.

Brock A. (2020b) 'Frack off': Towards an anarchist political ecology critique of corporate and state responses to anti-fracking resistance in the UK. *Political Geography* 82: 102246.

Brock A and Dunlap A. (2018) Normalising corporate counterinsurgency: Engineering consent, managing resistance and greening destruction around the Hambach coal mine and beyond. *Political Geography* 62: 33–47.

Brock A and Stephens-Griffith N. (2021). Policing environmental injustice. *IDS Bulletin*, Online First. https://doi.org/10.19088/1968-2021.130

Büscher B and Fletcher R. (2018) Under pressure: Conceptualising political ecologies of Green Wars. *Conservation and Society* 16: 105–113.

Camp JT and Heatherton C. (2016) *Policing the Planet: Why the Policing Crisis Led to Black Lives Matter*. New York: Verso Books.

Cavanagh CJ and Benjaminsen TA. (2017) Political ecology, variegated green economies, and the foreclosure of alternative sustainabilities. *Journal of Political Ecology* 24: 200–341.

Césaire A. (2001 [1955]) *Discourse on Colonialism*. New York: NYU Press.

Churchill W. (2002 [1988]) *Agents of Repression: The FBI's Secret Wars Against the Black Panther Party and the American Indian Movement*. Cambridge, MA: SouthEnd Press.

Crook M and Short D. (2014) Marx, Lemkin and the genocide-ecocide nexus. *The International Journal of Human Rights* 18: 298–319.

Crook M and Short D. (2021) Introduction. *Journal of Genocide Research* 23: 155–161.

Crook M, Short D and South N. (2018) Ecocide, genocide, capitalism and colonialism: Consequences for indigenous peoples and glocal ecosystems environments. *Theoretical Criminology* 22(3): 298–317.

Daggett CN. (2019) *The Birth of Energy: Fossil Fuels, Thermodynamics, and the Politics of Work*. Durham: Duke University Press.

Davidov V. (2012) From colonial primitivism to ecoprimitivism: Constructing the indigenous "savage" in South America. *Arcadia* 46: 467–487.

Davis R and Zannis M. (1973) *The Genocide Machine in Canada*. Montreal: Black Rose Books.

Del Gandio J and Nocella II AJ. (2014) *The Terrorization of Dissent: Corporate Repression, Legal Corruption, and the Animal Enterprise Terrorism Act*. Brooklyn: Lantern Books.

Duffy R. (2016) War, by Conservation. *Geoforum* 69: 238–248.

Dunlap A. (2014) Permanent War: Grids, Boomerangs, and Counterinsurgency. *Anarchist Studies* 22(2): 55–79.

Dunlap A. (2017) Wind Energy: Toward a "Sustainable Violence" in Oaxaca, Mexico. *NACLA* 49(4): 483-488.

Dunlap A. (2018) The 'solution' is now the 'problem:' Wind Energy, colonization and the 'Genocide-Ecocide Nexus' in the Isthmus of Tehuantepec, Oaxaca. *The International Journal of Human Rights* 42: 550–573.

Dunlap A. (2019a) 'Agro sí, mina NO!' the Tía Maria copper mine, state terrorism and social war by every means in the Tambo Valley, Peru. *Political Geography* 71: 10–25.

Dunlap A. (2019b) *Renewing Destruction: Wind Energy Development, Conflict and Resistance in a Latin American Context*. London: Rowman & Littlefield.

Dunlap A. (2020a) Bureaucratic land grabbing for infrastructural colonization: Renewable energy, L'Amassada and resistance in Southern France. *Human Geography* 13: 109–126.

Dunlap A. (2020b) Wind, coal, and copper: the politics of land grabbing, counterinsurgency, and the social engineering of extraction. *Globalizations* 17: 661–682.

Dunlap A. (2020c) The direction of ecological insurrections: political ecology comes to daggers with Fukuoka. *Journal of Political Ecology* 27(1): 988–1014.

Dunlap A. (2021a) Spreading 'green' infrastructural harm: Mapping conflicts and socioecological disruptions within the European Union's transnational energy grid. *Globalizations*: 1–25, https://doi.org/10.1080/14747731.14742021.11996518.

Dunlap A. (2021b) Book Review: The political economy of agrarian extractivism: Lessons from Bolivia by Ben McKay. *Journal of Agrarian Change*: 1–6.

Dunlap A and Correa-Arce M. (2021) 'Murderous Energy' in Oaxaca, Mexico: Wind factories, territorial struggle and social warfare. *Journal of Peasant Studies*: 1–27.

Duran E and Duran B. (1995) *Native American Postcolonial Psychology*. Suny Press.

Dunlap A and Fairhead J. (2014) The militarisation and marketisation of nature: An alternative lens to 'climate-conflict'. *Geopolitics* 19: 937–961.

Dunlap A and Jakobsen J. (2020) *The Violent Technologies of Extraction: Political Ecology, Critical Agrarian Studies and The Capitalist Worldeater*. London: Palgrave.

Dunlap A and Sullivan S. (2020) A faultline in neoliberal environmental governance scholarship? Or, why accumulation-by-alienation matters. *Environment and Planning E: Nature and Space* 3(2): 552–579.

EC. (2020) *Critical raw materials for strategic technologies and sectors in the EU—A foresight study*. Available at: https://ec.europa.eu/docsroom/documents/42881.

Elliott-Cooper A. (2021) Futures of Black resistance: Disruption, rebellion, abolition. In: *Black Resistance to British Policing*. Manchester University Press, 164–183.

Fairhead JR. (2018) Technology, inclusivity and the Rogue: Bats and the war against the 'invisible enemy'. *Conservation and Society* 16: 170–180.

Fairhead J, Leach M and Scoones I. (2012) Green Grabbing: a new appropriation of Nature? *Journal of Peasant Studies* 39(2): 237–261.

Federici S. (2009 [2004]) *Caliban and the Witch: Women, The Body and Primitive Accumulation*. New York: Autonomedia.

Fisk R. (2007) *The Great War for Civilisation: The Conquest of the Middle East*: Vintage.

Foucault M. (1980 [1972]) On popular justice: A discussion with maoists. In: Gordon C (ed) *Power/Knowledge: Selected Interviews & Other Writings, 1972–1977*. New York: Pantheon Books, 1–36.

Foucault M. (1995 [1977]) *Discipline and Punish: The Birth of the Prison*. New York: Random House.

Foucault M. (2002 [1973]) Truth and judicial forms. In: Faubion JD (ed) *Power: Essential Works of Foucault, 1954–1984, Volume 3*. London: Penguin, 1–89.

Foucault M. (2003 [1997]) *"Society Must Be Defended:" Lectures at the College De France 1975–1976*. New York: Picador.
Foucault M. (2007 [1978]) *Security, Territory, Population: Lectures at the College De France 1977–1978*. New York: Picador.
Franquesa J. (2018) *Power Struggles: Dignity, Value, and the Renewable Energy Frontier in Spain*. Bloomington: Indiana University Press.
Galeano E. (1997 [1973]) *Open Veins of Latin America: Five Centuries of the Pillage of a Continent*. London: Monthly Review Press.
Gelderloos P. (2013) *The Failure of Nonviolence: From Arab Spring to Occupy*. Seattle: Left Bank Books.
Gelderloos P. (2017) *Worshiping Power: An Anarchist View of Early State Formation*. Oakland: AK Press.
Gelderloos P. (2020) *Debunking the myths around nonviolent resistance*. Available at: https://roarmag.org/essays/chenoweth-stephan-nonviolence-myth/.
Gibson KJ. (2007) Bleeding Albina: A History of Community Disinvestment, 1940–2000. *Transforming Anthropology* 15(1): 3–25.
Granovsky-Larsen S and Paley D. (2019) *Organized Violence: Capitalist Warfare in Latin America*. Regina: University of Regina Press.
Granovsky-Larsen S and Santos L. (2021) From the war on terror to a war on territory: Corporate counterinsurgency at the Escobal mine and the Dakota Access Pipeline. *Canadian Journal of Latin American and Caribbean Studies/Revue canadiennedes études latino-américaines et caraïbes*, 1–25.
Guardian. (2021) *The Guardian view on the heat dome: Burning through the models*. Available at: https://www.theguardian.com/commentisfree/2021/jul/08/the-guardian-view-on-the-heat-dome-burning-through-the-models.
Gudynas E. (2009) Diez tesis urgentes sobre el nuevo extractivismo. In: Schuldt J, Acosta A, Barandiarán A, et al. (eds) *Extractivismo, política y sociedad*. Quito: Centro Andino de Acción Popular (CAAP) and Centro Latinoamericano de Ecología Social (CLAES).
Gudynas E. (2021) *Extractivism: Politics Economy and Ecology*. Black Point: Fernwood Publishing.
Hanaček K, Kröger M, Scheidel A, et al. (2022) On thin ice—The Arctic commodity extraction frontier and environmental conflicts. *Ecological Economics* 191: 1–15.
Hickel J. (2020) *Less Is More: How Degrowth Will Save the World*. London: Random House.
Higgins, P. (2012) *Earth Is Our Business: Changing The Rules of the Game*. London: Shepheard-Walwyn Publishers.
Hochmüller M and Müller M-M. (2017) Countering criminal insurgencies: Fighting gangs and building resilient communities in post-war. In: Moe LW and Müller M-M (eds) *Reconfiguring Intervention: Complexity, Resilience and the 'Local Turn' in Counterinsurgent Warfare*. London: Palgrave Macmillan.

Holloway, J. (2010) *Crack Capitalism*. London: Pluto.
Hönke J and Müller M-M. (2016) *The Global Making of Policing: Postcolonial Perspectives*. Routledge.
Huff A. (2021) Frictitious commodities: Virtuality, virtue and value in the carbon economy of repair. *Environment and Planning E: Nature and Space*: 25148486211015056.
Huff A and Orengo Y. (2020) Resource warfare, pacification and the spectacle of 'green' development: Logics of violence in engineering extraction in southern Madagascar. *Political Geography* 81: 102195.
Huggins MK. (1998) *Political Policing: The United States and Latin America*. Durham: Duke University Press.
Huggins MK, Haritos-Fatouros M and Zimbardo PG. (2002) *Violence Workers*. Berkeley: University of California Press.
Isenberg AC. (2000) *The Destruction of the Bison: An Enviornmental Hisotry. 1750–1920*. New York: Cambridge University Press.
John H. (2010) *Crack Capitalism*. London: Pluto.
Kallianos Y, Dunlap A and Dalakoglou D. (Forthcoming) Introducing infrastructural harm: Rethinking moral entanglements, spatio-temporal modalities, and resistance(s). *Globalizations*.
Kaur R. (2021) Nuclear necropower: The engineering of death conditions around a nuclear power plant in south India. *Political Geography* 85: 102315.
Kitson, F. (1971). *Low Intensity Operations*. London and Boston: Faber & Faber.
Koppes CR. (1985) Agent Orange and the official history of Vietnam. *JSTOR*, 131–135.
Kovel J. (2007 [2002]) *The Enemy of Nature: The End of Capitalism or the End of the World?* London: Zed Books.
Kovel J. (2007) *The Enemy of Nature: The End of Capitalism or the End of the World?* London: Zed books.
Kröger M. (2021) *Iron Will: Global Extractivism and Mining Resistance in Brazil and India*. Ann Arbor: University of Michigan Press.
Kröger M. (2022) *Extractivisms, Existences and Extinctions: Monoculture Plantations and Amazon Deforestation*. Routledge.
Lang M and Mokrani D. (2013 [2011]) *Beyond Development: Alternative Visions from Latin America*. Amsterdam: Transnational Institute.
Liboiron M. (2021) *Pollution is Colonialism*. Durham: Duke University Press.
Linebaugh P and Rediker M. (2013) *The Many-Headed Hydra: Sailors, Slaves, Commoners, and the Hidden History of the Revolutionary Atlantic*. New York: Beacon Press.
Lubbers E. (2012) *Secret Manoeuvres in the Dark: Corporate and Policy Spying on Activists*. London: Pluto.

Lunstrum E. (2014) Green militarization: Anti-poaching efforts and the spatial contours of Kruger national park. *Annals of the Association of American Geographers* 104: 816–832.

McCoy KE. (2005) Trained to torture? The human rights effects of military training at the school of the Americas. *Latin American Perspectives* 32: 47–64.

Menton M and Le Billon P. (2021) *Environmental Defenders: Deadly Struggles for Life and Territory*. London: Routledge.

Merchant C. (1983) *The Death of Nature: Women, Ecology, and the Scientific Revolution*. New York: Harper & Row.

Morris B. (2015) *Anthropology, Ecology, and Anarchism: A Brian Morris Reader*. Oakland: PM Press, 89–93.

Moses AD. (2008) Empire, colony, genocide: Conquest, occupation, and subaltern resistance in world history. In: *War and Genocide*. Oxford: Berghahn.

Moses D and Stone D. (2013) *Colonialism and Genocide*. London: Routledge.

Neocleous M. (2014) *War Power, Police Power*. Edinburgh: Edinburgh University Press.

Nixon R. (2011) *Slow Violence and the Environmentalism of the Poor*. Cambridge: Harvard University Press.

Oltermann P. (2021) *At least 58 dead in Germany as heavy rains bring catastrophic flooding*. Available at: https://www.theguardian.com/world/2021/jul/14/firefighter-drowns-and-army-deployed-amid-severe-flooding-in-germany.

Paley D. (2014) *Drug War Capitalism*. Oakland: AK Press.

Pellow DN. (2014) *Total Liberation: The Power and Promise of Animal Rights and the Radical Earth Movement*. Minneapolis: University of Minnesota Press.

Peluso NL and Vandergeest P. (2011) Political ecologies of war and forests: Counterinsurgencies and the making of national natures. *Annals of the Association of American Geographers* 101: 587–608.

Peluso NL and Watts M. (2001) *Violent Environments*. Ithaca: Cornell University Press.

Perlman F. 2010 [1983]. *Against His-story*. Against Leviathan Detroit: Red and Black Press.

Pulido L and De Lara J. 2018. Reimagining 'justice' in environmental justice: Radical ecologies, decolonial thought, and the Black Radical Tradition. *Environment and Planning E: Nature and Space* 1(1–2): 76–98.

Redfield P. (2005) Foucault in the tropics: Displacing the panopticon. In: Inda JX (ed) *Anthropologies of Modernity: Foucault, Governmentality, and Life Politics*. Oxford: Blackwell Publishing.

Rodgers D and O'Neill B. (2012) Infrastructural violence: Introduction to the special issue. *Ethnography* 13: 401–412.

Rodney W. (2009 [1972]) *How Europe Underdeveloped Africa*. Washington, DC: Howard University Press.

Rostow W. (1960) *The Stages of Economic Growth: A Non-Communist Manifesto*. Cambridge: Cambridge University Press.
Sakolsky RB and Koehnline J. (1993) *Gone to Croatan: Origins of North American Drop Out Culture*. New York: Autonomedia.
Sale K. (2000[1991]) *Dwellers in the Land: The Bioregional Vision*. University of Georgia Press.
Scheidel A, Del Bene D, Liu J, et al. (2020) Environmental conflicts and defenders: A global overview. *Global Environmental Change* 63: 1–12.
Schrader S. (2019) *Badges Without Borders: How Global Counterinsurgency Transformed American Policing*. Berkeley: University of California Press.
Scott JC. (1998) *Seeing Like a State: How Certain Schemes to Improve the Human Condition Have Failed*. New Haven: Yale University Press.
Scott JC. (2009) *The Art of Not Being Governed: An Anarchist History of Upland Southeast Asia*. New Haven: Yale University Press.
Scott JC. (2017) *Against the Grain: A Deep History of the Earliest States*. New Haven: Yale University Press.
Scott DN and Smith AA. (2016) Sacrifice zones in the green energy economy: The new climate refugees. *Transnational Law & Contemporary Problems* 26: 371.
Seigel M. (2018) *Violence Work: State Power and the Limits of Police*. Duke University Press.
Shapiro J and McNeish J-A. (2021) *Our Extractive Age: Expressions of Violence and Resistance*. Oxon: Routledge.
Short D. (2016) *Redefining Genocide: Settler Colonialism, Social Death and Ecocide*. London: Zed Books.
Siamanta ZC. (2017) Building a green economy of low carbon: the Greek post-crisis experience of photovoltaics and financial 'green grabbing'. *Journal of Political Ecology* 24: 258–276.
Siamanta ZC. (2019) Wind parks in post-crisis Greece: Neoliberalisation vis-à-vis green grabbing. *Environment and Planning E: Nature and Space* 2(2): 274–303.
Simpson M. (2019) Resource desiring machines: The production of settler colonial space, violence, and the making of a resource in the Athabasca tar sands. *Political Geography* 74: 1–12.
Smith G. (2017) *The War and Environment Reader*. Washington, DC: Just World Books.
Sovacool BK. (2021) Who are the victims of low-carbon transitions? Towards a political ecology of climate change mitigation. *Energy Research & Social Science* 73: 1–16.
Springer S, Locret M, Mateer J, et al. (2021) *Anarchist Political Ecology*. London: Rowman & Littlefield.

Sullivan S. (2017) What's ontology got to do with it? On nature and knowledge in a political ecology of the 'green economy'. *Journal of Political Ecology* 24: 217–242.

Tahir M. (2019) Violence work and the police order. *Public Culture* 31: 409–418.

Temper L, Avila S, Del Bene D, et al. (2020) Movements shaping climate futures: A systematic mapping of protests against fossil fuel and low-carbon energy projects. *Environmental Research Letters* 15: 1–23.

Thompson K. (2018) Prisons, policing, and pollution: Toward an abolitionist framework within environmental justice. *PomonaSenior Theses* 185. Claremont CA: Pomona College

Thomas W. (1994) *Scorched Earth: Military's Assault on the Environment*. Philadelphia: New Society Publishers.

Trebitch L. (2021) *Resilience, Disaster, Elusion: The Life and Times Below John Adams, Part I*. Tacoma: A Boulder on the Tracks.

USAID. (2011) *A Field Guide for USAID Democracy and Governance Officers: Assistance to Civilian Law Enforcement in Developing Countries*. Available at: http://pdf.usaid.gov/pdf_docs/Pnadu808.pdf.

Vergara-Camus L. (2014) *Land and Freedom: The MST, the Zapatistas and Peasant Alternatives to Neoliberalism*. Zed Books Ltd.

Verweijen J and Dunlap A. 2021. The evolving techniques of social engineering, land control and managing protest against extractivism: Introducing political (re)actions 'from above'. *Political Geography* 83 (1): 1–9.

Verweijen J and Marijnen E. (2018) The counterinsurgency/conservation nexus: Guerrilla livelihoods and the dynamics of conflict and violence in the Virunga National Park, Democratic Republic of the Congo. *The Journal of Peasant Studies* 45: 300–320.

Vidalou J-B. (2017) *Être forêts: habiter des territoires en lute*. Paris: Zones.

Warren S. (2019) Borders and the freedom to move. *Dialouges in Human Geography* 9: 223–225.

Williams K. (2007 [2004]) *Our Enemies in Blue: Police and Power in America*. Cambridge: South End Press.

Williams K, Munger W and Messersmith-Glavin L. (2013) *Life During Wartime: Resisting Counterinsurgency*. Edinburgh: AK Press.

Wolfe P. (2006) Settler colonialism and the elimination of the native. *Journal of Genocide Research* 8: 387–409.

PART I

Hydrocarbon Militarization

CHAPTER 2

A Postcolonial History of Accumulation by Contamination in the Gulf

Michael Hennessy Picard and Tina Beigi

Introduction

A central premise of the book argues that one of the driving forces behind a world colonized by heavy toxic waste, ecocide and climate catastrophe is the dominant political and economic order shaped by multiple systems of oppression such as colonial domination, white supremacy, patriarchy, and military-industrial complex. This chapter turns to the history of the Gulf to illustrate this enduring dynamic of socio-ecological catastrophe. We claim that a century of military aggression and industrial extraction contributed to the asymmetric distribution of energy and toxic waste between the West and the Middle East.

M. H. Picard (✉)
International Environmental Law, University of Edinburgh Law School, Edinburgh, UK
e-mail: m.picard@ed.ac.uk

T. Beigi
McGill University, Montreal, QC, Canada
e-mail: tina.mohammadbeygy@mail.mcgill.ca

A. Dunlap and A. Brock (eds.), *Enforcing Ecocide*,
https://doi.org/10.1007/978-3-030-99646-8_2

Already, a rich historiography shows the systematic relationship between military destruction and oil extraction, where depletion of ecosystems above ground facilitates extraction of natural resources underground (Ross, 2017; Bigger & Neimark, 2017; Bonneuil & Fressoz, 2016). Conquest and plunder enable imperial powers to satisfy their voracious energy demands, but also deplete the ecosystem of their enemies' territories in the process. Therefore, a study of capital accumulation must compose with its corollary dynamic of ecology contamination.

In the age of engine combustion born out of the world's first oil-based military conflict of 1914–1918, Western imperial interests applied "brute force technologies"—such as aerial bombing—to destructive capacity, deliberately targeting the natural environments sheltering the enemy (Ediger & Bowlus, 2019; Winegard, 2016; Barr, 2012). As the discovery of oil combustion accelerated the speed, power, and performance of warships and aircraft, so did control over vast reserves of oil become a strategic war aim for Western powers, which eventually led them to detach Arabia from the Ottoman enemy to secure their grip over the riches of the Gulf (Black, 2004). By artificially propping up militarized petro-states in the old Ottoman provinces, the Western Powers depleted existing ecosystems, which had formed the material basis of social arrangements in the Gulf. For example, British naval operations disrupted local pearl-diving fisheries in the Arabian Gulf (Bowen, 1951). Then, when Western rivalry for oil peaked in response to military needs of the Second World War, the Gulf region became militarized once more with aircraft, ammunition, and biochemical weaponry to protect the complex network of pipelines and oil tankers scattered across the land (Russell & Tucker, 2004). Military rule and capitalist restructuring degraded the local environment and alienated people (Neocleous, 2011; Harvey, 2005). For example, the British Royal Air Force bombarded nomadic Bedouins to clear the way for pipelines (Batatu, 1982): labor was divorced from its means of subsistence and self-sufficiency made redundant (Marks, 2011).

According to Ecomarxist dependency theory, the intoxication of postcolonial peoples and contamination of their modes of social reproduction is the driving force of profit accumulation at the heart of the metabolic rift between core and peripheral states (Foster et al., 2011; Foster, 2000). Pollution weakens the possibilities of subsistence, let alone resistance of postcolonial peripheries, whose ecologies are degraded for purposes of metropolitan value extraction (Gordillo, 2014). Building on Ecomarxist theory, this chapter elaborates on one striking feature of Western imperialism in the Gulf, which is the dynamic process of *accumulation by contamination.*[1] Accumulation by contamination encompasses the wartime act of polluting a territory to clear the way for its exploitation, as well as the act of accumulating wealth by simultaneously disseminating waste on the conquered territory after major hostilities have ceased. Accumulation by contamination in the Gulf manifests through drilling and bombing, contaminating land and people to reproduce capitalist modes of accumulation. Extractive plunder is impossible without military coercion, which relies on the displacement of indigenous communities through acts of pollution and contamination (Yates, 2011; Liboiron, 2021). Both kinds of pollution—war pollution from above and oil pollution down below—converge around the capitalist necessity to feed productivity gains of the combustion engine, which propels war machines and irrigates global markets. Before turning to the history of oil wars in the Gulf, the following section elaborates on the dialectics of contamination by the war machine and the extractive regime.

Accumulation by Contamination in the Postcolonial Era

Appropriation and contamination form a double helix in the history of warfare: just as Tartars in medieval times catapulted diseased bodies of dead warriors over city walls to conquer them, so does modern warfare use a wide range of toxic chemicals and ammunition to flush out combatants from fortified positions, such as in the 2004 and 2006 battles of Fallujah (Tessier, 2007). Like most animals using excreta to appropriate places, human warfare deploys sophisticated techniques of contamination to expropriate enemies and colonize resources (Serres, 2010). The symbolic order echoes biophysical contamination by reversing causality: in justification for war in successive Gulf wars, Western powers allege that the contaminated victims are the carriers of the disease. Metaphors of hygienic cleansing litter the legal and military literature, ordering to "drain the swamp", "search and sweep", "contain", "purge", and legally

[1] Although we amend "accumulation by contamination" to adapt it to an Ecomarxist framework, the concept is taken from F. Demaria (2016).

transplant "democracy" to territories under occupation (Lakoff, 1991). Weaponizing nature is justified by naturalizing warfare.

Our case study on the postcolonial history of the Gulf starts at the end of the First World War. The 1919 Versailles conference formally prohibited colonial annexation and ushered in a new "postcolonial era", which placed under a regime of international supervision territories "inhabited by peoples not yet able to stand by themselves under the strenuous conditions of the modern world", as defined by Art. 22 of the Covenant of the League of Nations.

Along with innovative tools of social and legal engineering, industrial efficiency in oil-run aerial and naval warfare allowed the British Empire, acting as a mandated power, to clear Mesopotamia for extraction. During wartime, shortly after the campaign against the Ottomans, the British bombarded the lands of the *fellahin* (traditional peasantry) in Kut, Amhara and Bagdad (Batatu, 1982), effectively expropriating them. This paved the way for "peacetime" oil extraction under the League of Nations mandate (1920–1932). Throughout the 1920s and early 1930s, military bombing threatened the livelihood of Bedouins, Assyrians, and Kurds, forced into accepting industrial drilling within artificially created State boundaries (Silverfarb & Khadduri, 1986; Thomas, 2003). In Kirkuk and Mosul, "immense palls of smoke rose heavenwards from the oilfields at frequent intervals" (Beeby-Thompson, 1961). Recalling the analogy of the "oil curse as the devil's excrement",[2] British extraction appropriated the Gulf by smearing it with excess fires and leaking pipelines, creating huge clouds of putrid smoke asphyxiating vegetation in the process (Ross, 2017). Beyond mere "negative externality", pollution is a colonial agent sweeping over land to clear the ground for further conquest and economic exploitation, long after the official cessation of hostilities (Colgan, 2013).

It is worth mentioning at this stage that the militarized petro-State formations propped up by Western legal engineering in turn used ecological degradation as a weapon against their indigenous population upon achieving nominal sovereignty. Accumulation by contamination is intertwined with endogenous State reproduction and shall not be solely attributed to regional relations of subordination to foreign imperial relations. Because of increasing regional State rivalry, Bahrain, Kuwait, Oman, Saudi Arabia, Yemen, Qatar, Iraq, and the United Arab Emirates have

[2] The quote is attributed to Venezuela Minister of mines Juan Pablo Perez Alfonzo in 1976 (see Tinker Salas, 2009).

suffered contamination by heavy toxic metal, and petroleum hydrocarbon. In the aftermath of the 1991 Gulf War, for example, oil field fires have chronically contaminated coasts, seas, soil, and air (Freije, 2015). As much as wealth from oil extraction is unevenly distributed, the pollution from oil wars is disproportionally dumped on vulnerable and marginalized communities in the region. Along the Tigris river, in Basra and in the Shatt-al-Arab, the prevalence of disease is especially high among the impoverished and malnourished, which are exposed to water sources contaminated with mercury, arsenic, lead, cobalt, cadmium, petroleum products, oil, soot from oil fires, and depleted uranium (Zolnikov, 2013).

Tied in with the importance of the war machine is the role of oil extraction for the system of accumulation by contamination. The environmental legacy of oil extraction in the Gulf reveals the extent to which the petroleum industry is responsible for global contamination levels. Oil cartels are linked to 71% of industrial greenhouse gas emissions since 1988 (Griffin & Heede, 2017). Over half of global industrial emissions since 1988 can be traced to twenty-five corporate and state producers (ibid.). The state producers are mostly located in the Gulf region, whereas the major corporate cartels, such as ExxonMobil, Shell, and BHP Billiton, are headquartered in the West. While oil wealth accumulates in vertically integrated companies, oil pollution spreads horizontally among marginalized communities. Major impacts of oil exploration, drilling, and extraction include deforestation, ecosystem destruction, chemical contamination of land and water, long-term harm to animal populations (particularly migratory birds and marine mammals), human health and safety risks for neighboring communities and oil industry workers, and displacement of communities (O'Rourke & Connolly, 2003; Mendelssohn, 2012).

Historically, the polluting effects of accumulation by contamination have been disproportionally experienced in the peripheries. Political ecologists call this plundering and degradation of poorer countries' natural resources, which cause an asymmetric global distribution of wealth and waste, an *ecological debt* (Paredis, 2009). Ecological debt is an indicator of the cumulative historical socio-ecological subsidy "paid" by the peripheries necessary to maintain the core's industrial techno mass (Warlenius et al., 2015). Research on total energy and material consumption shows that core regions within the world economy have significantly higher "metabolic" rates than peripheral regions, yet consistently displace their environmental load on the latter (Singh & Kennedy, 2015).

One country within the localized periphery of the Gulf has been particularly subjected to this ecological debt: Iraq. Iraq has become an extractive wasteland through a century-long accumulation of extractive policies for the benefit of Western corporations. Coercive military control over oil deposits contributed to the asymmetric stock distribution of energy for the West and toxic waste for the Iraqi soil, peoples, and cultures.

Iraq's Postcolonial History of Accumulation by Contamination

Iraq's sovereign debt rose in tandem with the ecological debt dumped on it by the West, revealing the inextricable relationship between the political ecology of oil extraction and the political economy of State building (Iraq Administration Reports 1914–1932, 1992). As Western extractive industries transformed the ecology of the Gulf region, so did Anglo-Saxon legal transplants transform rules of property and interest. By the time British-appointed King Faisal ascended the throne of Iraq in 1925, financial speculation was already funding the construction and maintenance of environmentally costly army bases, roads, railroads, pipelines, canals, residences, and embassies (Black, 2004). Such capital-intensive projects weighed heavy on the nascent State, which was to repay infrastructure costs with future oil revenue (Geoff & Phillip, 2011).

Rising military and infrastructure costs had to be reimbursed by collecting taxes. Those who resisted the taxing and drilling down below were subjected to military repression from above, under the euphemistic policy of "morale bombing", which was the weapon of choice of the British to squash Arab rebellion in Iraq (Cox, 1985). Bombing and drilling shaped and transformed the entire system of property rights in the Gulf, but also its ecology (Grove, 2019). With the full military support of the British Royal Air Force, foreign oil cartels and agricultural engineers restructured the land to guarantee extraction and tax revenue (Jones, 1977). The mechanized oil boom transformed social relations and fixed new boundaries, incorporated a local landed class into the global economy, while excluding and inflaming the resistance of the wider community, which became bound by indenture on the estates of a landed ruling minority.

Postwar Iraq was essentially controlled by a new military regime of panoptical vision to protect the new boundaries of oil concessions, which terrorized rural populations in the name of foreign oil extraction (Iraq Administration Reports 1914–1932, 1992). The British Royal Air Force became the bailiff of powerful oil men, revealing the symbiotic relationship between the military and the industry (Sluglett, 2007; Meilinger, 2017). Winston Churchill, then Colonial Secretary and Hugh Trenchard, founder of the Royal Air Force, waged a deadly and toxic bombing campaign, which satisfied the need to protect oil field exploration from Bedouin looting activity over a vast and remote territory. Striped of their means of survival, Bedouin tribes flocked to privately owned estates and to urban areas, looking for work.

In the 1920s, fumigating air raids turned the tribes into servitude on the estates of a landed ruling minority backed by the British military. One such onslaught was launched by the Royal Air Force in 1923–1924 in Southern Iraq, where the peasants and nomadic tribesmen from the Euphrates refused to pay up taxes to the tribal leaders responsible for collecting them. Later, Air Force operations raided the Kurds and Assyrians, confining them to an insular existence within the confines of State borders (Omissi, 1989).

While a select group of indigenous rulers became landed aristocrats, the British, US, French, and Dutch oil company shareholders monopolized underground property rights. From the beginning of Iraq's history, the uneven distribution of capital created ostentatious wealth on one side of the spectrum, and "superfluous" or "disposable" categories of population on the other. In the new Kingdom, the old Ottoman bureaucracy was replaced by British administrators: only 3.74% of civil servants were Arabs, the rest were members of the British imperial service (Ireland, 1970). Annual reports to the League of Nations demonstrate that Britain held the financial authority to control the flow of capital and establish sovereign debt repayment schemes over Iraq for the purpose of accumulation (Iraq Administration Reports 1914–1932, 1992).[3]

[3] See Report by H.M. Government to the Council of the League of Nations on *the Administration and Progress in Iraq during the Period 1920–1931*, Colonial, no 58, H.M.S.O., London, 1931; British embassy in Iraq, "Annual Report on Iraq for 1933". 28 March 1934, FO 371/17871, E2204/2204/93; Newton to foreign office, 10 June 1940 and 19 October 1940, FO 371/24556, E2198/E2913/203/93.

Extraction Rebellion

As bombing campaigns progressively dislodged the local population, the indignity of British tactics ignited armed resistance across the country. In June 1920, a coalition of disgruntled tribal sheikhs, religious dignitaries, and vociferous nationalists rebelled against the British policy of extraction and displacement. In "the Year of the Catastrophe" (*Am al-Nakba*), the "Awakening" (*Thawra*) was considered the catalyst of Arab nationalist sentiment against accumulation by contamination (Batatu, 1982).

In a coordinated effort to disrupt the enemy's objectives, nationalists targeted the new transportation routes and building sites. Raids on British lines of communication increased. Trains were looted. Petrol dumps blew up (Black, 2004). However justified the 1920 revolt may have been against the perceived expropriation without compensation, the endogenous violence inflicted an additional wave of pollution upon the land. Iraqis targeted the industrial infrastructure and military equipment imported by the occupying power, aggravating heavy metal land contamination.

Encouraged by the wave of nationalism, Iraqi delegates requested the election of a Convention and appealed for a united Arab Government elected by universal suffrage.[4] Instead, the Royal Air Force used "aerial policing" to level whole villages. "With most of the leaders under arrest or in exile, the tribes and towns of southern Iraq submitted to British authority" (Yaphe, 2010). In the end, airstrikes and military blockades succeeded in enclosing the agricultural workforce within landed estates modeled on English aristocratic domains (Thomas, 2003). Once the revolt had been crushed, the British High Commission established a network of military airbases across the Gulf, backed by a local central authority capable of protecting the flow of crude across large stretches of sand. Processing and transporting the swelling flow of crude generated a range of impacts on nearby ecosystems. In 1927, oil drilling in Kirkuk hit a gusher flowing uncontrollably for a week, pouring 95,000 barrels per day into the surrounding environment (Bamberg, 1994). Once struck, oil was then carried off by tankers on the Shatt-al-'Arab channel, which was deepened and widened to accommodate crude exports, disturbing in the

[4] Acting Civil Commissioner. Review of the Civil Administration of Mesopotamia to His Majesty's Government, Indian Office, December 3, 1920 [Cmd. 1061], 141.

process the "feeding and spawning grounds of important commercial fish and shrimp species" (Ross, 2017).

The cycle of coercive oil extraction and ecological impoverishment produced political instability in the Kingdom, which is revealed by the fact that fifty-eight governments succeeded each other between 1921 and the nationalist revolution of 1958 (Catherwood, 2005). By the time the British Mandate was over in 1932, vast tracts of Iraqi land had become militarized to protect the growing oil revenue of the State. The legacy of the British mandate in Iraq created an atmosphere of suspicion, betrayal, and revolt (Farouk-Sluglett & Sluglett, 1987). Soon enough, rival military factions fought for the control of the State.

The dazzling spectacle of oil extraction and bomb explosions created an endless appetite for power among rival factions, which claimed to offer a postcolonial alternative, while effectively replicating the hegemonic military structure of government. By using oil as a revenue stream for political repression instead of wealth redistribution, the Iraqi State mirrored the very process of accumulation by contamination established by British indirect rule (Dodge, 2003). While military officers controlled oil revenue in Baghdad, a landed aristocracy ruled over enclosed agricultural estates in the countryside. Because of this "great transformation" (Polanyi, 1944) where Iraqis became part of a market society, farmers were not only economically dispossessed from their land by accumulating sheikhs, they were also "contaminated" by new social arrangements. Although the entire society transformed, the shifting political economy of the Oil Kingdom weighed most heavily on the poor, such as small-scale farmers. Between 1932 and 1958, the State relegated the "superfluous categories" of peasants and farm laborers to the slums of urban centers, ravaged by water pollution, trachoma, and dysentery. While oil revenues kept coffers filled to the brim, the military regime was incapable of providing elementary social services (Dann, 1970). Following exogenous extraction by explosion, the endogenous exploitation of oil resources led to a dynamic of social exclusion and fragmentation.

The Saddam Years: War Pollution by Oil Accumulation

By 1958, Iraq had entered a new historical phase, characterized by the internal adoption of a violent cycle of accumulation by contamination. Since its cartel origins, the country had been an inherently weak client

State ruled by a military minority elite prone to corruption and authoritarian rule (Farouk-Sluglett & Sluglett, 1987). Under the new nationalist regime of 1958, the State's petroleum assets provided vast powers of patronage to a military elite (Tripp, 2007). Repeating the criminal foundations of the State, rival military factions fought for the control of its coercive apparatus (Tilly, 2017). The result was a succession of coups and countercoups between opposing kin-based alliances (1963, 1968) to control and redistribute oil revenue among their respective client networks (Tripp, 2007).

In 1968 the Baath Party ruthlessly emerged as the victor of the political struggle for the control of Iraq's extractive economy. Under the growing influence of Saddam Hussein, the nationalized oil industry (1972) became a "slush fund" for high officials within the State apparatus (Williams, 2009). In an unprecedented escalation of violence, Iraqi oil greased the process of accumulation by contamination.

Under the Baath regime of Saddam Hussein, Iraq's oil revenue stream reinforced the criminal foundations of the Iraqi State. Although Hussein's regime used the rent from the Iraq National Oil Company to fund industrialization and educational reforms—women literacy, for instance—, oil revenue was diverted to the acquisition of foreign military equipment. As we shall see, Iraq's internal capital accumulation ultimately led to regional ecological contamination at the expense of equal resource distribution among the population.

Hussein's fear of an internal coup was partly deflected by the projection of violence outward, during the 1980–1988 war against Iran, and in 1990–1991 against Kuwait. These wars fueled by oil revenue targeted industrial and military sites, armaments factories and oil refineries, which led to acute chemical pollution over the course of successive airstrikes. The endogenous process of accumulation would unleash a wave of military contamination, which peaked against the Kurdish people, collectively punished for siding with Iran. In 1988, the Kurds were targeted by artillery shells and airstrikes in the city of Halabja and gassed with nerve agents and mustard gas. As many as 5,000 Iraqi Kurds, mostly women and children, were killed. Decades after the attack, unexploded shells and residue from the gas that spread over the city still cause congenital defects (Kelly, 2007; Mlodoch, 2017).

The Iran–Iraq War

The Iran–Iraq war of 1980–1988 incurred as many as 1.5 million casualties (Ferretti, 1990). Belligerents on both sides used Western military equipment purchased with their national oil revenue. Environmental damage inflicted by the war is scattered and inconclusive, because of a general lack of concern for monitoring or clean-up (Walker, 1989). Some effects are known, such as the fact that ground battles and aerial bombardments caused extensive forest destruction and soil erosion. Tar and asphalt dumped on the coastal region between Abadan and the strait of Hormuz posed a great threat to already endangered species. Leaks from oil tankers in the Gulf are believed to be the cause (ibid.). The bombing of oil platforms polluted the Gulf, while sunken ships and bombed wrecks have contaminated the Shatt-al-Arab waterway, threatening its ecosystem and the fishing industry.

The impact of war on farmland was equally devastating: in Kermanshah, the conflict contaminated more than 300,000 hectares of irrigated farmland (UN Secretary-General, 1991). Millions of date palms and "5,000 hectares of orchards were destroyed, some 130,000 hectares of natural forest and 753,000 hectares of pasture land in the war-afflicted provinces were also rendered unusable (Amirahmadi, 1996). All five Iranian provinces impacted by the war appeared to be contaminated" by toxic materials emanating from chemical and biological weapons. The situation was compounded by soil compaction, flooding, and salinization where irrigation canals were destroyed (ibid.). On coastal strips and in mainland waterways, military waste destroyed the prawn-fishing industry and intoxicated the rural population. As a result of war, studies have shown a higher rate of disease incidence, such as eye infection, skin ailments, stomach illness and acute respiratory disease (ibid.). Since the end of the war, there has been an alarming increase in health-threatening insects and pests, in the Karoun river in the South-western province of Iran (Nasirian, & Irvine, 2017). Finally, unidentified minefields and unexploded war materials demonstrate the enduring sanitary impact of the Iran–Iraq war (UN Secretary-General, 1991).

The 1991 Gulf War

A mere two years after the end of the Iran–Iraq war (August 1988), the Gulf War (August 1990–February 1991) sparked a deadly combination of air power from above and oil sabotage down below. Due to the overwhelming fire storm of Western air power (Hallion, 1992; Sadiq & McCain, 1993; Smith, 2017), the Iraqi forces opened oil valves of the Sea Island pipeline, releasing oil from numerous tankers, oil lakes and fire trenches, as part of a scorched earth policy in a desperate retreat from Kuwait in 1991. The goal of the spill was to impede Coalition troops from attempting beach landings, but ultimately the spill could not reverse the outcome of war. Over 240 million gallons of crude oil were dumped into the Persian Gulf (Hawley, 1992). For the first time on a regional scale, oil pollution was used as a tactic of war and devastated the biodiversity of uninhabitable coastlines.

In a twisted reversal of the logic of accumulation by contamination, Iraq's military resorted to contamination by retreat. When the war aims of securing more oil wells are unsuccessful, sabotage during withdrawal becomes a policy of last resort. Oil spillage in the Persian Gulf tarred beaches and killed more than 25,000 birds, whereas oil spilled on land formed huge pools in lowlands, covering fertile croplands. They turned fields untillable, which led to food shortages. The fires released nearly half a billion tons of carbon dioxide, the leading cause of global warming, emissions greater than all but the eight largest polluting countries for 1991 that will remain in the atmosphere for more than a century. The oil that did not burn in the fires traveled by air in the form of nearly invisible droplets resulting in an oil mist or fog that poisoned trees and grazing sheep, contaminated fresh water supplies, and found refuge in the lungs of people and animals throughout the Gulf (Carr, 2007).

Following the Gulf War, Iraqi Shia in the South rebelled in March 1991. The uprising was crushed by the Iraqi Government, which launched a brutal campaign to drain the marshes of southern Mesopotamia and economically siege a previously self-sufficient population. "The state used hydrological infrastructure to divert water from the wetlands, permanently desiccating the area" (Ahram, 2015). Here again, the State coercively degraded the environment to expropriate peoples and restructure their land to better submit them to State discipline. Between 1991 and 1997, a system of dams, dikes, and canals was built to turn the wetlands into dry, salty lands. Once drained, the land would be leased to coopted tribal leaders for commercial agriculture and oil exploration, making survival contingent on cooperation with the State. The genocide and ecocide against the Marsh Arabs and their lands is an ongoing process. The latest contribution to this dynamic is the partnership between conservation NGOs and the oil industry on marsh "restoration projects". In adherence to neoliberal conservation practices, these projects have ensured the continuation of the Marsh Arabs displacement and social alienation from the lands (Priestley, 2020).

The 2003 Iraq War

The 2003 invasion of Iraq by the forces of the Coalition removed Hussein from power after his twenty-four-year rule. While mirroring previous practices of accumulation by contamination, the Iraq War displayed the use of specific weapons responsible for environmental and sanitary damage. The indiscriminate use of prohibited chemical weapons and toxic gases may have contributed to the high percentage of civilian casualties. Reports indicate that women and children mortality rates exploded 50-fold since the US invasion and bombardment campaigns (Burnham et al., 2006). During the November 2004 battle of Fallujah, the US army had recourse to highly toxic white phosphorus to clear the city of insurgents (Tessier, 2007). The chemical agent was also reported to have directly affected civilians in the densely populated areas of Nasariyah, Fallujah, and Baquba (ibid., 355).

In a 2005 report entitled "Assessment of environmental hot spots in Iraq", the United Nations Environment Programme estimated that industrial and military pollution contaminated ten sites with high levels of radioactive waste and forty-two sites with dioxin and depleted uranium (UNEP, 2005). Depleted Uranium (DU) used by Coalition forces in 2003 is a heavy metal particularly favored by the military industry for its penetrating properties of armored equipment. Previously used during the 1991 Gulf War, DU widely spreads in the air, soil, and water, particularly in dust storms over dry landscape. An estimated 250.000 to 300.000 small-caliber munitions were shot for every Iraqi insurgent killed in the Iraq War. When the hardened shell casings of ammunition explode, their toxic components contaminate soil and water (Al-Azzawi, 2006). "Between 1.000 and 2.000 tons of toxic and radioactive depleted uranium [...] have been used in Iraq by American and British forces during the war" (Levy & Sidel, 2008). The high prevalence of radioactive and toxic uranium in Iraqi soil and infrastructures constitutes a widespread and long-lasting threat to the health of the Iraqi population, which has been plagued by a high rate of cancers and birth defects (Fathi et al., 2013). Still today, decontamination of depleted uranium requires the removal of contaminated soil and its treatment as radioactive waste (Al-Sabbak et al., 2012).

The Occupying Power explicitly pleaded military necessity to justify the destruction of the enemy's environment. In 2003, on the grounds of "national security", the United States President Bush signed Executive Order 13303 granting all US entities operating in Iraq immunity from legal proceedings.[5] Legal immunity for military contamination enabled the US "shock doctrine" to establish a regime of accumulation for the corporate members and political allies of the Coalition. The Occupying Power siphoned 90% of the Development Fund for Iraq (made of frozen assets and oil revenue from the previous regime), by awarding 74% of contracts to US firms such as Bechtel (electricity), Halliburton (logistical support), Dyn-Corp, Vinnell and USIS (security firms and defense), Creative Associates (education), and Research Triangle Institute (local democracy). Only 2% of contracts were awarded to Iraqi companies. Refusing to hire Iraqi nationals for security reasons, the Department of Defense outsourced its labor tasks to private security and service companies such as Kellogg, Brown & Root, and Blackwater Worldwide. Shortly after the transfer of the Fund to the CPA President Bush signed Executive Order 13303 granting all US entities which were awarded payment under the Fund immunity from legal proceedings.[6] The Occupying Power also destroyed national public monopolies and replaced them with private extractive activities of foreign corporations. A dozen rounds of oil and gas licensing bids took place during the occupation campaign of

[5] Executive Order 13303, "Protecting the Development Fund of Iraq and Certain other Property in Which Iraq Has an Interest", Report on the National Emergency with Respect to the Development Fund for Iraq, Executive Order 13303, consistent with Section 401(c) of the National Emergencies Act, 50 U.S.C. 1641(c), and Section 204(c) of the International Emergency Economic Powers Act, 50 U.S.C. 1703(c), 22 May 2003.

[6] Coalition Provisional Authority Order Number 17 (Revised), "Status of the Coalition Provisional Authority, MNF – Iraq, Certain Missions and Personnel in Iraq", CPA/ORD/27 June 2004/17, Article 1, Section 4: "the multinational force, foreign liaison missions, their personnel, property, funds and assets and all international consultants shall be immune from Iraqi legal process".

2003–2011, awarding contracts to foreign investors and contractors, such as Halliburton, Baker Hughes, Weatherford International, and Schlumberger, which won the largest portion of the subcontracts to drill for oil, build wells and refurbish old equipment. In conclusion, the US-led process of accumulation by contamination further accentuated the asymmetric distribution of energy and toxic waste between corporate bodies and social metabolisms in Iraq.

The Rise of ISIL and Contamination by Repetition

By the time the US left Iraq in December 2011 in compliance with the terms of a bilateral *Status of Forces Agreement*,[7] the Pentagon had sold the Iraqi Defense Ministry $1.3 billion in tanks, helicopters, planes, and guided missiles (James & Nahory, 2013). The US also spent $1.4 billion of Iraqi treasury funds to finance the Ministry of Interior's secret prison program, train militias, and arm the new police force (Perito, 2007; Sedra, 2007). As a result, al-Maliki's Government of Iraq leveraged the new security apparatus to marginalize, arrest, and torture Sunni elected officials.[8] This power imbalance, sowing the seed of Sunni resentment, coincided with the growing instability in Syria, which ultimately led to the rise of the Islamic State of Iraq & the Levant (ISIL). Once the US Army departed, ISIL rose out of the rubble to become a profitable multinational oil business operation. ISIL was "adept at exploiting decades-old transnational gray markets for oil and arms trafficking" (Weiss & Hassan, 2015).

Because they stretch across desert land, pipelines are easily tapped into. Once tapped, the oil can be "bunkered" into tanks and sailed off into the Gulf. Resource extraction and armament depot looting provided ISIL with the material arsenal necessary to back its political claims (Malik, 2015). ISIL grew stronger by reviving old ties on a deregulated oil market (Weiss & Hassan, 2015). Between 2011 and 2016, ISIL orchestrated

[7] See "Agreement Between the United States of America and Republic of Iraq On the Withdrawal of United States Forces from Iraq and the Organization of Their Activities during Their Temporary Presence in Iraq", Mason, R. C. (2009, July) 'US-Iraq Withdrawal/Status of Forces Agreement: Issues for Congressional Oversight', Library of Congress, Washington DC. Congressional Research Service.

[8] Maliki purged Government ministries of their Sunni representatives, such as Vice President Tariq al-Hashemi and Sunni Finance Minister Rafi al-Issawi, who were charged with having links to terrorism (See Katzman, 2015).

attacks by funneling cash, arms, and oil through similar channels of undercover networks as those used by the Baath shadow State to smuggle oil out of Iraq during the US embargo of 1991–2003 (Natali, 2015).

By seizing oil wells and military equipment left behind by the American occupation, ISIL reproduced the accumulation by contamination of the occupying forces (Berger & Stern, 2015). ISIL acquired the extractive means to finance war-making, thereby adding another layer of rubble and dust to the ground they regained among the exploited people they pledged to protect: this pollution has come to be referred to as "the ISIL winter", which depicts the toxic fallout of three years of armed conflict (Almohsen, 2018). By detonating oil wells and torching sulfur plants across Iraqi provinces to fight against Government security forces, ISIL orchestrated environmental sabotage. "The burning of the [Mishraq Sulphur Plant] was a real case of using environmental damage as a weapon of war" (Zwijnenburgn & Postma, 2017).

Since many ISIL commanders had formerly served under the Baath regime of Hussein, they were familiar with the scorched earth tactics used during the retreat from Kuwait. The wreckage of twenty-five oil wells in Qayyarah provoked thick blinding smoke clouds stretching over tens of kilometers, turning people's skin and sheep's coats black from soot. This toxic legacy includes wide-scale cattle deaths, fields that no longer yield edible crops, and chronic breathing complications in children and the elderly. As a result, over 1,500 people were reportedly treated for suffocation in the Qayyarah, Makhmour, and Ijhala according to the Ministry of Health and the WHO (Weiss & Hassan, 2015). Heavily polluted waterways in the Basra region led to the collapse of agriculture and the displacement of entire communities from rural areas. As they flee, refugees in the city of Basra settle in severely polluted shanti towns, which pump their water in the Shatt-al-Arab river, now littered with debris, bacteria, chemicals, and salt (Guiu, 2020).

In the end, ISIL used oil both as a revenue stream and as an environmental weapon. The insurgency mirrored the very process of accumulation by contamination, which had been adopted by its enemy, the State. The fight by Government forces to regain control over ISIL-conquered territory, as well as the Coalition's systematic bombardment of Mosul, fueled another round of contamination, leaving behind a trail of blood and rubble in the ancient city (Knights & Mello, 2017).

Conclusion

A century of accumulation in the Gulf ultimately contaminated the human ecology of the region. This contamination was the product of the dynamic relationship between firebombing from above and oil drilling from below. First, wartime pollution contaminated modes of existence among pearl divers, fishers, farmers, and nomadic merchants throughout the twentieth century. Second, peacetime oil exploitation contaminated self-sufficient modes of social reproduction and replaced them with capitalist modes of accumulation, with far-reaching consequences for the Gulf's ecological conservation in the twenty-first century. Both kinds of contamination—war pollution from above and oil pollution down below—are products of capitalist accumulation, which propels war machines and irrigates global markets.

Over the course of a century, British "indirect rule" and American "regime change" secured oil concessions and military alliances, using war as a vehicle for fossil fuel accumulation. In turn, oil was harnessed as a political weapon for war making and State building, and was often consciously mobilized as a weapon of ecological warfare, especially by local actors, who ultimately replicated the hegemonic model of accumulation by contamination.

The Gulf is today one of the most militarized regions of the world, since oil extraction offered ceaseless occasions for military spending and war profiteering. As much as slaves working in plantations contributed to primary accumulation by dispossession in the colonial era (Beckert, 2015), planes over pipelines contributed to the hubris of accumulation by contamination in the postcolonial world.

With the discovery of new petroleum reserves on other continents and offshore, the oleaginous accumulation by contamination initiated a century ago in the Gulf has now colonized the planet. Our Ecomarxist study of the Gulf could find resonance elsewhere, in extractive regions of Latin America and Africa, which have come to symbolize the new geographies of dirty wars and mineral plunder. The primitive accumulation by contamination is in a process of constant relocation, revealing new actors. The Far East is now land-grabbing and drilling in the postcolonial world (Arboleda, 2020), while the West is pumping and fracking its own underground, with devastating effects on the quality of the water tables. As Michel Serres (2010) argued in his essay *Malfeasance: Appropriation Through Pollution?* globalization may paradoxically lead to worldwide dispossession of a polluted earth: *Res Nullius Mundus*.

Bibliography

Ahram, A. (2015). Development, counterinsurgency, and the destruction of the Iraqi marshes. *International Journal of Middle East Studies*, 47(3), 447–466.

Al-Azzawi, S. (2006). Depleted Uranium Radioactive Contamination in Iraq: An overview. *Global Research*, 1(4).

Almohsen, R. (2018, January 31). *The ISIS winter: The environmental impact of Middle East conflict*. The Open University. https://www.open.edu/openlearn/nature-environment/environmental-studies/the-isis-winter-the-environmental-impact-middle-east-conflict

Al-Sabbak, M., Ali, S. S., Savabi, O., Savabi, G., Dastgiri, S., & Savabieasfahani, M. (2012). Metal contamination and the epidemic of congenital birth defects in Iraqi cities. *Bulletin of Environmental Contamination and Toxicology*, 89(5), 937–944.

Amirahmadi, H. (1996). Iranian recovery from industrial devastation during war with Iraq. In J. Mitchell, *The long road to recovery: Community responses to industrial disaster*. New York: United Nations University Press.

Arboleda, M. (2020). *Planetary mine: Territories of extraction under late capitalism*. Verso.

Bamberg, J. H. (1994). *The History of the British Petroleum Company, The Anglo-Iranian Years 1928–1954*, vol. 2. Cambridge: Cambridge University Press, 157–60.

Barr, J. (2012). *A Line in the Sand: The Anglo-French Struggle for the Middle East, 1914–1948*. WW Norton & Company.

Batatu, H. (1982). *The old social classes and the revolutionary movements of Iraq: A study of Iraq's old landed and commercial classes and of its Communists, Baʿthists, and Free Officers*. Princeton University Press.

Beeby-Thompson, A. (1961). *Oil pioneer: Selected experiences and incidents associated with sixty years of world-wide petroleum exploration and oilfield development*. Sidgwick & Jackson.

Beckert, S. (2015). *Empire of cotton: A global history*. Vintage.

Berger, J. M., & Stern, J. (2015). *ISIS: The state of terror*. New York: HarperCollins.

Bigger, P., & Neimark, B. D. (2017). Weaponizing nature: The geopolitical ecology of the US Navys biofuel program. *Political Geography*, 60, 13–22.

Black, E. (2004). *Banking on Baghdad: Inside Iraq's 7,000-year history of war, profit, and conflict*. Wiley.

Bonneuil, C., & Fressoz, J. B. (2016). *The shock of the Anthropocene: The earth, history and us*. Verso Books.

Bowen, R. L. (1951). The pearl fisheries of the Persian Gulf. *Middle East Journal*, 5(2), 161–180.

Burnham, G., Lafta, R., Doocy, S., & Roberts, L. (2006). Mortality after the 2003 invasion of Iraq: A cross-sectional cluster sample survey. *The Lancet*, 368(9545), 1421–1428.

Carr, P. (2007). Shock and awe and the environment. *Peace Review*, 19(3), 335–342.

Catherwood, C. (2005). *Churchills Folly: How Winston Churchill Created Modern Iraq*. Basic Books.

Colgan, J. D. (2013). *Petro-aggression: When oil causes war*. Cambridge University Press.

Cox, J. L. (1985). A splendid training ground: The importance to the Royal Air Force of its role in Iraq, 1919–32. *The Journal of Imperial and Commonwealth History*, 13(2), 157–184.

Dann, U. (1970). *Iraq under Qassem: A political history, 1958–1963*. Praeger.

Demaria, F. (2016). Can the poor resist capital? Conflicts over accumulation by contamination at the ship breaking yard of Alang (India). *Nature, Economy and Society*, 273–304.

Dodge, T. (2003). *Inventing Iraq—The failure of nation-building and a history denied*. Oxford: Oxford University Press.

Ediger, V. Ş., & Bowlus, V. (2019). A farewell to King Coal: Geopolitics, energy security, and the transition to oil, 1898–1917. *The Historical Journal*, 62(2), 427–449.

Farouk-Sluglett, M., & Sluglett, P. (1987). *Iraq since 1958: From revolution to dictatorship*. KPI.

Fathi, R. A., Matti, L. Y., Al-Salih, H. S., & Godbold, D. (2013). Environmental pollution by depleted uranium in Iraq with special reference to Mosul and possible effects on cancer and birth defect rates. *Medicine, Conflict and Survival*, 29(1), 7–25.

Ferretti, M. J. (1990). The Iran-Iraq War: United Nations Resolution of Armed Conflict. *Villanova Law Review*, 35, 197.

Foster, J. B. (2000). *Marxs ecology: Materialism and nature*. NYU Press.

Foster, J. B., Clark, B., & York, R. (2011). *The ecological rift: Capitalisms war on the earth*. NYU Press.

Freije, A. M. (2015). Heavy metal, trace element and petroleum hydrocarbon pollution in the Arabian Gulf. *Journal of the Association of Arab Universities for Basic and Applied Sciences*, 17(1), 90–100.

Geoff, B., & Phillip, C. (2011). Budgetary and financial discontinuities: Iraq 1920–32. *Accounting History*, 21(3), 247–259.

Gordillo, G. R. (2014). *Rubble: The afterlife of destruction*. Duke University Press.

Griffin, P., & Heede, C. R. (2017). *The carbon major database*. CDP Carbon Majors Report. Retrieved February 12, 2021, from https://b8f65cb373b1b7b15febc70d8ead6ced550b4d987d7c03fcdd1d.ssl.cf3.rackcdn.com/cms/

reports/documents/000/002/327/original/Carbon-Majors-Report-2017.pdf

Grove, J. V. (2019). *Savage ecology: War and geopolitics at the end of the world.* Duke University Press.

Guiu, R. (2020, February). *When canals run dry: Displacement triggered by water stress in the south of Iraq*. Norwegian Refugee Council and the Internal Displacement Monitoring Center. Retrieved January 18, 2022, from https://www.nrc.no/resources/reports/when-canals-run-dry

Hallion, R. (1992). *Storm over Iraq: Air power and the Gulf War*. Smithsonian Books

Haraway, D. (2015). Anthropocene, capitalocene, plantationocene, chthulucene: Making Kin. *Environmental Humanities*, 6(1), 159–165.

Harvey, D. (2005). *The new imperialism: Accumulation by dispossession*. Oxford University Press.

Hawley, T. (1992). *Against the fires of hell: The environmental disaster of the Gulf War*. Harcourt Brace Jovanovich Publishers.

Iraq Administration Reports 1914–1932. (1992). Cambridge Archive Editions. Retrieved September 13, 2021, from https://www.cambridge.org/gb/academic/archive-editions/history/middle-east-history/iraq-administration-reports-19141932?format=WX&isbn=9781852073602

Ireland, P. W. (1970). *Iraq: A study in political development*. Russell & Russell.

James, P., & Nahory, C. (2013). *War and occupation in Iraq corruption, fraud and gross malfeasance*. Global Policy Forum. Retrieved February 20, 2021, from https://archive.globalpolicy.org/empire/us-military-expansion-and-intervention/iraq-8-14/37163.html

Jones, G. G. (1977). The British government and the oil companies 1912–1924, the search for an oil policy. *The Historical Journal*, 20(3), 647–672.

Josephson, P. (2002). *Industrialized nature: Brute force technology and the transformation of the natural world*. Island Press.

Kaldor, M., Karl, T. L., & Said, Y. (2007). *Oil wars*. Pluto Press.

Katzman, K. (2015). *Iraq: Politics, security, and US policy*. Vol. 22. Washington: Congressional Research Service.

Kelly, K. M. (1991). Declaring war on the environment: The failure of international environmental treaties during the person Gulf War. *American University International Law Review*, 7(4), 921–950.

Kelly, M. J. (2007). The Anfal trial against Saddam Hussein. *Journal of Genocide Research*, 9(2), 235–242.

Knights, M., & Mello, A. (2017). Defeat by annihilation: Mobility and attrition in the Islamic states defense of mosul. *CTC Sentinel*, 10(7), 1–7.

Lakoff, G. (1991). Metaphor and war: The metaphor system used to justify war in the Gulf. *Peace Research*, 25–32.

Leaman, G. (2004). Iraq, American empire, and the war on terrorism. *Metaphilosophy*, 35(3), 234–248.

Levy, B., & Sidel, V. (2008). The Iraq War. In B. Levy & V. Sidel (Eds.), *War & public health*. Oxford University Press.

Liboiron, M. (2021). *Pollution is colonialism*. Duke University Press.

Malik, S. (2015). The Isis papers: Behind death cult image lies a methodical bureaucracy. *The Guardian*. Retrieved March 19, 2021, from https://www.theguardian.com/world/2015/dec/07/isis-papers-guardian-syria-iraq-bureaucracy

Marks, S. (2011). Law and the production of superfluity. *Transnational Legal Theory*, 2(1), 1–24.

Mason, R. C. (2009, July 13). *US-Iraq withdrawal/Status of forces agreement: Issues for congressional oversight*. Library of Congress Washington DC Congressional Research Service. Retrieved August 19, 2021, from https://sgp.fas.org/crs/natsec/R40011.pdf

Meilinger, P. S. (2017). Trenchard and morale bombing: The evolution of Royal Air Force Doctrine before World War II. *The Journal of Military History*, 60(2), 193–220.

Mendelssohn, I. A., Andersen, G. L., Baltz, D. M., Caffey, R. M., & Carman, K. R. (2012). Oil impacts on coastal wetlands: Implications for the Mississippi River delta ecosystem after the deepwater horizon oil spill. *BioScience* 62(6), 562–574.

Mlodoch, K. (2017). The indelible Smell of apples: Poison Gas Survivors in halabja, Kurdistan-iraq, and their struggle for recognition. In B. Friedrich et al. (Eds.), *One hundred years of chemical warfare: Research, deployment, consequences* (pp. 349–362). Springer.

Moore, J. W. (2015). *Capitalism in the web of life: Ecology and the accumulation of capital*. Verso Books.

Nasirian, H., & Irvine, K. N. (2017). Odonata larvae as a bioindicator of metal contamination in aquatic environments: Application to ecologically important wetlands in Iran. *Environmental Monitoring and Assessment*, 189(9), 1–18.

Natali, D. (2015). The Islamic States Baathist Roots. *Al-Monitor*. Retrieved September 13, 2021, from https://www.al-monitor.com/originals/2015/04/baathists-behind-the-islamic-state.html

Neocleous, M. (2011). War on waste: Law, original accumulation and the violence of capital *Science & Society*, 75(4), 506–528.

Omissi, D. (1989). Britain, the Assyrians and the Iraq Levies, 1919–1932. *The Journal of Imperial and Commonwealth History*, 17(3), 301–322.

O'Rourke, D., & Connolly, S. (2003). Just oil? The distribution of environmental and social impacts of oil production and consumption. *Annual Review of Environment and Resources*, 28(1), 587–617.

Paredis, E. (2009). *The concept of ecological debt: Its meaning and applicability in international policy*. Cambridge: Academia Press.
Perito, R. (2007, February 7). *Reforming the Iraqi Interior Ministry, Police and Facilities Protection Service*. US Institute of Peace. Retrieved February 20, 2021, from https://www.usip.org/publications/2007/03/reforming-iraqi-interior-ministry-police-and-facilities-protection-service
Polanyi, K. (1944). *The great transformation*. Boston: Beacon press.
Priestley, C. (2020). We won't survive in a city. The marshes are our life: An analysis of ecologically induced genocide in the Iraqi marshes. *Journal of Genocide Research*, 1–23.
Ross, C. (2017). *Ecology and power in the age of empire: Europe and the transformation of the tropical world*. Oxford University Press.
Russell, E., & Tucker, R. (2004). *Natural enemy, natural ally: Toward an environmental history of warfare*. Oregon State University Press.
Sadiq, M., & McCain, J. (1993). *The Gulf War aftermath: An environmental tragedy*. Kluwer Academic Publishers.
Sedra, M. (2007). Security sector reform in Afghanistan and Iraq: Exposing a concept in crisis. *Journal of Peacebuilding & Development*, 3(2), 7–23.
Serres, M. (2010). *Malfeasance: Appropriation through pollution?* Stanford University Press.
Silverfarb, D., & Khadduri, M. (1986). *Britains informal empire in the Middle East: A case study of Iraq, 1929–1941*. Oxford University Press.
Singh, S., & Kennedy, C. (2015). Estimating future energy use and CO2 emissions of the world's cities. *Environmental Pollution*, 203, 271–278.
Sjostedt, B. (2020). *The role of multilateral environmental agreements: A reconciliatory approach to environmental protection in armed conflict*. Bloomsbury Publishing.
Sluglett, P. (2007). *Britain in Iraq: Contriving king and country, 1914–1932*. Columbia University Press.
Smith, G. (2017). *The war and environment reader*. Just World Books.
Tessier, J. D. (2007). Shake & (and) Bake: Dual-use chemicals, contexts, and the illegality of American white phosphorus attacks in Iraq. *Pierce Law Review*, 6, 323.
Thomas, M. (2003). Bedouin tribes and the imperial intelligence services in Syria, Iraq and Transjordan in the 1920s. *Journal of Contemporary History*, 38(4), 539–561.
Tilly, C. (2017). War making and state making as organized crime. In *Collective violence, contentious politics, and social change* (pp. 121–139). Routledge.
Tinker Salas, M. (2009). *The enduring legacy: Oil, culture, and society in Venezuela* (p. 237). Duke University Press.
Tripp, C. (2007). *A history of Iraq*. Cambridge University Press.

UN Secretary-General. (1991, December 24). *Irans reconstruction efforts in the wake of the conflict between the Islamic Republic of Iran and Iraq*. UN Secretary-General. Retrieved June 7, 2021, from https://digitallibrary.un.org/record/138790?ln=en

UNEP. (2005). *Assessment of environmental hot spots in Iraq*. United Nation Environment Program. Retrieved June 7, 2021, from https://www.unep.org/resources/report/assessment-environmental-hot-spots-iraq

UN Environment Technical Note. (2017). *Environmental issues in areas retaken from ISIL*. United Nation Environment Program. Retrieved June 7, 2021, from https://www.unep.org/resources/publication/environmental-issues-areas-retaken-isil-mosul-iraq-technical-note

Walker, A. (1989). Recessional and Gulf War impacts on port development and shipping in the Gulf States in the 1980s. *GeoJournal*, 18(3), 273–284.

Warlenius, R., Pierce, G., & Ramasar, V. (2015). Reversing the arrow of arrears: The concept of "ecological debt" and its value for environmental justice. *Global Environmental Change*, 30, 21–30.

Weiss, M., & Hassan, H. (2015). *ISIS: Inside the Army of Terror*. Regan Arts.

Williams, P. (2009). *Criminals, militias, and insurgents: Organized crime in Iraq*, US Army War College, Strategic Studies Institute.

Winegard, T. C. (2016). *The first world oil war*. University of Toronto Press.

Yaphe, J. (2010). Until they leave: Liberation, occupation, and insurgency in Iraq. In A. Baram, A. Rohde & R. Zeidel (Eds.), *Iraq between occupations*. Palgrave Macmillan.

Yates, M. (2011). The human-as-waste, the labor theory of value and disposability in contemporary capitalism. *Antipode*, 43(5), 1679–1695.

Zolnikov, T. R. (2013). The maladies of water and war: Addressing poor water quality in Iraq. *American Journal of Public Health*, 103(6), 980–987.

Zwijnenburg, W., & Postma, F. (2017). *Living under a Black Sky: Conflict pollution and environmental health concerns in Iraq*. PAX Netherlands. Retrieved November 20, 2021, from https://paxforpeace.nl/what-we-do/publications/living-under-a-black-sky

CHAPTER 3

Beyond Rentier State and Climate Conflict: Clashing Environmental Imaginaries and Ecological Oppression in Iran

Maziar Samiee

Introduction

In July 2021, water protests in the Khuzestan province turned into a nation-wide unrest that swept through Iran for nearly two weeks. Shooting unarmed protesters, security forces killed at least 8 people (see Fig. 3.1) and arrested hundreds, many of them from the Arab minority (Amnesty International, 2021). Although this has been the deadliest episode, Iran has been embroiled with major environmental protests for nearly two decades. Nevertheless, research on environmental justice in Iran is rather tenuous, remaining insignificant in environmental debates. The Environmental Justice Atlas (Temper et al., 2015) documents only six environmental conflicts for Iran, compared to 59 records for Turkey, of similar size and with similar ecological problems. Iran is also absent from

M. Samiee (✉)
University of Sussex, Brighton, UK
e-mail: M.Najdi-Samiee@sussex.ac.uk

A. Dunlap and A. Brock (eds.), *Enforcing Ecocide*,
https://doi.org/10.1007/978-3-030-99646-8_3

Fig. 3.1 Victims of ecological oppression in Iran include water protesters and environmentalists, own compilation

Davis and Burke's (2011) 'Environmental Imaginaries of the Middle East and North Africa' (MENA), which covers the historical ecology of 11 countries. This gap is partially due to the lack of academic freedom in Iran and high restrictions for scholars (Kinzelbach, 2020). Some prominent researchers of environmental studies in Iran have faced persecution and imprisonment (HRW, 2019) (see Fig. 3.1). Outside Iran, the research agenda is largely shaped by the orientalist gaze, or its main intellectual rival 'orientalism-in-reverse' (al-Azm, 1981), that each in their own way narrow the focus on topics such as Islamism as a reaction to 'Western' modernisation, essentialising, and exceptionalising the subject of the study along the way (Matin, 2013).

Iran holds 10% of global oil reserves and it is the second largest holder of natural gas reserves (OPEC, 2020). Petrodollars have been the main source of revenue for the Iranian government and often seen as a key element in formation and reproduction of it as a rentier state. Such a significant share of the world's fossil fuels makes the country central to any debate on climate mitigation and climate justice. Moreover, Iran faces numerous multifaceted ecological issues. It is ranked the fourth worst country in water crisis outlook (WRI, 2019), and water problems are projected to pose serious security challenges (Shahi, 2019) as many lakes, wetlands, and rivers are drying, resulting in serious groundwater depletion

(Madani, 2014). Excessive dam construction has created further problems (Yazdandoost, 2016). Several Iranian cities struggle with air pollution and dust storms, and Iran is consistently amongst the top 10 countries with highest CO2 emissions (IEA, 2021). The country has lost a third of its forested areas since the 1960s (Tahbaz, 2016) and its level of soil erosion is estimated to be four times higher than the global average (Amiraslani & Dragovich, 2011) with nearly 20% of the land at risk (UNDP, 2017). Consequently, land and marine ecosystems and biodiversity are under serious threat (Jowkar et al., 2016), with nearly 70% of the coral reefs in Persian Gulf considered lost (Bayani, 2016). Rising temperatures are expected to make vast areas of Iran uninhabitable (Pal & Eltahir, 2016). Land subsidence, flash floods, disruptions in water supply, increasing wildfires, and consecutive days of unbreathable air in major cities have already become the norm.

Ecological catastrophes in Iran have triggered organised and organic resistance that is met with state violence. Environmental NGOs face multiple constraints and have to manoeuvre shifting ideological red lines to avoid prosecution and closure (Fadaee, 2011). Environmental journalists also face persecution, comparable to that following coverage of military secrets (Schwartzstein, 2020). Repression and censorship make officially recognised channels of environmental grievances weak and ineffective. A recent crackdown on environmental defenders has led to the imprisonment of nine environmentalists for alleged espionage by the Islamic Revolutionary Guard Corps (IRGC) (HRW, 2019). One of them, a sociology professor and director of an environmental NGO Kavous Seyed Emami, died under suspicious circumstances while in solitary confinement (HRW, 2020). Lesser-known grassroots activists have been subject to similar treatments. Some argue that environmental tensions contribute to escalating wider political tensions in Iran (Sengupta, 2018).

This chapter demonstrates how state actions and the protection of particular political economic interests and power hierarchies are the root cause of accelerating trends of environmental crises in Iran. I highlight the role of the military in development strategies of hegemonic nationalist-Islamist factions, including the intertwined processes of securitisation and commodification or assetisation of nature they drive. Section 3.2 introduces the rentier state model and climate security framework—two predominant conceptualisations of nature in the region—along with relevant observations for Iran. These literatures are then critiqued in the following Sect. 3.3 where I discuss how such formalised models replace the agency of actors, namely the state, with an environmental determinism. Section 3.4 explores how a political ecology approach can avoid such determinism, and how the notion of environmental imaginary can help to historicise the human-nature relations and bring the state into analysis. In this context, I discuss how the hegemonic ecology of the Iranian state relies on the military and the market to clash with local environmental imaginaries and disfigure peoples' relations with their environment. Finally, the chapter concludes by summarising how environmental imaginary is central to oppression and resistance, and by sketching some future research directions.

A Rentier State Creating Its Own Climate Conflict?

Resource-rich countries, it is argued, lag in growth behind countries with less abundant resources (Sachs & Warner, 2001). There are various explanations for this supposedly economic curse. Rich natural resources can encourage voracious rent-seeking, inflaming corruption. Exporters of raw material could be vulnerable to volatility of international markets. And finally, high revenues from such exports appreciate the national currency, making other sectors of economy uncompetitive, the so-called *Dutch Disease* (Sala-i-Martin & Subramanian, 2003; Sachs & Warner, 2001). Some economists also emphasise that resource-backed state intervention in the economy leads to 'inefficient' allocation of resources (Amuzegar, 2008).

The *Political Dutch Disease* involves the consolidation of power by political elites through their grip on 'rentier sectors' which hinders the growth of 'productive sector' (Lam & Wantchekon, 1999). Such interdependent consolidation of rent-seeking economic interests and sociopolitical formation is a dynamic that reproduces itself and creates economic stagnation and political inertia, and gives rise to a 'rentier state' that harms 'development' (Mahdavy, 1970). This formation is argued to be durable since the ruling elite manages to reinvest the extracted resources back into its base and institutions (Smith, 2007). However, the rentier structure is not resilient and is prone to collapse in case of diminishing rents, as argued in the case of the Iranian Revolution in 1979 (Skocpol, 1982). The rentier state model establishes a research agenda that seeks to predict the level of democracy or authoritarianism based on the regimes' reliance on natural resources, as measured by share of oil and mineral exports in the economy (Ross, 2001; Wantchekon, 2002; Aslaksen, 2010). The ultimate curse of natural resources is said to be the susceptibility of such countries to violent conflicts and civil wars (Ross, 2015; Costello, 2016).

This formulation of the nature-state relations can directly link to the climate security perspective. The expanding body of literature on *climate conflict* (e.g. Hsiang & Burke, 2014) attributes instances of armed conflict and unrest to climate change. Although climactic events and trends are not necessarily seen as the primary factors of conflict, they act as catalysts or multipliers by exacerbating existing social, economic, political, and environmental tensions. Therefore, accelerating climate change is a determinant factor in heightened risks of conflict in future (Mach et al., 2019).

A synthesis of these two approaches suggests that a rentier state, in the age of climate change, is prone to slide in instability and conflict, including state violence. Iran, as an archetypal rentier state, and in fact as the first country that was originally conceptualised in this framework (Mahdavy, 1970), offers plenty of evidence for the entanglement of high oil revenues with lasting authoritarianism and chronic economic problems. Moreover, oil industries, water projects, deforestations, and other major drivers of ecological catastrophe in Iran are intertwined with preserving the ruling elites' interests, protected by the police and military, who themselves are amongst the perpetrators and beneficiaries of destructive economic activities.

For half a century (1960–2010) oil exports revenues have accounted for almost 20% of the total economic output in Iran, annually making up more than three quarters of all exports. Significantly, except for the Iran-Iraq war period (1980–1988), annual crude oil revenues—64 billion dollars a year between 2009 and 2018 (OPEC, 2014, 2020)—have been higher than the annual government budget (Mohaddes & Pesaran, 2013). A significant portion of this money funds multiple military, security, intelligence, judiciary, and ideological bodies that are unaccountable to the public. This comes at the expense of underfunded public needs (see Fig. 3.2).

Distribution of 'rents' expands beyond direct allocation of resources from the state budget. Parastatal conglomerates and foundations, including those controlled by the military, have a tight grip on Iranian economy. By order of the leader Khamenei, the Iranian constitution has been overturned and major national industries were privatised. This privatisation by and large falls on the petrochemical industries

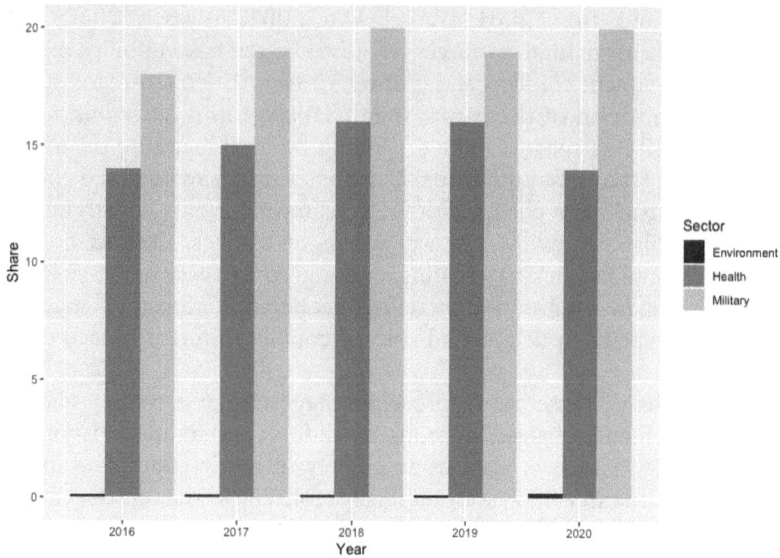

Fig. 3.2 Comparative share of military, health, and environmental spending as percentage of government budget, PBO, various years

(IPO, 2020). Privatised oil and gas companies are mostly acquired by pseudo-private parastatal foundations, through corrupt-ridden processes (Momeni & Haji Norouzi, 2018). Some of these conglomerates are affiliated with the IRGC which has had a rising economic presence (Forozan & Shahi, 2017) and many of these firms have been involved in various billion-dollar corruption scandals. Meanwhile, the same industries are responsible for soil saltification, groundwater reserve pollution, and marine pollution in the Persian Gulf and the Caspian Sea (Jafari, 2019a; Bayani, 2016). They also contribute to dust storms which lead to hazardous levels of particles in the air.

The rentier structure extends beyond the oil industry and similar cycles of extraction and exploitation contribute to other ecological catastrophes in Iran; from the mining and auto industries, monopolised by parastatal firms with impunity for their polluting activities, to waste management dominated by a 'rubbish mafia' (ISNA, 2020). Various types of land-grabbing and rapid transformations of thousands of acres of land, often by state-affiliated entities or individuals, generate billions of dollars for the perpetrators, including through using land as collateral for financial credit, while accelerating destructive trends such as deforestation and soil erosion (see Roudgarmi & Mahdiraji, 2020; Karji et al., 2019).

Perhaps the most pressing matter is water, that has brought Iran on the verge of becoming a 'water-bankrupt nation' (Madani et al., 2016). 'Excessive' construction of dams is a major issue. The total reservoir capacity is estimated around 40% higher than the total available freshwater resources (Yazdandoost, 2016). Despite the already oversized capacity, over a hundred more dams are currently in various stages of construction, along with dozens in planning and design stages. At least 60% of construction contracts are given to the parastatal and military conglomerates or their affiliates. At the top of these firms stands Khatam-ol-Anbia conglomerate, the core of Iran's Military-Subcontractor Complex. The Khatam 'base', under direct control of the IRGC's commander-in-chief, owns hundreds of firms in virtually all sectors of the economy. Almost a fifth of all dams under construction are directly contracted to Khatam or its satellites (see Appendix). Although dam construction may not provide much water for Iranian people, it sure is the breadwinner for entities like Khatam.

Ravaging air, water, and earth brings the state under fire. Resistance against ecological catastrophes has intensified with an increasing number of campaigns, peaceful protests, and riots. In recent years, such protests have transformed from single-issue campaigns into outright dissent against the state as a whole. In turn, the violent crackdown on protest further illustrates the polarisation of environmental politics in Iran along with the shrinking legitimacy of the Islamic Republic.

In July 2018, Iranian security forces used heavy machine guns against water protesters in the southern city of Khorramshahr (AP News, 2018). The port city was the scene of the first major battle in the Iran-Iraq war. Thus, Khormashahr and the wider Khuzestan province are destinations for 'war pilgrimage' (Bombardier, 2012) and prominent in Iranian narratives of the war as testimonies to righteous self-defence. In the place that was depicted as the stage for heroism, Iranian soldiers opened fire on their own population, protecting oil industries and dam construction businesses owned by the military. In another iconic event in Isfahan, farmers used the Friday prayers to protest for water, turning their backs to the prayers' imam, chanting 'turned against the enemy, embracing motherland' (BBC, 2018). In April 2019, during an unprecedented wave of flash floods affecting 26 out of all 31 provinces, flood victims repeatedly greeted officials and disaster relief troops with sticks and stones (DW, 2019). The water crisis is prominent, but grievances include other issues such as waste management, air quality, land-grabbing, and deforestation too.

Dynamics of political dissent in Iran changed in the late 2010s. It is marked by wider geographical spread and more diverse class composition, more explicit in its challenge to state authority, and reduced aversion to endure violence by dissenters (Shahi & Abdoh-Tabrizi, 2020). The transformation of environmental dissent cannot be seen in isolation from this trend. The effects of environmental crises and the consequent repression of protests are more severe in the areas inhabited by ethnic minorities of Iran or those with higher poverty rates (Hassaniyan, 2020; UNPO, 2018). Hundreds of Azeris who were arrested for peaceful demonstrations advocating the protection of lake Urmia (HRW, 2011) or dozens of Arabs and other locals who were arrested after forming a human chain in defence of Karun River (HRA, 2014) illustrate this. Indeed, some argue that worsening ecological crises could contribute to the intensification of resistance and repression (Shahi & Abdoh-Tabrizi, 2020; Waldman, 2018). In fact, the state's stubbornness in addressing ecological grievances—and its role in causing and exacerbating them—could be a key factor in the current severity of political dissent in Iran.

Through the lens of the rentier state and climate security, the Iranian state seems trapped in a vicious cycle of rent-seeking, ecological damage, and political dissent that eventually undermines its political and economic structures. But why would Iran avoid socio-ecological mitigation of the negative environmental impacts? Do the immediate economic benefits of dam construction for parastatal companies outweigh the strategic security costs of water protests for the entirety of the state? The Iranian state has been challenged domestically by the Green Movement in 2009 and waves of nation-wide protests in 2018, 2019, and 2021. Internationally, it has been exposed to costly wars in Syria and Iraq, and faces the new alliances of Israel with some Arab states, on top of the never-ending confrontations with the US. As these issues disproportionately affect areas with ethnic minorities, why is the Iranian state missing the chance to pre-empt the additional threat of 'eco-sectarianism', similar to what has been argued to have happened in Syria (Shahi & Vachkova, 2018)?

Rentierism and Security Without a State?

In one of the first studies of rentier states, Mahdavy (1970) sought to develop a theory with 'universal applicability'. Similarly, and around the same time, Falk (1971) sought to establish a universal system of political ecology and formulated a law for 'inverse relationship between the interval of time available for adaptive change and the likelihood and intensity of violent conflict' (Falk, 1971: 353 cited in Barnett, 2003). Since then, the rentier state and the climate security research programmes have largely avoided contingency of specific polities and transformations of power relations that involve state and nature.

The rentier state framework mixes up a variety of sociopolitical contexts and cannot capture their differences, or even appreciate the drastic changes in each of them. Iran has faced multiple rounds of international sanctions with shrinking oil revenues, but this has not broken the rentier structure. It is statistically dubious to state a correlation between resource abundance and underdevelopment or political instability and conflict when the so-called abundance is measured by share of raw material exports in the economy (Brunnschweiler & Bulte, 2008). State control over natural resources or the ownership by the ruling elite is presumed uncritically, whereas the key question is how the ruling elites have appropriated such resources and how they maintain their control. The state management of natural resources cannot be assumed as a purely economic rent maximising mission for a homogenous group of rulers and their allies, as development of property relations over natural resources is the main question in the first place. As such, appropriation and distribution of mineral and petroleum rents are not determinants of political struggles, rather they are distributed through multifaceted dynamics of political coalitions and adversaries (Di John, 2011). Primary reliance on econometrics methods makes much of the resource curse literature theoretically thin, and disconnected from wider theories in political economy, namely the staple thesis and theories of predatory state (Vahabi, 2018). This literature also fails to appreciate the breadth and depth of rentier relations in the MENA region, as it focuses on natural resources and cannot problematise other rent streams such as business deals with loyalists, foreign aid, and co-optation of the financial sector in the rentier structure (Malik, 2017).

Here lies the original sin of the rentier state model: preoccupation with rents from natural resources without wider conceptualisation of rent, which is an inherent feature of any market economy, and not necessarily limited to natural resources. Rentier state theory ignores that rents are enabled with the marketisation of nature, and not by abundance of resources per se. Therefore, it does not take into account how rentier structure expands well beyond extraction of raw material and includes industries, trade monopolies, and financial credits that are owned and managed by a multitude of private and public entities. As a result, the simple solution of privatisation to curtail the rentier problems often worsens the situation as it presupposes the blessings from market's invisible hand. But land-grab and water-grab in Iran are environmentally destructive rentier practices that are enabled not by their abundance, but with increasing marketisation. To frame oil rents in MENA economies as unearned and undeserved fortunes, all the while assuming that 'financial rents are a blessing that helps the market adjust to equilibrium... undermines what the anti-colonial national struggles were all about, which is the ownership of national resources' (Kadri, 2016: 127). In neoclassical economics income distribution is based on 'marginal productivity', hence rent extraction in a well-developed capitalist economy is impossible by definition (McGoey, 2017). Such a conceptualisation takes entrenchment of rentier structure and expansion of market mechanisms as seemingly separate processes. Therefore, from such a point of view, expansion of parastatal conglomerates in Iran should be either taken as evidence for dismantling the rentier structure or dismissed as perversion of the 'true' privatisation.

Shortcomings of the rentier state approach are mirrored in the climate security literature. These studies are especially influential amongst supranational bodies and western policy circles, but are often refuted. Historically, there has not been a correlation between outbreak of civil wars and climate disasters (Slettebak, 2012). Although statistical studies have produced limited and contradictory results for existence of a direct relation between climactic change and conflict (Selby, 2014), the climate security framework commits an 'epistemological slippage' to employ predictive natural sciences over imaginative and humanistic aspects of social life, creating a certain 'climate determinism' with simple answers for complex questions about the future of human societies and ecological issues (Hulme, 2011). Due to unfounded generalisation and embedded essentialisation, the politico-economic structures and power strategies of states and rival groups are overlooked, along with the relevant historical and international context (Selby & Hoffmann, 2014).

Iran has been insignificant in the climate security literature. But it is insightful to turn to Syria, the 'paradigmatic case' for this approach (Ide, 2018). Iran faces similar intertwined issues as Syria, such as of drought, internal migration, ethnic tensions, and authoritarianism. The Syrian civil war has propelled the climate conflict debate, but despite its ubiquity, no causal link could be established for climactic roots of the war and the argument only serves to depoliticise the war (ibid.; Selby et al., 2017). Crucially, calculated and shifting strategies of multiple states, Iran included, in utilising military and violence is replaced with an apolitical fatalism. Notably, Syria has not been studied as a failed rentier state, even though wider sociopolitical structures in Syria have been discussed in rentier framework (e.g. Beblawi, 1987; Hinnebusch, 2001; Daher, 2019). Likewise, other episodes of the Arab Spring, especially the uprisings in Egypt and Libya, and the 2013–2017 civil war in Iraq have not been studied as possible conjunctures of rentierism and environmental conflict, despite concurrence of heightened political instability and environmental disasters such as drought.

Importantly, neither rentier state theory nor climate security approaches prioritise the conceptualisation of the state, even though the state is the key enforcer of the property rights that give rise to rentierism, just as it is the unrivalled coordinator of environmental degradation through regulating and permitting catastrophic activities while framing the related issues as matters of security. This reduces politics to rigid rivalries and struggles over 'scarce' natural resources. Subjectivity and agency of collective actors and their historical legacies are replaced by a model of 'homo economicus' who operates based on 'rational choice'. But a rational state would not undermine its own security by driving ecological disasters, or would it? As mainstream economics argues, a rational actor optimises present and future earnings and costs. The state's rationality is myopic, suited to calculate short-term risks and benefits but unable to grapple with long-term consequences. Said differently, the state is trapped in a political inertia shaped by economic expectations. Either way, this methodological individualism focuses on choices determined by structural forces and objective conditions, at the expense of overlooking the evolutionary social relations (Cramer, 2002). Even when collective action of social groups in relation to state institutions is discussed, it is in terms of pregiven identities, as in the case of Syria where special access to economic rents is mainly framed in static 'sectarian' terms (see Haddad, 2012 for a critique).

Rentier state theory and climate conflict framework fail even by the positivist standards. Their abstract approaches cannot adapt to each concrete context so they keep changing the standards of assessment. In climate security both scarcity and abundance are blamed (Selby, 2014). The rentierism literature regards such states as fragile and over-powerful at the same time (e.g. Schwarz, 2008b) and is simultaneously puzzled by endurance and fragility within them. While it suffers from deep economism it regards political developments in rentier states to be purely ideological and disconnected from economic issues.

Climate security focuses on adaptation strategies for 'individual' entities, thus avoiding a wider systemic approach that engages with centrality of political oppression and violence that shape environmental conditions (Besthorn & McMillen, 2002). The marketisation of nature and state militarisation are often neglected in the climate security approach, or even advanced through 'green' economy policy proposals (Dunlap & Fairhead, 2014). Similarly, rentier state framings accentuate the determining effect of natural resources without exploring dynamics and strategies of states that cannot be traced back to exploitation of nature. Hence the rentier state framework ultimately adheres to the general depoliticisation in macroeconomics (Swanson, 2008). Depoliticisation and environmental determinism ultimately remove direct responsibility. Taking ecological catastrophes as a 'pure form of misfortune' absolves states and supranational bodies from bearing any responsibility for creating resilience against disasters. This framework can be employed in technical-managerial settings and is supposedly intact from power struggles (Mason, 2014). Preoccupation with climate conflict and climate refugees also diverts attentions from key question of restructuring global economy without magnifying existing inequalities (Hartmann, 2010). Similarly, rentier state provides justification for despotism and corruption as endemic and incurable condition of MENA countries, predetermined by nature. From the Saudi oil minister who said 'All in all, I wish we had discovered water' to the Iranian leader Khamenei who said 'this great God-given blessing has caused our country much economic, political, and social collapse', MENA rulers are first in line to decry the ills of massive windfalls and rentier states they control.

Ecological Oppression and State Environmental Imaginary

In a stark contrast to rentier state and climate security theories, political ecologists examine the interactions of ecosystems with politico-economic structures through analysis of dynamic social property relations that are shaped by and in turn structure power struggles at different global, national, and regional scales (Peet et al., 2010). The notion of *environmental imaginary* (Peet & Watts, 1996), a powerful yet somewhat overlooked idea (Nesbitt & Weiner, 2001), can help to conceptualise politics of nature in the context of hegemonic struggles and strategies beyond mere contestation over extraction and exploitation of inanimate resources.

In *Liberation Ecologies*, Peet and Watts (1996: 268) introduce environmental imaginary as 'a way of imagining nature, including visions of those forms of social and individual practice which are ethically proper and morally right with regard to nature'. Mitchel emphasises that the environmental imaginary 'is more than just a work of imagination' (in Davis & Burke, 2011: 267) and should be studied in conjuncture of natural forces, tools, and social technologies. Elaboration of environmental imaginary can further illuminate the relation of state to nature. As Bookchin (1982: 94) says 'the State is not merely a constellation of bureaucratic and coercive institutions. It is also a state of mind, an instilled mentality for ordering reality'. Such mentality involves peoples and places. Nationhood, albeit aiming for the creation of a common identity in a unified territory, does not only separate from the outside world, it excludes internally as well. Rural populations or nomadic people are framed as 'backward' or 'traditional', and the mission of states is to modernise, civilise, and absorb them into their own political economy (Neumann, 2004). Governing people includes controlling their environment and a mission to reshape their socio-ecological relationships. Defining the boundaries of this governance, mapping the territory, and mapping the state-society divide is intertwined with dismantling and enclosing the commons (ibid.). Spatial imaginary of the state is not solely shaped by geopolitical considerations, it also relies on an environmental imaginary that divides the territory to urban and rural areas, acreage and idle land, civility and wilderness, society, and nature. At the core of the developmental mission of the state is the question of how to separate, re-connect, and transform the compartmentalised territory.

While environmental imaginaries of people(s) have a high degree of locality, the state is concerned with an environmental imaginary that establishes and maintains a totality. The attempted totality of the state comes with an uneven inclusion of localities. As such, *ecological oppression* is the enforcement of specific environmental imaginaries, which is a necessary feature of the modern state (Scott, 1998). By imposing its vision of ecological relationships over its citizens and territory, the state conquers rival environmental imaginaries and undermines 'ecological wholeness' (Bookchin, 1982: 28).

In this light, when the environmental imaginary of the modern Iranian state is examined, it no longer appears as a rentier state hastily walking into its demise of climate conflict, but rather that its commitment to a hegemonic project rationalises the environmental costs and security consequences. Insistence on food self-sufficiency is a principle that shapes the water policy in Iran. A humongous 92% of water in Iran is used in agriculture, inefficiency of which is a secret to no-one (Madani et al., 2016). In Syria it is similar: there, the Baathist ideology's preoccupation with agricultural self-sufficiency created water scarcity (Barnes, 2009), later combined with the expansion of market policies in agriculture (Dahi & Munif, 2012), which contributed to erosion of its adaptive capacity (Ide, 2018). In Iran and Syria, not deterministic structural drivers but wider geopolitical considerations and the state's understanding of its territory and people justify such inefficient and environmentally disastrous operations.

The mercantilist food anxiety (see Foucault et al.'s second lecture, 2009) that leads to an isolationist self-sufficiency policy in Iran is in line with a national environmental imaginary and the emotional tone of predominant Iranian nationalism that represents itself as a defensive—thus moral—identity in which Iran is perceived to be a victim of its unique position at the crossroads of global events. Yet, located on the Iranian Plateau and protected by natural boundaries, Iranian civilisation is capable of surviving, adapting, and absorbing the uncivilised foes in its rich culture, thus enduring and outliving all the enemies. Geographical features provide an 'appearance of continuity' that helps to present a nation as a 'natural' grouping of people (Bloom, 1990).

Iran's supposedly defensive nationalism seeks to justify centralisation of power and militarised rule over people and places through highlighting perceived natural and international threats. It is best demonstrated in current prevalence of an ancient inscription in Persepolis. The inscription, from Darius I, the ruler of the First Persian Empire at its peak, is his prayer for Iran to be protected from 'the enemy, the drought, and the lies'. The inscription combines geopolitics, environmental imaginary, and ethics. It is a concise nationalist manifest listing the threats to security, territory, and population. Nature inspires other dimensions of Iranian nationalism including a celebration of diversity of indigenous peoples and ecosystems, as a case of unity in diversity (e.g. Ahmadi, 2005). But this perception of Iran as a country that encompasses varied landscapes and environments establishes the environmental imaginary of a uniquely diverse place, thus distinguished from supposedly uniformly arid Arab rivals (Abe, 2013).

The roots of Iranian nationalism go back to the dawn of the twentieth century. Rising capitalist relations and colonial powers posed increasingly complex challenges to the Iranian government and society. The 1906 Constitutional Revolution sought to establish a democratic and somewhat egalitarian state. Shortly after its establishment, the National Assembly passed a bill to devolve some of the state powers to local councils. This ambition never materialised. Ultimately Reza Shah emerged as the chief architect of the modern Iranian state. Internal instability and foreign interventions paved the way for him as a military leader who promised to restore order and strengthen the central government (Ghods, 1991a). One of his main modernisation policies was to centralise the army, cutting its reliance on warrior tribesmen. His military consumed a third of the national budget. He quashed rebels and autonomous regions, attempted to settle nomadic people in villages, and created trade monopolies and industries often run by military officers (Ghods, 1991b). These policies altered local economies and ecologies in favour of a centralised model. Prominence of the military and ambition for development continued and expanded under his son, Mohammadreza Shah. He embarked on forming one of the largest armies in the world. During his reign the army extended well beyond the immediate military function, as creation of Knowledge Corps, Hygiene Corps, and Construction Corps militarised education, health, and development. The armed forces also helped to disseminate the Shah's nationalist ideology (Looney, 1988). After the 1979 revolution the conventional army was coupled with the IRGC that is more ideologically driven. While a division of labour amongst the state bodies used to shape the strategies for transformation of the populace and their space, today the IRGC embodies ecological oppression in Iran on its own. It fuses militarisation and marketisation of nature in one as it has a direct role in repressing dissent along with expansive extractive economic activities, including in oil and water management. The Iranian military, praised for patriarchal protection of the motherland, has been central in authoritarian national integration: with its role in development projects, both as contractor and enforcer, it has irreversibly disfigured peoples' relation with their environment.

The hegemonic ecology of the Iranian state, at odds with local environmental imaginaries, is not static. The current hegemonic ecology has gradually shifted from a strong developmental state to a conglomerate state. While the Shah's developmental state was primarily concerned with transformation of people and nature to make them economically more productive, the current Islamic Republic's conglomerate state is primarily interested in making profits from transformations of people, nature, and even the state itself. This is evidenced in privatisation of nationalised industries and natural resources by overturning article 44 of the constitution, and by near absolute liquidation of land, including through increasing financialisation and use of land as collateral.

Khamenei has absolute power by law and directly controls some of the biggest parastatal companies. Nevertheless, current economic relations are not centralised around any one entity, and they do not form an overarching harmonious structure, but rather consist of wide and dispersed patrimonial networks that control various bodies of the government (Harris, 2013). This multipolar and multilevel structure creates responsibility bumpers, mechanisms of defusing disputes and grievances, or at least exhausting them without resolution, and ultimately denies people any decision-making powers. Iran's economic structures are rooted in a balancing act between reincorporation of some elements of pre-revolutionary Iran, managing internal disputes of diverging ruling factions, and responding to pressures from a shrinking social base. Integral evolvement of these structures protects and expands the vested interests of the agents of the proclaimed totality that prevents any devolution of power and undermines ecological wholeness.

The trajectory of the rural policies is illustrative of the hegemonic ecology in Iran. The rural population in Iran is simultaneously 'denigrated as backward' and 'idealized as the soul of the nation' (Ehsani, 2006: 93), and the state's relation to them is primarily defined by food self-sufficiency ambitions, a task that the farmers have not been given the tools and freedoms to carry out (ibid.). They are encouraged to grow a handful of crops that the state redeems strategic and secures with fixed prices.

The intended self-sufficiency does not materialise, but deviation from endemic produce and increasing pressure on water resources empowers the state bodies at the expense of deteriorating local networks and their environment.

Immediately after the 1979 revolution, peasant movements in some regions expropriated land from large landlords and established farmers' councils, but soon the new regime crushed them along with their urban socialist allies (see Zahedi, 2019). To reorganise the pre-revolution agricultural administration, multiple initiatives were created: Islamic Village Councils, Centres of Services for Rural and Nomadic People, Communal Cooperatives designed to restore the pre-land reform traditions, and Construction Jihad. These institutions failed to achieve their stated objectives, partially due to competition instead of cooperation, and partially because the planning remained top-down and did not include rural population in decision-making (Shakoori, 2001). Even the communal farming that intended to revive local social technologies was an archetypal and abstract model of traditional setting, without attention to actual regional particularities. In this setting the Construction Jihad enjoyed massive funds and failed to provide the basic rural needs (ibid.).

Construction Jihad, closely linked with the military, is one of the main constructors of dams in Iran. These dams are planned and built without consulting the population in affected areas. In one study, 'residents believed that state authorities intentionally blocked their access to information by excluding them from meetings and dam construction jobs' (Hoominfar & Radel, 2020: 14). This completes the cycle of ecological oppression; a government body affiliated with the military, designed to centrally plan and execute an abstract model of traditional farming, uses its exclusive access to resources to pursue construction projects that do not meet intended targets but irreversibly severe peoples' relation with their environment, and military violence forces people to live in this new reality.

But local communities do not accept the hegemonic ecology of dams as development (ibid.). Oppression breeds resistance and the hegemonic ecology in Iran faces counter-hegemonic environmental imaginaries. Environmental imaginary is a 'primary site of contestation; critical social movements have at their core environmental imaginaries at odds with hegemonic conceptions' (Peet & Watts, 1996: 263). As such, nature is not just a stock of resources that externally defined political rivals fight over, but it is itself a factor in shaping alternative identities, knowledges, ontologies, and politics. It can become a shelter from or a barricade against the state. In regions with significant demands for autonomy nature plays a significant role in imagining the nation. Kurdish people see themselves to have 'No Friend but the Mountains' (Boochani, 2018). The Aras River, which is a natural boundary between Iran and the Republic of Azerbaijan, has special symbolism for Iranian Turks/Azeris. Urban youth use ecotours to escape from the conservative Islamist laws. Nature becomes a space of transgression where they can sing, dance, and build a collective, relatively safe from the invasive Islamic state apparatus (Fadaee, 2018).

Nature has also served as a platform to mount offensives against the Iranian state. In 1920, the Socialist Soviet Republic of Iran came into being in Gilan province. The short-lived republic, aiming to spread the revolution to all of Iran, was rooted in a guerrilla movement called 'Jungle Movement', as partisans used forests of northern Iran as their hiding place (see Jafari, 2020). Decades later another guerrilla movement used the same forests to wage war against the Shah. Although militarily unsuccessful, their perceived bravery gained them significant popularity and fuelled the revolutionary fire through their imagery in arts and popular culture. Forests and nature are central in that revolutionary imagery (Vahabzadeh, 2015).

Heightened political dissent in Iran is increasingly intertwined with ecological grievances. But the oppositional forces often reduce these issues to troubles created by a uniquely corrupt rentier Islamic Republic and are yet to develop a counter-hegemonic ecology. Lack of engagement with this 'primary site of contestation' weakens their attempts in articulating an alternative. Even more important, without an alternative environmental imaginary that embraces decentralisation of power, local democracy, and liberation ecology, sheer transformation of political structures fails to stop the practices of ecological oppression.

Conclusion: Freedom from Resource Curse and Climate Conflict

This chapter demonstrated that ecological oppression is a crucial feature of the modern Iranian state. Abstract and ahistorical theories are insufficient for the study of ecological catastrophes and environmental grievances as these processes are not driven by natural and structural forces, but by a hegemonic ecology and violent imposition of a specific environmental imaginary upon peoples and places.

Environmental imaginary has primarily been applied to the context of global capitalism, colonialism, and in juxtaposing Western and/or modern versus Oriental and/or indigenous environmental imaginaries (e.g. Peet, 1985; Peet & Watts, 1996; Hartmann, 2010; Davis & Burke, 2011; Hoffmann, 2018). Exploring transformations of hegemonic ecologies of states in the postcolonial era can advance this area of research, as strategies and competing environmental imaginaries in each polity are not reducible to such broader dynamics. The idea of hegemonic ecology could be developed and refined further to problematise the state in this context.

Such a perspective is particularly relevant for the Middle East and North Africa, where politics is largely discussed in orientalist terms, or orientalism-in-reverse. Contrary to prevalent explanation of MENA authoritarianism as rooted in oil, one of the major democratic movements of Iran was shaped around nationalisation of oil. Although Mossadegh's premiership was cut short with the 1953 coup, the nationalisation movement has inspired other democratic struggles in Iran and beyond. Likewise, mass oil strikes that were a decisive episode of the 1979 revolution demonstrate a possibility of democratic mobilisation around oil (Jafari, 2019).

It is not in oil's nature to create authoritarianism or democracy, but how it is embedded in various political strategies decides its effects. Natural resources, abundant or scarce, cannot determine political stability or conflict on their own, and this is the same for climate trends, be it drought or a temperate weather. It is the ecological oppression and the undermining of local and democratic society-nature relations that leads to crises. Breaking the spell of resource curse and calming the climate conflict require an ecology of freedom.

Appendix

Share of parastatal companies from dams under construction in Iran

Firm	*Affiliation*	*Number of contracted dams*
Khatam-ol-Anbia	IRGC	18
Ab o Khak	Construction Jihad and Mostazafan Foundation	6
Jihad Nasr	" " " " "	6
Sabir	Shasta	5

(continued)

(continued)

Firm	*Affiliation*	*Number of contracted dams*
Jihad Tose-eh	Construction Jihad	4
Other parastatal	Various	13
Other sectors	Various	38
Unknown		16
Total		106

Aggregated data from Iranian Water Resources Management Company (www.wrm.ir), and official database for companies' registry (www.rrk.ir), as of August 2020

Bibliography

Amnesty. 2021. 'Iran's Security Forces Crush Protests with Ruthless Force'. https://www.amnesty.org/en/latest/news/2021/07/iran-security-forces-use-live-ammunition-and-birdshot-to-crush-khuzestan-protests/.

Abe, Satoshi. 2013. 'Conceptions of Nature in Iran: Science, Nationalism, and Heteroglossia'. *Journal of Anthropological Research* 69 (2): 201–23.

Ahmadi, Hamid. 2005. 'Unity Within Diversity: Foundations and Dynamics of National Identity in Iran'. *Critique: Critical Middle Eastern Studies* 14 (1): 127–47.

al-'Azm, Sadik Jalal. 1981. 'Orientalism and Orientalism in Reverse'. In *Khamsin 8*, 5–26. London: Ithaca Press. Reproduced in A. L. Macfie, ed., *Orientalism: A Reader*, NYU Press.

Amiraslani, Farshad, and Deirdre Dragovich. 2011. 'Combating Desertification in Iran Over the Last 50 Years: An Overview of Changing Approaches'. *Journal of Environmental Management* 92 (1): 1–13.

Amuzegar, Jahangir. 2008. 'Iran's Oil as a Blessing and a Curse'. *Brown Journal of World Affairs* 15 (1) (Fall/Winter).

AP News. 2018, July 1. 'Gunfire, Clashes amid Iran Protests Over Water Scarcity'.

Aslaksen, Silje. 2010. 'Oil and Democracy: More Than a Cross-Country Correlation?' *Journal of Peace Research* 47 (4): 421–31.

Barnes, Jessica. 2009. 'Managing the Waters of Ba'th Country: The Politics of Water Scarcity in Syria'. *Geopolitics* 14 (3): 510–30.

Barnett, Jon. 2003. 'Security and Climate Change'. *Global Environmental Change* 13 (1): 7–17.

Bayani, Niloufar. 2016. 'Ecology and Environmental Challenges of the Persian Gulf'. *Iranian Studies* 49 (6): 1047–63.

BBC. 2018, March. کشاورزان اصفهان نماز جمعه‌را بر هم زدند. https://www.bbc.com/persian/iran-43434396.

Beblawi, Hazem. 1987. 'The Rentier State in the Arab World'. *Arab Studies Quarterly* 9 (4): 383–98.

Besthorn, Fred H., and Diane Pearson McMillen. 2002, June. 'The Oppression of Women and Nature: Ecofeminism as a Framework for an Expanded Ecological Social Work'. *Families in Society* 83 (3): 221–32.

Blaikie, Piers, and Harold Brookfield, eds. 1987. *Land Degradation and Society*, 1st ed. Routledge.

Bloom, William. 1990. *Personal Identity, National Identity and International Relations* (Cambridge Studies in International Relations). Cambridge: Cambridge University Press.

Bombardier, Alice. 2012. 'War Painting and Pilgrimage in Iran'. *Visual Anthropology* 25 (1–2): 148–66.

Boochani, Behrouz. 2018. *No Friend but the Mountains*. Pan Macmillan.

Bookchin, Murray. 1982. 'The Ecology of Freedom: The Emergence and Dissolution of Hierarchy'. Cheshire Books, Palo Alto.

Brunnschweiler, C. N., and E. H. Bulte. 2008. 'ECONOMICS: Linking Natural Resources to Slow Growth and More Conflict'. *Science* 320 (5876): 616–17.

Bunker, Stephen G. 1985. *Underdeveloping the Amazon: Extraction, Unequal Exchange, and the Failure of the Modern State*. University of Illinois Press.

Costello, Matthew. 2016. 'Rentierism and Political Violence'. *Sociology Compass* 10 (3): 208–17.

Cramer, C. 2002. 'Homo Economicus Goes to War: Methodological Individualism, Rational Choice and the Political Economy of War'. *World Development* 30 (11): 1845–64.

Daher, Joseph. 2019. *Syria After the Uprisings: The Political Economy of State Resilience*. London: Pluto Press.

Dahi, Omar S., and Yasser Munif. 2012. 'Revolts in Syria: Tracking the Convergence Between Authoritarianism and Neoliberalism'. *Journal of Asian and African Studies* 47 (4): 323–32.

Davis, Diana K., and Edmund Burke. 2011. *Environmental Imaginaries of the Middle East and North Africa*. Athens: Ohio University Press.

Di John, Jonathan. 2011. 'Is There Really a Resource Curse? A Critical Survey of Theory and Evidence'. *Global Governance: A Review of Multilateralism and International Organizations* 17 (2): 167–84.

Dunlap, Alexander, and James Fairhead. 2014. 'The Militarisation and Marketisation of Nature: An Alternative Lens to "Climate-Conflict"'. *Geopolitics* 19 (4): 937–61.

DW. 2019, April 4. '"اعتراضات سیل‌زدگان؛ هشدار پلیس و وزیر کشور به "اخلالگران نظم'. Accessed 23 March 2021. https://www.dw.com/fa-ir/iran/a-48201500.

Ehsani, Kaveh. 2006. 'Rural Society and Agricultural Development in Post-revolution Iran: The First Two Decades'. *Critique: Critical Middle Eastern Studies* 15 (1): 79–96.

Fadaee, Simin. 2011. 'Environmental Movements in Iran: Application of the New Social Movement Theory in the Non-European Context'. *Social Change* 41 (1): 79–96.

———. 2018. 'Ecotours and Politics of Fun in Iran: From Contested State–Society Relations to Emancipatory Nature–Society Relations'. *The Sociological Review* 66 (6): 1276–91.

Falk, Richard. 1971. *This Endangered Planet: Prospects and Proposals for Human Survival*. New York: Random House.

Feitelson, Eran, and Amit Tubi. 2017. 'A Main Driver or an Intermediate Variable? Climate Change, Water and Security in the Middle East'. *Global Environmental Change* 44 (May): 39–48.

Forozan, Hesam, and Afshin Shahi. 2017. 'The Military and the State in Iran: The Economic Rise of the Revolutionary Guards'. *The Middle East Journal* 71 (1): 67–86.

Foucault, Michel, Michel Senellart, François Ewald, and Alessandro Fontana. 2009. *Security, Territory, Population: Lectures at the Collège de France, 1977–1978*. New York, NY: Picador/Palgrave Macmillan.

Ghods, M. Reza. 1991a. 'Iranian Nationalism and Reza Shah'. *Middle Eastern Studies* 27 (1): 35–45.

———.1991b. 'Government and Society in Iran, 1926–34'. *Middle Eastern Studies* 27 (2): 219–30.

Haddad, Bassam. 2012. 'Syria, the Arab Uprisings, and the Political Economy of Authoritarian Resilience'. *Interface: A Journal for and About Social Movements* 4 (1): 113–30.

Harris, Kevan. 2013. 'The Rise of the Subcontractor State: Politics Of Pseudo-Privatization in the Islamic Republic of Iran'. *International Journal of Middle East Studies* 45 (1): 45–70.

Hartmann, Betsy. 2010. 'Rethinking Climate Refugees and Climate Conflict: Rhetoric, Reality and the Politics of Policy Discourse'. *Journal of International Development* 22 (2): 233–46.

Hassaniyan, Allan. 2020. Environmentalism in Iranian Kurdistan: Causes and Conditions for Its Securitisation'. *Conflict, Security & Development* 20 (3): 355–378.

Hinnebusch, Raymond. 2001. *Syria: Revolution from Above*. London, UK and New York: Routledge.

Hoffmann, Clemens. 2018. 'Environmental Determinism as Orientalism: The Geo-Political Ecology of Crisis in the Middle East'. *Journal of Historical Sociology* 31 (1): 94–104.

Hoominfar, Elham, and Claudia Radel. 2020. 'Contested Dam Development in Iran: A Case Study of the Exercise of State Power Over Local People'. *Sustainability* 12 (13): 5476.

Hosseini, Vahid, and Hossein Shahbazi. 2016. 'Urban Air Pollution in Iran'. *Iranian Studies* 49 (6): 1029–46.

HRA. 2014. 'حمله نیروهای انتظامی و لباس شخصی به حامیان کارون'. https://www.hra-news.org/2014/hranews/b-294/.

HRW. 2011. 'Iran: Allow Peaceful Protests Over Lake's Destruction'. https://www.hrw.org/news/2011/09/10/iran-allow-peaceful-protests-over-lakes-destruction.

HRW. 2019. 'Iran: Environmentalists Sentenced'. https://www.hrw.org/news/2019/11/22/iran-environmentalists-sentenced.

HRW. 2020. 'Iran: Environmentalists' Unjust Sentences Upheld'. https://www.hrw.org/news/2020/02/19/iran-environmentalists-unjust-sentences-upheld.

Hsiang, Solomon M., and Marshall Burke. 2014. 'Climate, Conflict, and Social Stability: What Does the Evidence Say?' *Climatic Change* 123 (1): 39–55.

Hulme, Mike. 2011. 'Reducing the Future to Climate: A Story of Climate Determinism and Reductionism'. *Osiris* 26 (1): 245–66.

Ide, Tobias. 2018. 'Climate War in the Middle East? Drought, the Syrian Civil War and the State of Climate-Conflict Research'. *Current Climate Change Reports* 4 (4): 347–54.

IEA. 2021. *Global Energy Review 2021*. Paris: IEA.

IPO. 2020. Niafam.com. n.d. 'عملکرد تفصیلی سالیانه'. Accessed 10 March 2021. سالیانه عملکرد تفصیلی.

ISNA. 2020. 'مافیای زباله یکی از عوامل آلودگی محیط زیست'. https://www.isna.ir/news/99042820686/مافیای-زباله-یکی-از-عوامل-آلودگی-محیط-زیست.

Jafari, Peyman. 2019a. 'Iran's Petrochemical Industry: A Disaster Zone of Precarity and Pollution'. *Toxic News*. https://toxicnews.org/2019a/11/18/irans-petrochemical-industry-a-disaster-zone-of-precarity-and-pollution.

———. 2019b. 'Linkages of Oil and Politics: Oil Strikes and Dual Power in the Iranian Revolution'. *Labor History* 60 (1): 24–43.

———. 2020. 'When the Red Flag Flew Over Iran'. *Jacobin*. https://jacobinmag.com/2020/11/soviet-socialist-republic-iran-gilan-iran.

Jowkar, Houman, Stéphane Ostrowski, Morad Tahbaz, and Peter Zahler. 2016. 'The Conservation of Biodiversity in Iran: Threats, Challenges and Hopes'. *Iranian Studies* 49 (6): 1065–77.

Kadri, Ali. 2016. 'The Infeasibility of Revolution in Syria'. In *The Unmaking of Arab Socialism*, 117–58. Anthem Press.

Karji, Ali, Asregedew Woldesenbet, Mostafa Khanzadi, and Mohammadsoroush Tafazzoli. 2019. 'Assessment of Social Sustainability Indicators in Mass Housing Construction: A Case Study of Mehr Housing Project'. *Sustainable Cities and Society* 50 (October): 101697.

Kinzelbach, Katrin. 2020. 'Free Universities: Putting the Academic Freedom Index into Action'. GPPI.

Lam, Ricky, and Leonard Wantchekon. 1999. 'Dictatorships as a Political Dutch Disease'. Working Papers 795, Economic Growth Center, Yale University.

Looney, Robert E. 1988. 'The Role of Military Expenditures in Pre-revolutionary Iran's Economic Decline'. *Iranian Studies* 21 (3/4): 52–83.

Mach, Katharine J, Caroline M. Kraan, W. Neil Adger, Halvard Buhaug, Marshall Burke, James D. Fearon, Christopher B. Field. et al., 2019. Climate as a Risk Factor for Armed Conflict. *Nature* 571 (7764): 193–197.

Madani, Kaveh, Amir AghaKouchak, and Ali Mirchi. 2016. 'Iran's Socio-Economic Drought: Challenges of a Water-Bankrupt Nation'. *Iranian Studies* 49 (6): 997–1016.

Madani, Kaveh. 2014. 'Water Management in Iran: What Is Causing the Looming Crisis?' *Journal of Environmental Studies and Sciences* 4 (4): 315–28.

Mahdavy, Hossein. 1970. 'The Patterns and Problems of Economic Development in a Rentier Stare: The Case of Iran'. In M. A. Cook (Ed.), *Studies in Economic History of the Middle East*. London: Oxford University Press.

Malik, Adeel. 2017. 'Rethinking the Rentier Curse'. *International Development Policy | Revue internationale de politique de développement*, no. 7 (February).

Mason, Michael. 2014. 'Climate Insecurity in (Post)Conflict Areas: The Biopolitics of United Nations Vulnerability Assessments'. *Geopolitics* 19 (4): 806–28.

Matin, Kamran. 2013. *Recasting Iranian Modernity: International Relations and Social Change*. Routledge.

McGoey, Linsey. 2017. 'The Elusive Rentier Rich: Piketty's Data Battles and the Power of Absent Evidence'. *Science, Technology, & Human Values* 42 (2): 257–79.

Migdal, Joel S., ed. 2004. *Boundaries and Belonging: States and Societies in the Struggle to Shape Identities and Local Practices*. Cambridge: Cambridge University Press.

Mohaddes, Kamiar, and M. Hashem Pesaran. 2013. *One Hundred Years of Oil Income and the Iranian Economy: A Curse or a Blessing?* Rochester, NY: Social Science Research Network.

Momeni, Farshad, and Shima Haji Nourouzi. 2018. 'اقتصادی سیاسی خصوصی سازی همراه با فساد'. فصلنامه پژوهشهای اقتصادی 17 (67): 221–52.

MPORG. 2020. 'لایحه بودجه سال 1399 کل کشور'. Accessed 10 March 2021. https://www.mporg.ir/Portal/View/Page.aspx?PageId=37cd6f7f-f233-4584-9c0c-fbd09ccfb3ba.

Nesbitt, J. Todd, and Daniel Weiner. 2001. 'Conflicting Environmental Imaginaries and the Politics of Nature in Central Appalachia'. *Geoforum* 32 (3): 333–49.

Neumann, Rod. 2004. 'Moral and Discursive Geographies in the War for Biodiversity in Africa'. *Political Geography* 23(7): 813–37.

OPEC. 2014. 'OPEC Bulletin 2014'. n.d. Accessed 12 March 2021. https://www.opec.org/opec_web/en/publications/2746.htm.

OPEC. 2020. 'Annual Statistical Bulletin'. Accessed 30 March 2021. https://www.opec.org/opec_web/en/publications/202.htm.

Pal, Jeremy S., and Elfatih A. B. Eltahir. 2016. 'Future Temperature in Southwest Asia Projected to Exceed a Threshold for Human Adaptability'. *Nature Climate Change* 6 (2): 197–200.

Peet, Richard. 1985. 'The Social Origins of Environmental Determinism'. *Annals of the Association of American Geographers* 75 (3): 309–33.

Peet, Richard, and Micheal Watts. 1996. *Liberation Ecologies: Environment, Development, Social Movements*. London: Routledge.

Peet, Richard, Paul Robbins, and Michael Watts, eds. 2010. *Global Political Ecology*. Routledge.

Ross, Michael L. 1999. 'The Political Economy of the Resource Curse'. *World Politics* 51 (2): 297–322.

Ross, Michael L. 2001. 'Does Oil Hinder Democracy?' *World Politics* 53 (3): 325–61.

Ross, Michael L. 2015. 'What Have We Learned About the Resource Curse?' *Annual Review of Political Science* 18 (1): 239–59.

Roudgarmi, Pezhman, and Mohammad Taghi Amoozadeh Mahdiraji. 2020. 'Current Challenges of Laws for Preservation of Forest and Rangeland, Iran'. *Land Use Policy* 99 (December): 105002.

Sachs, Jeffrey D., and Andrew M. Warner. 2001. 'The Curse of Natural Resources'. *European Economic Review, 15th Annual Congress of the European Economic Association* 45 (4): 827–38.

Sala-i-Martin, Xavier, and Arvind Subramanian. 2003. *Addressing the Natural Resource Curse: An Illustration from Nigeria*. SSRN Scholarly Paper ID 420318. Rochester, NY: Social Science Research Network.

Schwartzstein, Peter. 2020. 'The Authoritarian War on Environmental Journalism'. https://tcf.org/content/report/authoritarian-war-environmental-journalism/.

Schwarz, Rolf. 2008a. 'From Rentier State to Failed State: War and the De-Formation of the State in Iraq'. *A Contrario* 5 (1): 102–13.

———. 2008b. 'The Political Economy of State-Formation in the Arab Middle East: Rentier States, Economic Reform, and Democratization'. *Review of International Political Economy* 15 (4): 599–621.

Scott, James C. 1998. *Seeing Like a State: How Certain Schemes to Improve the Human Condition Have Failed*. New Haven and London: Yale University Press.

Selby, Jan. 2014. 'Positivist Climate Conflict Research: A Critique'. *Geopolitics* 19 (4): 829–56.

Selby, Jan, Omar S. Dahi, Christiane Fröhlich, and Mike Hulme. 2017. 'Climate Change and the Syrian Civil War Revisited'. *Political Geography* 60 (September): 232–44.

Selby, Jan, and Clemens Hoffmann. 2014. 'Rethinking Climate Change, Conflict and Security'. *Geopolitics* 19 (4): 747–56.

Sengupta, Somini. 2018. 'Warming, Water Crisis, Then Unrest: How Iran Fits an Alarming Pattern'. *The New York Times.*

Shahi, Afshin, and Ehsan Abdoh-Tabrizi. 2020. 'Iran's 2019–2020 Demonstrations: The Changing Dynamics of Political Protests in Iran'. *Asian Affairs* 51 (1): 1–41.

Shahi, Afshin, and Maya Vachkova. 2018. 'Eco-Sectarianism: From Ecological Disasters to Sectarian Violence in Syria'. *Asian Affairs* 49 (3): 449–67.

Shahi, Afshin. 'Drought: The Achilles heel of the Islamic Republic of Iran'. *Asian Affairs* 50 (1) (2019): 18–39.

Shakoori, Ali. 2001. *The State and Rural Development in Post-revolutionary Iran.* New York and Hampshire, UK: Palgrave.

Shambayati, Hootan. 1994. 'The Rentier State, Interest Groups, and the Paradox of Autonomy: State and Business in Turkey and Iran'. *Comparative Politics* 26 (3).

Skocpol, Theda. 1982. 'Rentier state and Shi'a Islam in the Iranian Revolution'. *Theory and Society* 11 (3): 265–283.

Slettebak, Rune. 2012. 'Don't Blame the Weather! Climate-Related Natural Disasters and Civil Conflict'. *Journal of Peace Research* 49 (1): 163–76.

Smith, Benjamin. 2007. *Hard Times in the Lands of Plenty.* Ithaca, NY: Cornell University Press.

Swanson, Jacinda. 2008. 'Economic Common Sense and the Depoliticization of the Economic'. *Political Research Quarterly* 61 (1): 56–67.

Tahbaz, Morad. 2016. 'Environmental Challenges in Today's Iran'. *Iranian Studies* 49 (6): 943–961.

Temper, Leah, del Bene, Daniela, and Martinez-Alier, Joan. 2015. 'Mapping the Frontiers and Front Lines of Global Environmental Justice: The EJAtlas'. *Journal of Political Ecology* 22: 255–78.

UNDP. 2017. 'Iran and UNDP Strengthen Efforts to Combat Desertification'. n.d. UNDP in Iran. Accessed 23 February 2021. https://www.ir.undp.org/content/iran/en/home/presscenter/articles/2017/06/17/iran-and-undp-strengthen-efforts-to-combat-desertification.html.

UNPO. 2018. 'UNPO's Alternative UPR Report on Iran Denounces Environmental Issues and Human Rights Violations'. https://unpo.org/article/20552.

Vahabi, Mehrdad. 2018. 'The Resource Curse Literature as Seen Through the Appropriability Lens: A Critical Survey'. *Public Choice* 175 (3): 393–428.

Vahabzadeh, Peyman. 2015. 'Rebellious Action and the "Guerrilla Poetry": Dialectics of Art and Life in the 1970s Iran'. In K. Talattof (Ed.), *Persian Language, Literature and Culture: New Leaves, Fresh Looks; In Honor of Ahmad Karimi-Hakkak*, 103–22. London and New York: Routledge.

Waldman, Scott. 2018. 'Climate Change May Have Helped Spark Iran's Protests'. *Scientific American*. Accessed 23 March 2021. https://www.scientificamerican.com/article/climate-change-may-have-helped-spark-iran-rsquo-s-protests/.

Wantchekon, Leonard. 2002. 'Why Do Resource Dependent Countries Have Authoritarian Governments'. *Journal of African Finance and Economic Development* 5 (2): 57–77.

WRI. 2019. 'Aqueduct 3.0 Country Rankings'. *World Resources Institute*. https://www.wri.org/resources/data-sets/aqueduct-30-country-rankings.

Yazdandoost, Farhad. 2016. 'Dams, Drought and Water Shortage in Today's Iran'. *Iranian Studies* 49, (6): 1017–28.

Zahedi, Arya. 2019. 'The Iranian Revolution at the Twilight of the Worker's Council'. Commune (blog). https://communemag.com/the-iranian-revolution-at-the-twilight-of-the-workers-council/.

CHAPTER 4

Policing Indigenous Land Defense and Climate Activism: Learnings from the Frontlines of Pipeline Resistance in Canada

Jen Gobby and Lucy Everett

Introduction

As we sit at our laptops writing this chapter, 300+ wildfires are burning throughout British Columbia. In the Prairie provinces, farmers are calling for emergency relief as crops fail amid heat waves and intense drought. The global climate system reels and careens, increasingly out of balance, yet the Canadian federal and provincial governments are holding fast to an economic system addicted to fossil fuels and extractive economics. Canada

J. Gobby (✉)
Concordia University, Montreal, QC, Canada
e-mail: jengobby@gmail.com

L. Everett
McGill University, Montreal, QC, Canada
e-mail: lucy.everett@mail.mcgill.ca

A. Dunlap and A. Brock (eds.), *Enforcing Ecocide*,
https://doi.org/10.1007/978-3-030-99646-8_4

is not only failing to meaningfully reduce greenhouse gas emissions but is actively building new oil and gas pipelines to transport bitumen from the Alberta Tar Sands to the Pacific coast. This glaring failure of the Canadian government to respond to the climate crisis and transition away from fossil fuel extraction is being met by fierce and concerted resistance from Indigenous land defenders and increasing numbers of climate activists. Grassroots resistance has been thwarting these corporate interests, in many cases successful in delaying and even stopping proposed projects (see Gobby et al. 2021; Temper et al. 2020). Meanwhile, there has been an increase in the surveillance and criminalization of resistance efforts. This is especially so for Indigenous movements, who are disproportionately policed as they defend their communities, their rights, and their territories (Crosby & Monaghan 2018). Policing serves to actively protect the status quo which is driving ecological and climate catastrophe and is starkly at odds with Canada's climate targets and reconciliation promises.

In this chapter we examine how the violent criminalization of Indigenous land defenders is employed by the state as a tool to repress the fights against extractivism and the fossil fuel hegemony in Canada. We explore these dynamics through the lens of multiple forms of power exerted by the police and the corresponding, counter-power cultivated by frontline communities and grassroots social movements. Through this analysis, we argue that the policing of anti-pipeline movements has been serving to actively enforce the violation of Indigenous rights and continuation of Canada's fossil fuel dependency, and its contribution the climate crisis.

We come to the writing of this chapter deeply invested in, and grateful to, the work of resistance movements. Jen Gobby is a settler climate organizer and a postdoctoral researcher based in unceded Abenaki territory in rural Quebec. Lucy Everett is a student, researcher, and activist of mixed white settler (British, Scottish, Mennonite) and Red River Metis descent, currently based on unceded and illegally occupied Coast Salish territories in so-called British Columbia. Both of us have been involved with anti-pipeline efforts and have, to varying degrees, experienced the policing of the movements that we are part of. We aim to do research and writing that contributes to movement efforts while helping to expose the colonial violence and injustice inherent in Canadian politics and economic policies.

This chapter is based on a mixed methodology that includes reviewing recent literature on policing, surveillance, and criminalization of environmental and Indigenous movements in Canada. It further draws on the findings from recent research projects each of us have, separately, led in collaboration with Indigenous land defenders resisting pipeline developments. Between 2016 and 2018, Jen conducted more than 50 interviews and several focus groups with climate activists and land defenders involved with anti-pipeline organizing across Canada. Through 2020–21, Lucy conducted 15 interviews with land defenders, economists, public finance experts, and government officials about the financial accounting practices involved in the TransMountain Expansion (TMX) pipeline project. This project was conducted in collaboration with the Tiny House Warriors, an Indigenous-led group resisting the TMX pipeline. In this chapter, we draw on existing literature, our own experiences in anti-pipeline organizing, and on the interviews we carried out with activists, land defenders, and other experts.

In the next section, we describe the settler colonial context in which the racist, violent policing of Indigenous movement is taking place. We then draw on Rodríguez et al.'s (2017) Conflict Transformation framework to explore four forms of power—*discursive, institutional, relational, and material*—that are being wielded by the security state in Canada in the policing and criminalization of Indigenous movements. Finally, we describe the ways these forms of power are being wielded by frontline communities and grassroots movement as *counter power* against extractive industries and the settler colonial state.

The Context

Ongoing Settler Colonialism

Settler colonialism is a form of colonialism[1] that established, built and currently maintains the Canadian state. Settler colonialism "functions through the replacement of Indigenous populations with an invasive settler society that, over time, develops a distinctive identity and sovereignty" (Barker and Lowman n.d., see also Wolfe 2006; Veracini 2011). With settler colonialism, the colony is not just a place to extract resources to send back home; instead the colony becomes home to the

[1] Colonialism generally refers to the exertion of control over territory or resources outside the official boundaries of a state or empire (Barker and Lowman 2015).

settlers (Waldron 2018). Making home for colonizers requires the elimination of Indigenous Peoples and their relationships with the land (Wolfe 2006; Shoemaker 2015; Veracini 2011, 2014).

Since Europeans began colonizing the lands now known as Canada, the extraction and exploitation of natural resources has been the basis of the state's economic functioning (Simpson 2019). From as far back as the fur trade right up to this current moment of tar sands expansion, extractive industries have been damaging social and ecological systems (Chodos 1973; Simpson 2019; Huseman and Short 2012), meanwhile generating wealth and power for settlers. Settler colonial policies and structures of dispossession, elimination, and assimilation have been developed to overcome the obstacle that Indigenous Peoples and their rights pose to the settlers' access to lands and resources required to build the nation state and economy. Canada's extractive economy has been and still is premised on the dispossession of land and resources from Indigenous peoples (Bernauer et al. 2018).

Dene scholar Glen Coulthard (2014) makes clear that the primary purpose of settler colonialism is access to land, for settlement and for capital accumulation. Nishnaabeg scholar Leanne Betasamosake Simpson writes that "the Canadian state has always been primarily interested in acquiring the … rights to [Indigenous] land for settlement and for the extraction of resources" (2017: 42). Settler colonial society requires the elimination of Indigenous societies and Indigeneity (Wolfe 2006; Crosby 2021) in order to access lands and resources to exploit them for capital accumulation and distribution among settlers.

The Truth and Reconciliation Commission (TRC) Report, released in 2015, starkly exposed many to the huge suffering that has come directly from historical and ongoing colonial relations between the Canadian state and Indigenous Peoples, naming the historical treatment of First Nations, Inuit, and Metis peoples by the Canadian state as "cultural genocide" (TRC 2015: 1). Almost two decades prior, the 1996 Royal Commission on Aboriginal People explicitly laid out Canada's imposition of a colonial relationship on Indigenous Peoples, which includes.

> residential schools, forcible relocation, the imposed Band Council system, institution of a pass system, germ warfare, outlawing of ceremonies such as the potlatch and traditional activities such as fishing, failed treaty processes, and other forced assimilation policies. Currently, it takes the form of the imposition of foreign governance systems legislated through the Indian

> Act and state sanctioned appropriation of Indigenous lands and resources. (Walia 2012: 241; see also Dusault and Erasmus 1996)

The impacts of ongoing settler colonialism include systemic social, economic, and health inequalities between Indigenous Peoples and settler Canadians (see Manuel and Derrickson 2017: 78; TRC 2015: 146–147). In 2017, the United Nations Committee on the Elimination of Racial Discrimination urged the Canadian government to address what it found were persistent violations of Indigenous rights. In its 2018 *World Report*, Human Rights Watch states that the Canadian government "has yet to pay adequate attention to systemic poverty, housing, water, sanitation, healthcare, and education problems in Indigenous communities" (Human Rights Watch 2018: n.p.).

The lands and waters that Canadian settlers and the crown now claim as their own were not freely handed over by Indigenous Peoples. These territories were never ceded, neither by treaty nor war. They were taken through force and coercion. In the cases where treaties were signed, the land was often subsequently seized through transparent violations of treaty agreements (McFarlane and Schabus 2017). In various ways, Canada's claim to state control over these lands and waters "rest[s] on fragile claims and legal fictions grounded in the likes of the Doctrine of Discovery and terra nullius and thus is ripe for contestation on a variety of sociolegal and political fronts" (Crosby 2021: 5; see also Reid 2010 and Manuel and Derrickson 2015).

As one Land Defender we interviewed explained

> Racism is at the foundation of Canada's claim to sovereignty, of having power over us. A Canadian state shouldn't exist. But because of racism, because of the Doctrine of Discovery, they claim the right to exist. [Canada] exists because of racism.[2]

Canadian governments' claims to Indigenous lands and to natural resources thus rest on racist, colonial, and deeply contested grounds. These claims have been resisted continuously since Europeans first arrived, as documented in works such as Gord Hill's 500 years of Resistance (2010). Indigenous rights and resistance pose a significant threat to both the ideational and material dimensions of the settler colonial project, and

[2] JG Interview #38, Mik'maw, July 28, 2017.

for this and other reasons, Indigenous movements are subject to disproportionally high levels of surveillance, police violence, and criminalization (Crosby and Monaghan 2018).

The Systemic Racism in the Police and Judicial Systems

Indigenous land defenders face more risk of arrest and criminalization than settler activists; they are monitored more closely, violently, and frequently on mere suspicion rather than reasonable or probable grounds (Monaghan and Walby 2017). Indigenous communities defending their lands and waters face particularly high rates of violent repression, criminalization, and surveillance (see Nikiforuk 2019; Monaghan and Walby 2017). This reflects a broader trend globally. The UN reported in April 2019 that Indigenous peoples are facing an escalation in criminalization and violent repression especially when exercising and defending their rights to their territories and natural resources.[3]

This is part of a broader reality in Canada whereby racialized populations, particularly Black and Indigenous peoples, are at greater risk of police violence (Monaghan and Walby 2017) and are overrepresented in Canadian incarceration rates (Owusu-Bempah and Wortley 2014; see also Maynard 2017). For example, 27% of the incarcerated people in Canada are Indigenous, despite constituting only 4% of Canada's population (Brake 2018). This trend is also gendered: a report released by Canada's Correctional Investigator showed that Indigenous women (and Two-Spirit people) represent 37% of all incarcerated women and 50% of all female maximum-security inmates (Brake 2018).

As the late Arthur Manuel, Secwepemc thinker and political leader, wrote:

> For Indigenous people in Canada, the colonial law enforcer meets us at every turn... If we try to keep resource extractors from moving onto our lands, injunctions against us are quickly awarded and the police swoop in for mass arrests. Canadian jails are full of our young men and women. That is colonialism. That is oppression. (Manuel and Derrickson 2017: 73)

Kanahus Manuel, daughter of Arthur Manuel and one of the founders of Tiny House Warriors, explained in an interview how her hair "stands up

[3] https://www.un.org/press/en/2019/hr5433.doc.htm.

... to think of how cruel Canada and the crown [are] ... over some civil disobedience. That shouldn't ever lead you to jail time. But as Natives, we know we are targeted with jail time at a far greater rate than the other races here in Canada".[4] This insight was mirrored by an Anishinaabe land defender, who explained:

> When it comes to Indigenous Peoples defending their own territory, there is a history of law enforcement escalating quickly. This is not a new thing; this has been happening for a long time. It is a racist system playing out ... white supremacy that is keeping Indigenous Peoples in fear because either you're going to get shot, the military is going to be called, or you're going to be facing life in prison. So, it's not the same story when it comes to a bunch of white people in kayaks blocking the freighters.... For Indigenous Peoples it's our lives that are at stake when it comes to defending the land and the water.[5]

Underlying this systemic racism in policing in Canada are the logics and interests of settler colonialism and the incessant drive to "shor[e] up access to territories for state formation, settlement, and capitalist development" (Howe and Monaghan 2018: 332, see also Coulthard 2014). To protect such interests, the Canadian security state has developed and deployed "elaborate bureaucratic mechanisms, policing tactics and technologies, and systems of security governance that operate within the contours of the eliminatory logic of settler colonialism" (Crosby 2021: 5).

Although the repression of Indigenous resistance is continuous throughout Canada's history, the ways Indigenous communities and movements are policed ve changed over time. As Crosby and Monaghan (2018) have shown, using extensive data from Access to Information requests, recent trends in surveillance and policing of movements have been developed in response to ongoing and mounting community and movement resistance to oil and gas pipelines and other extractive industry development over the last two decades. Crosby and Monaghan refer to this new dynamic of policing as the *security state*, which they describe as a "sprawling array of national security and policing agencies, industry and corporate partners, and public bureaucracies that are increasingly integrated through surveillance, intelligence databanks,

[4] LE Interview #41, Kanahus Manuel, September 3, 2020.

[5] JG Focus Group Interview #2, Anishinaabe, September 28, 2017.

and institutional partnerships in efforts to pre-empt or disrupt potential threats" (Crosby and Monaghan 2018: 3). For example, in 2007, Indigenous and Northern Affairs Canada (INAC) launched a "hot spot reporting system" to monitor "native unrest" (Manuel and Derrickson 2017: 226). In 2017, the RCMP's Community-Industry Response Group (C-IRG) was created in British Columbia to oversee policing related to the energy industry specifically. In 2012, in response to the Idle No More movement, the Royal Canadian Mounted Police (RCMP) set up Project SITKA to monitor and track Indigenous protests and activists (Nikiforuk 2019). These characterize the current policing efforts in Canada to quell Indigenous decent and facilitate and enforce the expansion of the fossil fuel industry, placing the full force of the state behind industry to maintain the extractive, carbon polluting status quo.

Policing Resistance Movements in Canada

The 2018 report from the IPCC stated that addressing the mounting climate crisis requires limiting global warming to 1.5 °C, the temperature increase limit necessary to avert the dangerous destabilization of ecological and social systems—which will itself require "rapid, far-reaching and unprecedented changes in all aspects of society". Not only is the Canadian government failing to take this kind of meaningful action to address the climate crisis, they are actively repressing—through surveillance, injunctions, criminalization, and violence—the very people and communities that are standing up to defend their waters, lands, and rights and to fight for a habitable earth (Ceric 2020; Nikiforuk 2019; Monaghan and Walby 2017).

This increase in the criminalization of Indigenous land defense has been playing out in multiple fights against new oil and gas pipelines including the Northern Gateway, Line 3, Line 9, KeystoneXL, Energy East, Coastal Gas Link, TransMountain, and other pipelines over the last decade.[6] And as Indigenous Nations and communities across the country have been joining forces with each other and with settler environmental and other social justice movements, the collective power to resist extractivism in Canada has been building (Gobby 2020). People have set up

[6] Northern Gateway, KeystoneXL, and Energy Energy projects were successfully stopped. The resistance to Line 9 was not successful. Line 3, Coastal Gas Link, and TransMountain are under construction but continuing to be resisted.

blockades and long-term protest camps, launched international boycotts, manually shut down pipelines, brought industry to court, and taken to the street en masse to draw attention to injustice and to amplify the voices and demands of those resisting (Gobby et al. 2021). In some cases, community and movement efforts have led to delaying and even halting projects as well as to legislative victories, new legal precedents, and nurturing relationships between people and the land as communities form to defend life (Black et al. 2014; Gosine and Teelucksingh 2008; Gobby 2020, 2021). Crobsy and Monahan (2018: 5) contend that the policing and criminalization of these movements is a response to the "growing momentum in the politics of Indigenous self-determination, meaningful efforts at settler solidarities, and the abilities to disrupt the status quo. As agencies that serve to protect the status quo of settler colonialism, they are also responding to the need to protect the wealth accumulated through the exploitation of Indigenous lands" (Crosby and Monaghan 2018: 5).

In virtually every single one of these pipeline fights, Indigenous land defenders face discriminatory and violent treatment by police and private security forces. As one example, there have been three rounds of mass arrests of Indigenous land defenders and allies protesting TMX[7] in Tsleil-Waututh Territory, on so-called Burnaby Mountain, a site of resistance to TMX along the pipeline's terminal spread. The first round of arrests in 2014 included an 11-year-old girl that was placed in police custody for allegedly breaching an injunction zone. Protestors reported that Indigenous people were treated more violently, both physically and verbally, by the RCMP as compared to the treatment of settlers present (Spiegel 2021). Similarly, it was reported that during the spring and summer of 2018 at the gates of the TMX construction site, Indigenous and settler protestors were arrested in very different manners. While arresting non-Indigenous protestors, RCMP officers were careful to adhere to the 5-step process and some cases RCMP officers even pleaded with protestors to leave so they did not have to arrest them. This is in stark contrast to the manner in which RCMP officers violently attacked several Indige nous land defenders, physically assaulting them before arresting them, even breaking the wrist of one of the land defenders[8] (Hermes 2021).

[7] The TransMountain pipeline is a proposed twinning of an existing pipeline that brings bitumen oil from the Alberta tar sands to the ports of Vancouver for export. Despite fierce and ongoing protest for over a decade now, the project is being built.

[8] LE Interview #41, Kanahus Manuel, Sept, 3, 2020.

In August 2018 a group of Indigenous women held a news conference outside of the provincial courthouse in Vancouver to address the targeting of Indigenous women by the RCMP. One woman, Crystal Smith, reported being "violently arrested" in front of her children. She explained, "I had bruises on my wrists. I had bruises on my upper body from the force that they used on me… It is here that we see the difference between Indigenous bodies and non-Indigenous bodies… It is here that we see the difference of attitude that the RCMP has" (quoted in Brake 2018).

In another prominent pipeline fight, also in British Columbia, land defenders of the Wet'suwet'en Nation set up the Unist'ot'en resistance community on their traditional territory through which multiple oil and gas pipelines[9] have been proposed. In February 2019, the RCMP "moved in with a large force on all … camps, forcibly removing the Wet'suwet'en and their supporters and occupying the territory so that pipeline crews could commence work" (Dhillon and Parish 2019: n.p.). Notes from an RCMP strategy session, accessed by *The Guardian* newspaper, showed that Canadian police had been prepared to use lethal force and shoot Indigenous land defenders, receiving instructions from commanders to "use as much violence … as you want" (Dhillon and Parish 2019: n.p.). This willingness to use lethal violence to push pipelines through Indigenous territory makes clear that police are being employed to ensure and enforce the interests of extractive economics.

This violent repression of Indigenous lifeways and resistance movements is deeply in conflict with the Canadian government's rhetoric of reconciliation, including their recent commitments to respect Indigenous rights as affirmed by the United Nations Declaration on the Rights of Indigenous Peoples (UNDRIP). Additionally, this deployment of state force against those standing up to protect the environment and future generations is deeply at odds with the urgent need to address the climate crisis. The policing of anti-pipeline movements has been serving to *actively enforce the violation of Indigenous rights and continuation of Canada's fossil fuel dependency, and its contribution the climate crisis*. In short, the Canadian government is enforcing ecocide.

[9] These proposed pipelines are part of an 'energy corridor' aiming to connect the Alberta tar sands and shale gas extraction projects with ports and LNG processing terminals in Kitimat and Prince Rupert on BC's west coast.

Exploring the Conflict Transformation Framework

In the following sections, we draw on Iokine Rodriguez and colleagues' (2017) Conflict Transformation framework, which offers three forms of power: *discursive power, institutional power, and relational power*. By power, we are referring to the ability of different social actors or groups of actors to influence others, the course of events, and outcomes. We add to Rodriguez et al.'s framework by offering a fourth form of power: *material power*, which we conceptualize as including both the power wielded by financial and other resources as well as the power of physical force.

We engage with this framework to help unpack and tease apart the multiple forms of power the security state in Canada is using against communities and movements in order to enforce pipeline expansion. Such power analysis aids us in identifying and understanding different strategies being used. This understanding can then inform movement counter strategies, rendering them more effective in applying power, and indeed, better *enforcing* Indigenous rights and the defense of lands, water, and climate. Exploring power as a multidimensional phenomenon, rather than a singular force, can help identify where and how we can most effectively defend life in the face of ecocide.

Discursive Power

> The field of security governance is constituted by an evolving set of discursive and social practices regarding imagined and perceived threats (Crosby 2021: 2)

Power can be wielded through words, framings, concepts; by the ways notions of "normal", "acceptable", "justified", "threat", etc., are constructed (Rodríguez et al. 's 2017). In recent trends in the policing of Indigenous movements, we see a clear use of discursive power to delegitimize Indigenous resistance, and to paint resistance as threatening the safety and wellbeing of settler Canadians and their interests. This strategy is commonly employed by pro-extractive industry state and non-state actors to portray community leaders and activists as obstacles to development, as posing risks to national security, as undermining Canadian values, and as mobilizing disruptive and violent events.

As evidenced in RCMP reports and other documents, Indigenous grievances are framed by the security state as matters of perception and assertions of self-determination, seen as "irrational, criminal, and a potential threat to the material and immaterial interests of settler society" (Crosby 2021: 8). Within police reports, Indigenous political grievances are dismissed "within stereotypical tropes of Native greed, of acting out of individual self-interest" and as marginal within Indigenous communities, whereby the grievance and resulting resistance is only being expressed by a small faction within a community (Crosby 2021: 13). As Crosby makes clear, "security officials deploy designations such as'factions' or 'splinter groups' to marginalize and delegitimize those community members seen as acting outside the authority of the band council system" (2021: 13). This approach demonstrates a concerted strategy to divide and pacify different Indigenous communities.

The security state and industry also employ narratives that create false impressions of support for energy infrastructure projects. This entails misleading maneuvers and questionable rationales to claim social license and community approval, which serve to fashion the police-military interventions as "objective" and "neutral", "conceal[ing] their racialized and prejudicial characters" (Howe and Monaghan 2018: 331). This hides the ways that the RCMP's risk management tools (and "risk" constructions) are, in actuality, antagonizing and attempting to delegitimize Indigenous and environmental movements (Howe and Monaghan 2018).

Perhaps the most pervasive and powerful discursive maneuver in the recent policing of Indigenous movements has been the framing resistance to oil and gas pipelines within narratives of the "war on terror". This evocation of the war on terror has increased the security state's power to monitor and police Indigenous resistance, for example, expanding Canadian Security Intelligence Service (CSIS)'s mandate to use covert actions and tactics, even ones that violate the Canadian charter of rights and freedoms (Moore 2015). Along with other countries, security agencies in Canada are now classifying environmental activities as domestic terrorist threats, bringing anti-terrorism legislation and rhetoric to realm of civil disobedience (Le Billon and Carter 2012) and associating Indigenous land defense with terrorism, extremism, and violence. Yet, there is little to no evidence of Indigenous movements using or planning to use violent tactics (Le Billon and Carter 2012; Crosby and Monaghan 2018).

As a Kanien'kehá ka land defender explained in an interview,

> There is a conflict between companies and Indigenous Peoples ... to the point that ... if you're a native then you are a terrorist. If you look at how policing is used against activists and environmentalists, it's fucking disgusting. [They justify it by] painting us as wack jobs ... crazy people, not like everybody else. They try to make us out like we don't know what we are talking about, like we're not part of society[10]

This blurring of the very real lines between land defense, civil disobedience, and terrorism leads to an implicit equivalence between people defending their lands and communities against unwanted extractive development and violent terrorism aimed at injuring and killing people (Monaghan & Walby 2017). This conflation helps justify the extensive policing and criminalization of Indigenous movements that challenge extractive capitalism (Crosby and Monaghan 2018: 3; see also Moore 2015).

The war on terror brought an expansion of the definition of terrorism to include any interference with Canada's economic or financial stability and interference with "critical infrastructure" (Moore 2015). Indeed, special branches have emerged in Public Safety Canada and the RCMP that center around critical infrastructure (Monaghan and Walby 2017; Crosby 2021). This further justifies domestic surveillance and the repression of Indigenous land defense (Spice 2018; Crosby and Monaghan 2018; see also Granovsky-Larsen and Santos 2021). Oil and gas pipelines have become labeled as critical infrastructure, which has rendered resistance to pipelines as domestic terrorism. As one of many examples, the protection of critical infrastructure was evoked repeatedly in the orders to invade and dismantle the Wet'suwet'en resistance camps, citing the CGL pipeline as critical to the wellbeing of Canadians and land defenders and their supporters cast as terrorists (Spice 2018).

We see here how language is used to paint oil and gas pipelines in the public eye not as profit seeking, climate disrupting industrial projects but as critical to national wellbeing and normalized as unavoidable common sense (Spice 2018). The development of oil and gas and the invasion of Indigenous lands is normalized and framed as being in the best interest of Canada. It then becomes harder to imagine decolonial and ecologically viable futures (Spice 2018). The settler state is shaping narratives around extractive projects as part of natural advancement of the state

[10] JG Interview #7, Kanien'kehá ka, June 5, 2017.

while masking violence caused to Indigenous land and bodies (Spice 2018).

The security state, we can see, is creating concepts, narratives, and definitions that serve extractive settler colonial interests, while justifying the monitoring, categorization, and profiling of Indigenous resistance. This "proliferation of mechanisms and technologies to protect critical infrastructure – or render it resilient – present new dynamics of racializing surveillance that targets Indigenous peoples as national security threats" (Crosby 2021: 9). This is discursive power at play in Canada to serve extractive interests and settler colonialism.

Institutional Power

> Laws are created to protect people that own things, ownership, privacy. The rules are very vague, and they permit officers to arrest anyone in almost any situation.[11]

Discursive power is not just about influencing public opinion and maintaining colonial legitimacy in the hearts and minds of Canadians, it does much more than that. Certain concepts—such as national interest and critical infrastructure—justify the building of pipelines and become institutionalized. They are legal tools to allow for the state to render certain things legal or illegal, and then to mobilize massive resources for surveillance and policing, and to employ physical force against those posing threats to the ability to build pipelines. Concepts become legal instruments, determining what the state can do to people, determining whose lives and interests are protected, and who is subject to violent arrests and imprisonment.

Discursive power, then, undergirds *institutional power*, a form of power that wielded through engagement with formal institutions within economic, legal, and political systems such that one can more effectively influence outcomes (Rodríguez et al. 's 2017). Institutional power is exerted through the Canadian constitution, which claims colonial jurisdiction over stolen Indigenous lands. It is exerted through laws and regulations that make the forcing of pipelines through Indigenous territories "legal". It is exerted through bodies like the National Energy Board

[11] JG Focus Group Interview #2, September 28, 2017.

(NEB), whose studies and consultations assess and advise decision-makers about potential social and environmental impacts of proposed pipelines and other projects. It is exerted through government decision-making to approve projects. It is exerted through the courts of law that grant injunctions to allow for the police to remove land defenders and their blockades that impede the construction of pipelines. Courtrooms are "routinely places of asserting settler-colonial power over Indigenous laws and values, denying Indigenous claims, controlling Indigenous bodies and defining – with colonial law – what is relevant or irrelevant about Indigenous land, governance systems and life" (Spiegel 2021: 2). Institutional power is exerted through the courts that render judgments and sentencing, and then it is exerted through the prison system where land defenders serve sentences.

This list is by no means exhaustive of the vast web of colonial institutions in Canada that are involved with the development of pipelines and the repression of those who oppose them. All these institutional arrangements are important to understand, but in this section, we focus in on one specific tool of institutional power: court injunctions. These have been a tool of institutional power used frequently to protect corporate interests against Indigenous land defenders and others resisting pipelines in Canada.

An injunction is court order issued by a judge in response to an application filed by a party involved in a lawsuit and is meant to protect that party's interests or rights while a case is under judicial review (Ceric 2020). Through an injunction, one or more of the parties involved in a trial is ordered to do or refrain from doing a specific action or actions. An injunction granted by the BC Supreme Court in 2018 ordered land defenders on Burnaby Mountain to refrain from obstructing an entrance to a TMX facility, destroying signage or fencing around TMX sites, or coming within 5 m of TMX property (Hermes 2021).

For an injunction to be granted, the plaintiff must show that without it, they are likely to suffer irreparable harm, meaning harm that cannot be financially compensated. The plaintiff must also show that the injunction is in the public interest and that the injunction's benefit to the plaintiff outweighs its burden on the defendant, that the injunction is in the public interest. With injunctions associated with extractive projects, the applicant is generally the corporate proponent of the project, the standard for getting an injunction is minimal and often based on the alleged irreparable harm associated with impeding the progress of extractive projects (Ceric 2020).

The failure to comply results in a charge of *contempt of court*. When land defenders and allies violate the terms of injunction, for example by refusing to leave a blockade, the courts can issue an enforcement order empowering police to arrest alleged "contemnors" (Ceric 2020). Those arrested under an injunction are not charged under the criminal code, which would include provisions for the accused's defense, with the reasons for and circumstances around the protest activities taken into account by the judge. By arresting people through injunctions, such protections under the Criminal Code are circumvented (Ceric 2020; Hermes 2021).

Court injunctions effectively constitute privately bought deterrence mechanisms that serve to criminalize resistance "a-priori" (Brock 2020). Court injunctions stifle and repress resistance in general, but are a tool used specifically against Indigenous rights and resistance (Ceric 2020). In the Yellowhead Institute report, Kruse and Robinson (2019) explain that injunctions are regularly used against First Nations, often in the context of resistance blockades, "to circumvent [the Nations'] ability to assert Aboriginal rights/title and treaty rights in relation to Crown and corporate development and projects" (n.p.). As Kanahus Manuel, a Secwepemc land defender deeply involved with the resistance to Trans-Mountain pipeline explained: "it's very biased, these injunctions, they really side with the corporations".[12]

Her father, Arthur Manuel (2017: 215) explained that injunctions are being used as a "legal billy club", through which the "assertion of [Indigenous] rights on the ground is instantly criminalized by the Canadian state". Put otherwise, injunction is a "blunt instrument in opposition to Indigenous law" (Pasternak and King 2019: 29). Injunctions are among the legal weapons of institutional power to facilitate corporate "access to resources and lands and easing the operation of extractive capitalism" (Ceric 2020: n.p.). Court injunctions are essential for enforcing ecocide and the trampling of Indigenous rights and self-determination.

[12] LE Interview #41, Kanahus Manuel, Sept, 3, 2020.

Relational Power

> Power is better conceptualized as a relationship than as a static entity.... Power constitutes a relationship. (Collins and Bilge 2016: 28)

Intersectional feminist scholars have long argued that power is inherently relational and Foucault conceived of power in relational terms as well, moving through networks, flowing or shared between institutions or people (1971). Through Rodriguez et al.'s framework, we see *relational power* as built through networks—of people, organizations, and state institutions. Such networks facilitate the collaboration, the sharing of information and resources between different actors to work together toward common objectives. Relational power is built through increasing the collaboration between actors (Rodríguez et al. 's 2017). Although relational power can refer to a very broad range of social phenomena, here we focus in on one specific form of relational power being built in Canada; the ways that government and law enforcement agencies and private sectors have come to work together more and more closely over the last decade to monitor, police, and criminalize Indigenous movements that are resisting extractive projects in Canada. We describe the ways that the security state is building *power* through *relational* means, including partnerships (see Brock 2020) and networks of intelligence sharing (see chapter 10, this volume). Examples include the Critical Infrastructure Partners (CIP), Critical Infrastructure Intelligence Team (CIIT), Energy and Utilities Sector Network (EUSN), and Aboriginal Joint Intelligence Group (JIG), and others, as described below. By offering these examples, we illustrate how the partnerships, coordination, and sharing of intelligence is building the power of the security state to enforce extractivism.

Public Safety Canada's website explains Critical Infrastructure Partners (CIP):

> Strengthening the resilience of critical infrastructure (CI) requires *collaborative work among all partners and stakeholders.* Public Safety Canada works closely with CI stakeholders, including federal departments and agencies, provinces and territories, owners and operators, the research and development community and international counterparts. Building on

> this approach, Public Safety Canada works with its partners to share information, manage risks and reduce CI vulnerabilities.[13]

In 2015, when Bill C-51 was passed, oil and gas infrastructures (including pipelines) became defined as critical infrastructure and through this designation, resistance to fossil fuels became categorized as domestic terrorism, sparking a marked increase in coordination between fossil fuel industries and law enforcement agencies, including through CIP (Spice 2018). As an example of this increased cooperation in policing, the surveillance of activists and land defenders protesting the Northern Gateway pipeline involved RCMP, CSIS, NEB, energy corporations, and private security firms and involved widespread sharing of personal information between these agencies (Monaghan and Walby 2017).

The Critical Infrastructure Intelligence Team (CIIT) collaborates with a wide variety of agencies including Public Safety Canada, CSIS, Integrated Terrorism Assessment Centre, provincial government agencies, private sector stakeholders, international partners, Transport Canada, Natural Resource Canada, Finance Canada, and Bank of Canada. This long list demonstrates the incredible extent of cooperation—including those inside and outside conventional security intelligence domains—that is happening in order to protect pipelines through the widespread surveillance of activists and land defenders (Monaghan and Walby 2017). Through these *relations of cooperation* across such a wide diversity of institutions, the security state can access and mobilize more resources and other kinds of support for the surveillance and repression efforts than law enforcement agencies would on their own.

Another example is Canada's National Strategy for Critical Infrastructure which has "solidified a decades-long effort to establish corporations as policing partners" (Crosby 2021: 3), effectively allowing corporations to actively participate in monitoring and profiling Indigenous land defenders and communities (Crosby 2021). This is to the extent that security clearance is given to private sector personnel to access classified intelligence and there has been a marked increase of private sector actors on intelligence distribution lists (Crosby 2021: 6). Furthermore, the energy industry has the ability to consult regularly with national security and to upload their own incident reports directly to RCMP databases,

[13] https://www.publicsafety.gc.ca/cnt/ntnl-scrt/crtcl-nfrstrctr/crtcl-nfrstrtr-prtnrs-en.aspx, emphasis added.

allowing for privately collected intelligence to be aggregated into RCMP national security threat assessments. In these and other ways, the energy industry actors have become "deputized" in the field of national security and CIP-oriented surveillance (Monaghan and Walby 2017). These relationships between state and industry are nurtured in other ways too, for example with pipeline companies sponsoring network meetings or providing food for such meetings (Monaghan and Walby 2017). Here we see how increasing collaboration between government, law enforcement, and industry, is working together to build power to increase their efficiency and effectiveness in enforcing extractivism and criminalizing land defense.

This intelligence sharing that facilitates the enforcement of extractivism is formally institutionalized through the Energy and Utilities Sector Network (EUSN) which is made up of government departments, law enforcement agencies, and CI owners/operators. EUSN meetings "facilitate intelligence transfers between security agencies and private sector corporations, and the latter now enjoy a privileged position to influence the perception and labelling of threats" (Crosby 2021: 6). As corporations are empowered to define the "threats" that justify policing and as the sharing of intelligence is facilitated by EUSN for increased efficiency, the relational power of the security state is strengthened. As industry comes to benefit from the force of the law and as law enforcement comes to benefit from corporate resources, the power of both is increased through the relations they've formed.

Though the security state monitors and polices a wide range of movements, this relational power being built through interagency and industry collaboration is being wielded specifically against *Indigenous* movements and communities. For example, the RCMP has an operational working group, the Aboriginal Joint Intelligence Group (JIG), to "develop a 'national approach to Aboriginal disturbances' (Crosby 2021: 9) and to collect, analyze, disseminate intelligence on tensions/conflicts within Aboriginal communities and surrounding areas as they may escalate to civil disobedience and unrest" (Crosby 2021: 9). They use open-source intelligence and gather information from internal and external partners to produce weekly reports distributed to about *450 partners* including government agencies, law enforcement, and private sector. They also produce and distribute an annual *Aboriginal Communities of Concern Strategic Intelligence Report* to "identify communities that are considered highly volatile" and "individuals who are causes of concern to public safety" (Crosby 2021: 9–10). As such a long list of actors access such intelligence (through these vast networks of interagency and industry relations being forged) the power of the security state to enforce extractivism is strengthened. Another coordinated effort targeting Indigenous movements was SITKA, a secret investigation, during 2014–2015, led by the National Intelligence Coordination Centre (NICC), a branch of the Federal Policing unit of the RCMP that serves as an *"intelligence sharing hub"* connecting federal, regional, and local arms of the RCMP (Howe and Monaghan 2018: 332–333). The purpose of SITKA was to monitor and track Indigenous protests and activists (Nikiforuk 2019). Many have argued that SITKA was undertaken without reasonable grounds, and in violation of Charter-protected rights to free expression and privacy and that Project SITKA is part of a "trend of erosion of Canadians' freedoms and expression and assembly" (CJFE 2016[14]). Again, the capacity of the security state to monitor and police movements is strengthened through the intelligence sharing and other coordination, in this case among the different arms of the RCMP.

In all we have described above, we see *relational power* at work through the coordination, collaboration, and intelligence sharing among a vast network of public and private agencies working together to protect extractive interests. It is also important to note (and is well documented in the literature on counterinsurgency) that the security state also pays close attention to the relational power of communities and movements, strategically aiming to disrupt coalition and alliance building within and across movements (see Brock and Dunlap 2018; Brock 2020; Dunlap 2020). For example, SITKA used a Public Order Profile Scale (POPS) to identify networks, risk potential, growth of protest, and "fueling factors" that are required for a successful protest to grow. They associate risk with successful Indigenous protest and success with increase the '*connectivity* between Indigenous movements and their allies (Howe and Monaghan 2018: 338). Indeed, the point of POPS is to "evaluate the *connectedness* of sites of protest to larger social movements and/or other allied groups" (Howe and Monaghan 2018: 338). This allows the state to strategically target individuals or groups that are considered to be "central, organizational, nodes" (Howe and Monaghan 2018: 329). Violence toward

[14] https://www.cjfe.org/cjfe_condemns_project_sitka_targeted_surveillance_of_indigenous_land_defenders.

these central individuals is often "punitive and spectacular" (Howe and Monaghan 2018: 330).

It becomes clear that the security state understands relational power, as they work to increase their own relational power though elaborate networks of state and industry actors and through strategically aiming to disrupt the solidarity and alliances that build movement and communities' relational power, as they aim to disrupt our collective abilities to challenge extractivism and settler colonialism.

Material Power

We build on Rodríguez et al.'s (2017) framework of 3 forms of power by adding a fourth form: *Material Power,* which we conceptualize as including both physical force and financial resources that enable the enforcement of extractivism and settler colonialism. We ground our understanding of this type of power in Historical Materialism, based on Marxist theory that emphasizes how social ideas and institutions stem from material economic base, such as land and raw materials or "resources" (see Coulthard 2014; Carroll and Sarker 2016). Employing this lens to the analysis of extractive capitalism on stolen Indigenous lands, we see the material power of the settler state as the consequence of centuries of concentration of capital in the industrialized global north as the result of global imperialism and colonialism; these processes necessarily involved the plundering of Indigenous territories, destruction of communities, and theft of resources around the world, among other horrors (Coulthard 2014; Kallis 2018).

Given the internationally recognized sovereignty of Indigenous Nations, settler colonial states like Canada employ a specific "domestic" iteration of the military industrial complex to provide the material means (capital) for fossil fuel extraction. Pseudo-militaristic institutions like the RCMP were initially created to protect the illegitimate property claims to Indigenous lands of the early settler state (see Simpsons 2019), and therefore only exist because of accumulated colonial settlement and state expenditures rooted in the stolen wealth that they enforce access to for state interests. To this day, in many cases, this state military capital is mobilized to steal further wealth from unceded Indigenous lands without their consent in service of ecocidal industry and resource extraction-driven capital accumulation. In the case of the RCMP occupations of Secwepemc and Wet'suwet'en territory to enforce the illegal construction of the TMX and CGL pipelines, the sovereignty of these nations over their unceded traditional territories implores us to view this violence as a crisis of international military occupation. Despite its "fairy-tale" wishes, Canada is not the only sovereign at the table: When they send in the RCMP, "they're sending in an external military force. We call them police, but they're operating on territory that on the map should not be marked as Canada, so that basically means they're a Canadian military force on an offensive".[15]

The institutions of occupation and surveillance that are described in this chapter are "embedded within a complex of state-sponsored violence upon which Canada was founded – practices with which the police are historically complicit" (Hargreaves 2017: 34–35). Criminalization is a tool weaponized against Indigenous land defenders despite their rightful title to the territories on which it occurs (Verweijen and Dunlap 2021). The material power of the state which enables this violent criminalization to repress Indigenous resistance to extractive projects is starkly illustrated by the creation of the BC RCMP Community-Industry Response Group (C-IRG) in 2017. C-IRG officers and private security contractors show up to patrol resistance camps in unceded Secwepemc and Wet'suwet'en territory daily, sometimes twice a day.[16] Over the last year, C-IRG has been the main deployment at the Fairy Creek Old Growth land defense blockades on Pacheedaht lands, with 480 arrests to date in service of private logging corporation Teal Jones—all while wildfires and harmful smoke are consuming much of the province.

Additionally, one of the key tools the Canadian state has used to suppress Indigenous resistance is manufactured systemic generational poverty on reserves, despite the fiduciary duties outlined in the constitution (Green 2014: 230–231). Given that Indigenous peoples have been confined to 0.2% of their traditional territories on reserves, many cannot access their wealth of resources such as berries, fish, and game anymore. This is either due to settlement/private property claims backed by the force of the state or because the food, medicines, and waters have been poisoned over decades of industrial activity, with the accumulation of toxins rendering traditional foods unsafe to consume. This has forced

[15] LE Interview #42, D.T. Cochrane 2020, September 9 2020. See also Arthur Manuel and Derrickson (2017).

[16] LE Interview #41, Kanahus Manuel, Sept, 3, 2020.

Indigenous peoples to rely, often to a significant degree, on the market economy for their livelihoods, which can be largely inaccessible given how much of their wealth has been stolen in the form of land.[17] Conditions on-reserve were dubbed the "fourth world" by George Manuel due to the difference in living conditions and HDI rankings compared to the rest of Canada (Manuel and Derrickson 2015: 169, 221). On-reserve programs have been underfunded as long as they have existed, leading to the crises seen in reserve communities currently such as overcrowded and unsafe housing, high rates of physical and mental health epidemics, among a vast array of other persistent consequences of settler colonialism. Court battles trying to get injunctions against oil and gas energy conglomerates are simply too expensive for band councils to risk fighting, given their capital constraints and knowing the bias of the courts and the likelihood they will lose.[18]

These two components of accumulating stolen wealth for the security state plus the decades of underfunding Indigenous communities have contributed to the serious and real material problems faced by many Indigenous peoples, creating a sort of "carrot and sticks" dialectic of material power to uphold the interests of the ecocidal settler status quo. The "carrots" are the cash settlement agreements such as modern treaty frameworks that offer extensive funding to Indian Act mandated Band & Council (rather than the rightful collective title-holders of the Nation) in exchange for the extinguishment of their title rights, and industry Impact Benefit Agreements (IBAs) or Mutual Benefit Agreements (MBAs) offering funding for the right-of-way for a project (like TMX). These coercive agreements offer communities financial resources that are urgently needed to solve material problems on-reserve like housing and health crises. But this funding comes with the multiple consequences associated with ongoing fossil fuel extraction, including but not limited to sexual violence associated with man camps contributing to the MMIWG2S genocide, and toxification of lands and waters necessary for cultural survival.[19] The "stick"—if communities refuse to allow the violence of extractivism on their home territories despite the material

[17] LE Interview #41, Kanahus Manuel, September 3, 2020.

[18] LE Interview #41, Kanahus Manuel, September 3, 2020.

[19] Secwepmecul'ecw Assembly 2017, Women's Declaration Against TransMountain Man Camps. Women's Declaration | secwepemculecw.

benefits of doing so—is police violence, criminalization, and physical harm to Indigenous peoples and lands. This violence is both real and threatened, and it cannot be separated from the historical development of the Canadian state. Ultimately, the utility for the settler state of the "stick" of violence against Indigenous peoples is dependent upon the continuous flow of material resources to the bureaucracy of the state complex that enables multi-billion-dollar expenditures on policing, RCMP, CSIS, and other institutions of colonial violence.

> The real problem today in British Columbia is that the government is not even putting forward a serious proposal. They know that most Indigenous groups will never accept what they are offering, but at the same time they know that if Indigenous people protest, they can simply get an injunction and then an enforcement order and then they can use armed force to push Indigenous peoples to the side while they carry on with their development without an agreement. (Manuel and Derrickson 2017: 183)

Another example of state violence mobilized against Indigenous land defenders during TMX's attempted construction was the peaceful AIG Insurance occupation by Indigenous Youth group Braided Warriors in so-called Vancouver on February 19, 2021. It was caught on videos and live streams as police started beating the youth who were singing and holding ceremony. One officer violently pulled the ponytail of one youth; elders and youth were thrown onto the ground and stomped on while others were thrown onto broken glass strewn on the concrete from the AIG lobby windows that the cops had shattered minutes before. Everyone was violently removed from the lobby of the building and had belongings and medicines confiscated for hours behind the police line remaining outside. I (co-author LE) had shown up to support the youth organizers. In my case, the "stick" was literal. It was the blunt end of a police baton to the back of my head. This left me with a concussion. Many of the other Indigenous youth required acute and ongoing medical attention for the injuries they sustained during that action, some of whom were arrested and held in jail for hours despite requiring medical attention.

The state would not be capable of deploying the physical force that they do if they did not have the extensive funding, or financial power, to do so. Moreover, the threat of bodily harm from police brutality or the financial/legal implications of disobeying police injunctions enables the further concentration of material wealth for the settler state through the suppression of resistance in communities suffering from its violent business-as-usual approach to state sovereignty. This is ***material power*** at work in the policing of Indigenous movements.

As various anti-pipeline activists we have talked to make clear: "The state apparatus has done a wonderful job maintaining the complete monopoly of violence and bringing in the heavy guns".[20] This is used "to discourage protesters. That is certain".[21] It "closes the space for open expression and the ability to denounce unwanted projects. This is an enormous injustice.[22] "What they actually want to do is to criminalize political dissent. They want to create such a condition of fear that even people who are acting under moral necessity or democratic justification think twice before they engage in their action. It amounts to an attempt to criminalize Indigeneity itself".[23]

Conclusion

> No amount of state terror can convince us that we do not have the right to protect our homelands (Braided Warriors 2021).

As we have shown, over the last decade, there has been a new concerted approach to policing social movements in Canada. Discouraging dissent and criminalizing Indigenous resistance is done in the service of extractive, colonial capitalism and it appears to be intended by the state to deter direct actions that interfere with the bottom line of the extractive industry. This is profoundly unjust, and it is ecocidal. As the climate crisis intensifies threatening much of life on Earth, the Canadian economy remains locked into fossil fuel extraction and exploitation, and government, industry, and law enforcement are colluding to try to make sure new pipelines are built. That is the bad news. However, despite these efforts, the good news is that frontline communities and social movements are strategically building and wielding discursive, institutional, relational, and material power against these violent status-quo interests. In terms of ***discursive***

[20] JG Interview #11, June 7, 2017.

[21] JG Interview #17, June 1, 2017.

[22] JG Interview #18, June 15, 2017.

[23] JG Focus Group Interview #2, September 28, 2017.

power, Indigenous movements and others have been actively engaging in sophisticated communications media campaigns to reshape the conversation around extractivism in Canada. This disturbs the old *jobs versus environmental* narrative, drawing attention to the root causes of climate change, targeting colonial capitalism and offering up transformative visions of Land Back and Indigenous self-determination as the solutions to the climate and inequality crises (see Coulthard 2014; Simpson 2017). In one important discursive move, Freda Huson, spokesperson for the Unist'ot'en camp, flips the script on critical infrastructure, explaining.

> So industry and government always talk about critical infrastructure, and their critical infrastructure is making money, and using destructive projects to make that money, and they go by any means necessary to make that happen .... For us, our critical infrastructure is the clean drinking water, and the very water that the salmon spawn in.... That salmon is our food source; it's our main staple food. That's one of our critical infrastructures. And there's berries that are our critical infrastructure, because the berries not only feed us, they also feed the bears, and the salmon. And each and every one of those are all connected, and without each other, we wouldn't survive on this planet (Huson quoted in Spice 2018: 40).

Here, Freda Huson offers a transformative reframing of the concept of critical infrastructure, conceived from within a Wet'suwet'en law and cosmology, pointing toward a radically different value system.

As for *institutional power*, frontline communities have been engaging in, and in some instances winning, legal cases. They have been working with international bodies to advocate for their rights. For example, Arthur Manual worked hard to gain international recognition of Indigenous propriety rights, "for Aboriginal title and rights [to be] recognized at the international financial level as a real proprietary interest, just as Indigenous rights to self-determination are recognized at the UN and in international human rights bodies" (Manuel and Derrickson 2017: 190). His daughter Kanahus continues the fight against the TMX pipeline. She hosted a Human Rights Symposium, bringing "experts together to show people that we do have experts on our side, legal experts, human rights experts and advocates on our side that stand with us, that legitimize that we do, to really protect us".[24] The Wet'suwet'en hereditary

[24] LE Interview #41, Kanahus Manuel, Sept, 3, 2020.

chiefs submitted a formal request to UN to monitor the RCMP, the Canadian government, and Coastal GasLink's actions on their land. But perhaps most powerfully, Indigenous communities are exercising institutional power by enacting their own laws and governance systems and powerfully assuming authority for sovereign decision-making on their territories. In many cases, this has involved communities successfully issuing eviction notices to pipeline companies (Gobby et al. 2021).

Relational power too has been building within anti-pipeline movements and through wider networks of solidarity (Gobby et al. 2020). One manifestation of this is the development of formal alliances between different Indigenous Nations and communities facing unwanted pipelines. The leadership at Unist'ot'en helped forge the Sacred Fire Network to coordinate and share information and resources between many Indigenous front lines in BC. And the fight against the Energy East pipeline inspired the forging of the Treaty Alliance Against Tar Sands Expansion. Similarly, the fight against the Northern Gateway pipeline was successful in no small part because of The Yinka Dene Alliance which constituted an unprecedented coming together of many different First Nations across BC.

Another show of *relational power* is the solidarity blockades that erupted across Canada in 2020, under the banner #ShutDownCanada, in solidarity with Wet'suwet'en land defenders resisting the Coastal Gas Link pipeline in BC in face of violent police repression. By blockading rail lines, port entrances, and ferry terminals, these coordinated resistance efforts brought economic activity to a halt, showing "rather forcefully the power that non-elites have to stop economic power in its tracks" (Shantz 2020: n.p.). We see here the vast networks of frontline struggles joining forces, centering Indigenous struggles, and coalescing around transformative goals such as land restitution (#LandBack) and Indigenous self-determination (Pasternak and King. 2019).

And indeed, the Indigenous led anti-pipeline movements are wielding incredible *material power*: as the example about solidarity blockades above shows, solidarity between peoples and social movements constitutes material threat and wields material power against extractive interests and the colonial state. So many of the resources sought by extractive companies are on unceded Indigenous territories, and communities' repeated commitments to erect blockades and place their bodies in the way of the pipelines is a real financial liability for said companies. And this risk and liability simply grows as more people join in support. The greater and stronger the networks of solidarity, the greater the material threat to the oppressive exploitative status-quo interests. Indeed, the #shutdowncanada solidarity blockades mentioned above stranded hundreds of millions of dollars in goods and stalled some cross-border trade (Forrest 2020; see also Diabo 2020).

Blockades force the meeting of ecocidal extractivism and decolonial resistance; they are zones of conflict, and conflict is expensive. From an investment standpoint, zones of civil unrest and conflict are laden with many financial risks and uncertainties. Indigenous resistance is posing a heavy financial risk and loss to companies trying to build pipelines. Consequently, investors are pulling out, as the case of Kinder Morgan withdrawing from TMX shows. Indigenous resistance and land defense embody their own type of material counter-power by interfering with the ecocidal/genocidal processes of capital accumulation. Kanahus Manuel explains that.

> We say that one of the biggest forms of resistance is living on the land. And one of the biggest things is that we want to create risk and uncertainty for the TransMountain pipeline... getting out there and using different civil disobedience tactics to stop construction...we know that as long as we can deter construction, it costs them money... If you were to calculate us - the Shut Down Canada, the Land Back at Six Nations, the Atlantic fishing dispute happening right now, the TransMountain pipeline - those are the liabilities, right there! How much are we costing Canada? [It's] the amount of money that they put in for their police, their C-IRG, their CSIS agents, and all of this money that they put to suppress us, because if you keep on calculating that, if that doesn't stop, you're going to really be able to see the amount of risk that we're creating in this country.[25]

We include this very brief overview of the discursive, institutional, relational, and material power being enacted from below to make clear that communities and movements are by no means passive victims of state repression and the enforcement of ecocide. They are fierce warriors; they are brilliant strategic thinkers who are expanding networks of communities resisting extractivism and envisioning and enacting flourishing alternatives. As Anne Spice (2018) makes clear, Indigenous resistance to infrastructure protects Indigenous relations against the violence of settler

[25] LE Interview #41, Kanahus Manuel, Sept, 3, 2020.

colonial invasion and thus opens alternatives to living in good relation to territories; these are not just spaces of negation but also spaces of radical possibility under Indigenous leadership and jurisdiction.

In the chapter we have demonstrated the profound harms and maintenance of settler colonial ecocide, but also the ardent resistance against it. We encourage anyone reading this to find a way to wield your own power—through the ***discourses*** you attend to, learn from, and share; through the ***institutions*** you can influence through your work, your vote, and your advocacy; through your ***relations and networks***, by joining movements and enacting solidarity with those taking direct action; and through offering ***material*** support to the folks on the frontlines—in financial or physical form. It is not possible to be a passive bystander to all that is going on, nor an objective, dispassionate researcher. Frontline communities and social movements are building power through tireless and incredibly risky work but to tip the balance of power, resistance needs to be mainstreamed. This moment in history calls everyone to join the frontline, like our lives depend on it.

Acknowledgements We'd like to acknowledge and express gratitude for the help of Simona Bobrow and Rachel Ivey in conducting the literature review for this chapter.

Bibliography

Barker, A. and E.B. Lowman. n.d. "Settler Colonialism." Global Social Theory. https://globalsocialtheory.org/concepts/settler-colonialism/.

Barker, A. and E.B. Lowman. 2015. *Settler: Identity and Colonialism in 21st Century Canada*. Halifax, NS: Fernwood Publishing

Bernauer, W., Heller, H., & Kulchyski, P. 2018." From Wallmapu to Nunatsiavut: The Criminalization of Indigenous Resistance." *Monthly Review*, 69(8): 33–40.

Black, T., S. D'Arcy, and T. Weis (eds.). 2014. *A Line in the Tar Sands: Struggles for Environmental Justice*. Oakland, CA: PM Press.

Brake, J. 2018. Women Speak Out Against Criminalization of Land Defenders, Water Protectors. APTN News, August, 3. https://www.aptnnews.ca/national-news/women-speak-out-against-criminalization-of-land-defenders-water-protectors/

Brock, A ., and A. Dunlap. 2018. "Normalising Corporate Counterinsurgency: Engineering Consent, Managing Resistance and Greening Destruction

Around the Hambach Coal Mine and Beyond." *Political Geography* 62(1): 33–47.

Brock, A. 2020. "'Frack Off': Towards an Anarchist Political Ecology Critique of Corporate and State Responses to Anti-fracking Resistance in the UK." Political Geography https://www.sciencedirect.com/science/article/pii/S0962629819301192?dgcid=author

Carroll, W. K., and K. Sarker, K. (Eds.). 2016. A World to Win: Contemporary Social Movements and Counter-Hegemony. ARP Books.

Ceric, I. (2020). Beyond Contempt: Injunctions, Land Defense, and the Criminalization of Indigenous Resistance. *South Atlantic Quarterly*, *119*(2), 353–369.

Chodos, R., 1973. *The CPR: A Century of Corporate Welfare*. Toronto: James Lorimer & Company.

Cochrane, D.T. Interview, September 9 2020.

Collins, P.H., and S. Bilge. 2016. *Intersectionality*. Hoboken, NJ: Wiley.

Coulthard, G.S. 2014. *Red Skin, White Masks: Rejecting the Colonial Politics of Recognition*. Minneapolis, MN: University of Minnesota Press.

Crosby, A. 2021. The Racialized Logics of Settler Colonial Policing: Indigenous 'Communities of Concern'and Critical Infrastructure in Canada. *Settler Colonial Studies*, 1–20. https://www.tandfonline.com/doi/abs/https://doi.org/10.1080/2201473X.2021.1884426?journalCode=rset20.

Crosby, A., and J. Monaghan. 2018. *Policing Indigenous Movements: Dissent and the Security State*. Halifax: Fernwood Publishing.

Dhillon J, and W. Parrish W. 2020. Exclusive: Canada Police Prepared to Shoot Indigenous Activists, Documents Show. *The Guardian*, December 20. https://www.theguardian.com/world/2019/dec/20/canada-indigenous-land-defenders-police-documents

Diabo, R., 2020. Canada's colonial web entraps Wet'suwet'en nation. *Indigenous Policy* J. 31 (1).

Dunlap A. (2020) Wind, Coal, and Copper: The Politics of Land Grabbing, Counterinsurgency, and the Social Engineering of Extraction. *Globalizations* 17(4): 661–682.

Dusault, R., and G. Erasmus. 1996. Royal Commission on Aboriginal Peoples. People to People, Nation to Nation: Highlights from the Report of the Royal Commission on Aboriginal Peoples. Ottawa, ON: Minister of Supply and Services Canada.

Forrest. M. 2020. Trudeau's Next Political Crisis: Rail blockades. Politico. https://www.politico.com/news/2020/02/21/trudeaus-rail-blockades-116599.

Foucault, M. 1971. "Orders of Discourse." *Social Science Information* 10(2): 7–30.

Gobby, J., 2020. More Powerful Together: Conversations with Climate Activists and Land Defenders. Halifax: Fernwood Publishing.

Gobby, J., L. Temper, M. Burke, and N. von Ellenrieder. 2021. Resistance as governance: Transformative Strategies Forged on the Frontlines of Extractivism in Canada. The Extractive Industries and Society.

Gosine, A., and C. Teelucksingh. 2008. *Environmental Justice and Racism in Canada: An Introduction.* Toronto, ON: Emond Montgomery

Granovsky-Larsen, S., and L. Santos. 2021. From the War on Terror to a War on Territory: Corporate Counterinsurgency at the Escobal Mine and the Dakota Access Pipeline. Canadian Journal of Latin American and Caribbean Studies/Revue canadienne des études latino-américaines et caraïbes: 1–25.

Green, J. 2014. I*ndivisible: Indigenous Human Rights.* Winnipeg, MB: Fernwood Publishing

Hargreaves, A. 2017. *Violence Against Indigenous Women: Literature, Activism, Resistance.* Waterloo, ON: Wilfred Laurier University Press

Hermes, K. 2021. How Canada Is Targeting Indigenous Resistance to TMX. Briarpatch, January 19. https://briarpatchmagazine.com/articles/view/how-canada-is-targeting-indigenous-resistance-to-tmx-trans-mountain-pipeline

Hill, G. 2010. *500 Years of Resistance.* Halifax: Fernwood Publishing, .

Howe, Miles, and Jeffrey Monaghan. 2018. "Strategic Incapacitation of Indigenous Dissent: Crowd Theories, Risk Management, and Settler Colonial Policing." *The Canadian Journal of Sociology/Cahiers Canadiens De Sociologie* 43(4): 325–348. *JSTOR*, www.jstor.org/stable/26573404. Accessed 30 April 2021.

Human Rights Watch. 2018. World Report 2018.

Huseman, J., and D. Short. 2012. "'A Slow Industrial Genocide': Tar Sands and the Indigenous Peoples of Northern Alberta." *The International Journal of Human Rights* 16(1): 216–237.

IPCC. 2018. Summary for Policymakers. In Global Warming of 1.5°C. An IPCC Special Report on the Impacts of Global Warming of 1.5°C Above Pre-Industrial Levels and Related Global Greenhouse Gas Emission Pathways, in the Context of Strengthening the Global Response to the Threat of Climate Change, Sustainable Development, and Efforts to Eradicate Poverty, ed. V. Masson Delmotte, P. Zhai, H. Pörtner et al. World Meteorological Organization, Geneva.

Kallis, G. 2018. Degrowth. Newcastle Upon Tyne: Agenda.

Kruse, M, and C. Robinson. 2019. Injunctions By First Nations: Results of A National Study, November 14. Yellowhead Institute. https://yellowheadinstitute.org/2019/11/14/injunctions-by-first-nations-results-of-a-national-study/.

Le Billon, P., and A. Carter. 2012. Securing Alberta's Tar Sands: Resistance and Criminalization on a New Energy Frontier. In *Natural Resources and Social Conflict*, 170–192. London: Palgrave Macmillan.

Manuel, A. and Grand Chief R.M. Derrickson. 2015. Unsettling Canada: A National Wake-Up Call. Toronto, ON: Between the Lines

Manuel, A., and Grand Chief Derrickson. 2017. Reconciliation Manifesto: Recovering the Land, Rebuilding the Economy. Toronto, ON: James Lorimer

Manuel, K. Interview, September 30 2020.

Maynard, R. 2017. Policing Black Lives: State Violence in Canada from Slavery to the Present. Halifax, NS: Fernwood Publishing.

Merriam Webster. 2020. Historical Materialism. https://www.merriam-webster.com/dictionary/historical%20materialism

Monaghan, J., and K. Walby. 2017. "Surveillance of Environmental Movements in Canada: Critical Infrastructure Protection and the Petro-Security Apparatus." *Contemporary Justice Review* 20(1): 51–70. https://www.tandfonline.com/doi/full/, https://doi.org/10.1080/10282580.2016.1262770

McFarlane, P., and N. Schabus (eds.). 2017. Whose Land Is It Anyway? A Manual for Decolonization. Vancouver, BC: Federation of Post-Secondary Educators of BC.

Moore, J. 2015. In the National Interest? Criminalization of Land and Environmental Defenders in the Americas. Mining Watch.

Nikiforuk, A. 2019. When Indigenous Assert Rights, Canada Sends Militarized Police. https://thetyee.ca/Analysis/2019/01/17/Indigenous-Rights-Canada-Militarized-Police/

Owusu-Bempah, A., and S. Wortley. 2014. "Race, Crime, and Criminal Justice in Canada." The Oxford Handbook of Ethnicity, Crime, and Immigration, 321–359

Pasternak, S., and H. King. 2019. Land Back: A Yellowhead Institute Red Paper. https://redpaper.yellowheadinstitute.org/wp-content/uploads/2019/10/red-paper-report-final.pdf

Public Safety Canada. 2021. https://www.publicsafety.gc.ca/cnt/rsrcs/pblctns/srtg-crtcl-nfrstrctr/index-en.aspx.

Reid, J. (2010). "The Doctrine of Discovery and Canadian Law." *The Canadian Journal of Native Studies* 30(2): 335–359.

Rodríguez, I., M. Inturias, and J. Robledo. 2017. Transforming socio-environmental conflicts: A framework for analysis.

Simpson M. 2019. "Resource Desiring Machines: The Production of Settler Colonial Space, Violence, and the Making of a Resource in the Athabasca Tar Sands." *Political Geography* 74: 1–12.

Simpson, L.B. 2017. As We Have Always Done: Indigenous Freedom Through Radical Resistance. Minneapolis, MN: University of Minnesota Press.

Shantz, J. 2020. "'Shut Down Canada'—From a Politics of Appeal to a Politics of Disruption." The Straight. https://www.straight.com/news/1372156/jeff-shantz-shut-down-canada-politics-appeal-politics-disruption.

Shoemaker, N. 2015. "A Typology of Colonialism." *Perspectives on History* 53(7): 29–30.

Spice, A. 2018. "Fighting Invasive Infrastructures: Indigenous Relations Against Pipelines." *Environment and Society* 9(1): 40–56.

Spiegel, Samuel J. 2021. Climate Injustice, Criminalisation of Land Protection and Anti-colonial Solidarity: Courtroom Ethnography in an Age of Fossil Fuel Violence." *Political Geography* 84: 102298. https://doi.org/10.1016/j.polgeo.2020.102298.

https://www.sciencedirect.com/science/article/pii/S0962629820303619

Temper, L., Avila, S., Del Bene, D., Gobby, J., Kosoy, N., Le Billon, P., ... & Walter, M. (2020). "Movements Shaping Climate Futures: A Systematic Mapping of Protests Against Fossil Fuel and Low-Carbon Energy Projects." *Environmental Research Letters*, 15(12): 123004.

TRC (Truth and Reconciliation Commission of Canada). 2015. Final Report of the Truth and Reconciliation Commission of Canada: Honouring the Truth, Reconciling for the Future.

Veracini, L. 2011. "Introducing: Settler Colonial Studies." *Settler Colonial Studies* 1(1): 1–12.

Veracini, L. 2014. "Understanding Colonialism and Settler Colonialism as Distinct Formations." *Interventions* 16(5): 615–633.

Verweijen, J., and A. Dunlap, A. 2021. The Evolving Techniques of the Social Engineering of Extraction: Introducing Political (Re) actions 'from Above'in Large-Scale Mining and Energy Projects.

Walia, H. 2012. "Decolonizing Together: Moving Beyond a Politics of Solidarity toward a Practice of Decolonization." In E. Shragge, A. Choudry, and J. Hanley (eds.), Organize! Building from the Local for Global Justice. Oakland, CA: PM Press

Waldron, I. 2018. There's Something in the Water: Environmental Racism in Indigenous and Black Communities. Halifax, NS: Fernwood Publishing

Winter, J. 2021. 'War in the Woods': Activists Blockade Vancouver Island in Bid to Save Ancient Trees. *The Guardian*, April 9. https://www.theguardian.com/environment/2021/apr/09/canada-logging-old-growth-trees-vancouver-island

Wolfe, P. 2006. "Settler Colonialism and the Elimination of the Native." Journal of Genocide Research, 8(4): 387–409.

Woods, A. 2016. What Is Historical Materialism?" https://www.marxist.com/an-introduction-to-historical-materialism/part-one-what-is-history.htm

PART II

Enforcing Extraction

CHAPTER 5

Global Britain and London's Mega-Mining Corporations: Colonial Ecocide, Extractive Zones, and Frontiers of Martial Mining

Daniel Selwyn

Introduction

Spreading radioactive isotopes between both polar regions, the first atomic bomb test in Alamogordo, New Mexico, by the United States' (US) army is being proclaimed by geological scientists as the inception of a new epoch in the 4.6 billion-year history of planet Earth (Connor, 2015).[1] Initiated as a settler colonial project on unceded Indigenous land—from scientific compounds on San Ildefonso Pueblo territory to

[1] Notable critiques of the 'Anthropocene' epoch have located the origins of irreversible climate change in the organisation of economic production under capitalism as a 'world-ecology,' as opposed to being inherent to the human species: a Capitalocene (Moore, 2015). Decolonial critiques extend this by centring the experiences and epistemologies of Black, Indigenous and other racialised peoples to unsettle the anthropocentrism, Eurocentrism, and whiteness in contemporary narratives of ecological collapse, which erase the enduring colonisation and racialisation integral to accumulation: a Racial Capitalocene (Gill, 2021a, b; Vergès, 2017; Yussof, 2018).

D. Selwyn (✉)
London Mining Network, London, UK
e-mail: dselwyn@protonmail.com

A. Dunlap and A. Brock (eds.), *Enforcing Ecocide*,
https://doi.org/10.1007/978-3-030-99646-8_5

detonations adjoining Mescalero Apache reservations—ensuing decades of nuclear warfare, in which eight nations alone are responsible for 2,056 nuclear test explosions, have proliferated cancers and incinerated sacred lands from Pacific atolls to Aboriginal Australia and the Algerian Sahara (Churchill, 2003; Arms Control Association, 2020). These geographies have been rendered as sacrifice zones by global military powers, uninhabitable on temporal scales several orders of magnitude more than the 10,000-year lifespan of atmospheric carbon dioxide emissions. The world's foremost nuclear powers—Russia, the US, the United Kingdom (UK), France, and Israel—maintain 97% of the 15,913 nukes on alert, with 'Global Britain' announcing a 40% stockpile increase after departing the European Union (Grove, 2019: 42; Reif and Bugos, 2021). Before attaining weapons grade, however, uranium is mined as ore, a process which not only releases radon gases and daughter products to settle in the lungs of miners and nearby communities but pollutes soils and waterways, as extracting each tonne produces at least five hundred times its equivalent in toxic waste; an 'ecological holocaust' from inception (Churchill, 2003: 120). Once mined, uranium is processed into yellowcake, a concentrated powder, before being converted into hexafluoride and then enriched and pressed into fuel for projectiles, explosives, and atomic weapons (Hecht, 2012: 59). The apocalyptic climax of this process is witnessed by the bombings of Japanese cities Hiroshima and Nagasaki in August 1945, which extinguished up to 340,000 human lives (Campaign for Nuclear Disarmament, 2021). Led by the US with British and Canadian support, the Manhattan Project procured radioactive minerals from Navajo reservations and the Shinkolobwe mine in the Congo, then under Belgian colonial rule (Churchill, 2003; Hecht, 2012). By the time Shinkolobwe reached this global significance, Congo was Africa's most profitable colony, with mushrooming rubber plantations satiating appetites for automobiles across imperial metropoles following the advent of the inflatable tyre. This was crucially structured by a holocaust which killed over 10 million Congolese people through exhaustion, starvation, and sadistic forms of labour bondage (Hochschild, 1998). Congo's independence was postponed until 1960, and its essence remained illusory. When the Democratic Republic of Congo (DRC)) elected Patrice Lumumba as its first Prime Minister, his anti-imperialist and Pan-Africanist resource nationalism proved intolerable to British, American, and Belgian secret services, who colluded to assassinate him and usher in decades of repression and looting under military general Mobutu Sese Soko. (Corera, 2013; Zeilig,

2008). By the end of the century, Congo was the epicentre of the world's deadliest conflict since 1945, involving at least seven African countries and 5.4 million deaths (Reyntjens, 2004).

Tracing the spectacular violence of nuclear annihilation to the slower, but no less deadly, violence of industrial resource extraction and labour exploitation, this chapter uncovers multi-dimensional militarised processes enforcing ecological catastrophe: from war and policing to the armed force necessary to appropriate land and extract minerals, which, in turn, materialise the imperial violence of global military powers. In doing so, it proposes martial mining[2] as a concept that inheres the intimate and interdependent relationship between the arms trade, industrial resource extraction, and widespread ecological degradation integral to the operations and technologies of racial capitalism on local to global scales. This chapter proceeds by outlining overlapping spatial and temporal constellations of martial mining. First, it examines colonial conquests and ecocides over the long sixteenth century that enabled the emergence of London's mega-mining corporations during Europe's industrialisation in the nineteenth century. Outlining a theoretical and historical context, this section also emphasises how global landscapes, atmospheres, and biophysical processes have been transformed by ongoing mass extinction events, including Indigenous genocides and chattel slavery, which structure racial capitalism. Second, it reads militarised extractive zones almost 12,000 kms apart alongside each other: the Grasberg mine integral to Indonesia's military occupation of West Papua and massacres of striking mineworkers on the South African platinum belt in Marikana. Despite having specific historical and material contexts, shared social and ecological relations of corporate imperialist plunder and neo-colonial state governance will be highlighted. Finally, this chapter explores glacial, abyssal, and planetary frontiers of mining and military activity that increasingly accompany contemporary extractive zones and enduring colonial ecocides. Connecting seemingly disparate geographies and temporalities of warfare and extraction, martial mining describes a fundamentally colonial and capitalist planetary relation that organises the accumulation of land, minerals, and profit from the depths of oceans to celestial bodies in space.

[2] To my knowledge, this conjugation was first introduced and developed in Selwyn (2020).

Colonial Ecocide and Extractive Zones: Becoming a Mining Giant

Mining giants make strenuous efforts to portray themselves as benign technological institutions, which herald the arrival of modernity and progress around the world. In reports adorned with photos of wind turbines, electric vehicles, and solar panels, Rio Tinto champions itself as a 'pioneer' of materials 'essential for a low-carbon future' and 'human progress,' while Anglo American similarly brands itself as 'essential to modern life' by 'leading in sustainable mining' (Selwyn, 2020: 24). Not only do these narratives displace communities resisting mining outside categories of modernity, humanity, and sustainability, they also erase legacies of social and ecological violence which enabled their transformation into multi-billion-dollar transnational corporations (ibid.). This interacts with other strategies of 'corporate counterinsurgency,' from social development programmes and security provision to psychological operations and media manipulation (Brock and Dunlap, 2018; Dunlap 2020). Fundamentally unsettling these distortions, Gómez-Barris (2017: xvi) defines the 'extractive zone' as a 'colonial paradigm, worldview and technology that reduces, constrains and converts life into commodities for capital accumulation,' which is maintained through carceral regimes of surveillance and bordering, and martial regimes of dispossession and warfare. Exploring the rise and reach of London's mining giants Anglo American, BHP, Glencore, and Rio Tinto will reveal that extractive zones are integrally structured by colonial violence against people and nature. Functioning as a 'method of genocide,' the (legal) crime of ecocide—including soil degradation, water contamination, habitat destruction, and biodiversity loss—holds particular significance to martial mining (Short, 2016: 6).

Racial Capitalism, Mass Extinctions, and Resource Wars

After bubonic plagues, unrelenting military conflict, and widespread peasants' revolts decimated the economic surplus of feudal Europe's ruling class, colonial frontiers became the 'organising principle of metropolitan wealth' (Patel and Moore, 2018: 14). The aristocracies of Portugal and Castile began funding expeditions to the coasts of West Africa in 1441, established sugar-slave plantations in Madeira a decade later, and invaded the shores of the Americas in 1492 after Muslims and Jews had been expelled from the Iberian peninsula after seven centuries of Islamic rule in continental Europe (McKittrick, 2015; King, 2019). This corresponds with another accentuated ruling class strategy, as capitalist modes of production and ownership emerged in a European context already saturated by technologies of racialisation (Robinson, 2000). Importantly, feudal Europe's marginal others—Jewish, Irish, Gypsy, Roma, and Traveller communities, among others—experienced stigma and dehumanisation, from political exclusion to wholesale expulsions, before the colonial encounter (ibid.). With the globalisation of Europe's biome, at least 54 million Indigenous people in the Americas—and an incalculable number of languages and cosmologies—were exterminated by pathogens, forced labour, and exhaustion (Yussof, 2018; Grove, 2019). This colonial catastrophe is inscribed in the Earth's geological structure by a period of global climate cooling in 1610,[3] emphasising racial capitalism's emergence as a world-system at the nexus of 'genocide-ecocide' (Short, 2016; Gill, 2021a, b).

Constructed as a 'region of plunder, discovery, raw resources, taming and classification,' the 'New World' stabilised a 'geopolitical tradition of conquest, colonisation and extraction' (Grove, 2019: 38). Mining is integral rather than incidental to this tradition. Although mining and metallurgy have been practised for millennia, Mumford (2010/1934: 84) correlates the amassing of raw materials necessary for new modes of warfare with the rise of modern states, which are martial by definition. War made Europe's industrialisation both 'necessary and possible,' initiating a feedback loop between mass production, accelerating resource demands, and overseas conquests (Satia, 2018: 6). This is emphasised by a tenfold expansion in the continent's armies between 1530 and 1710, a comprehensive 'military revolution' (ibid.). Marx's assertion that slavery gave 'value to the colonies,' which in turn 'created world trade,' is critical to note here (Agozino, 2014). The 'unique twin birth' of liberalism and racial slavery (Losurdo, 2011: 35) is encapsulated in Britain's rise to global dominance 'on the business model of chattel enslavement,' as nearly 13 million Africans were trafficked across the Atlantic to labour on plantations and mines (Beckles, 2018). Moreover, one of the largest loans in history (equivalent to £300 billion today) compensated British slaveowners in 1833 for their 'loss of property,' while compelling enslaved

[3] As land formerly cultivated by Indigenous civilisations was gradually reclaimed by wilder arboreal nature, more carbon dioxide was sequestered from the atmosphere, causing global temperatures to drop (Yussof, 2018).

people to continue working in unpaid 'apprenticeship' (Manjapra, 2018). The 'complex historical afterlives' of these inverted reparations—still being repaid by British taxpayers until 2015—inheres in railways, factories, gunboats, and steamships, which made new generations of colonial infrastructures and enterprises possible, including a network of mines from Jamaica to Australia, where the world's largest mining company began its accumulation and ascendancy (Yussof, 2018: 42).

Involving at least 500 frontier massacres, Britain's colonial settlement of Australia appropriated land from Aboriginal inhabitants exclusively for white possession (Dovey, 2017; Moreton-Robinson, 2015). Concurrent with genocidal gold rushes in Victoria, a syndicate of European investors incorporated the Broken Hill Proprietary (BHP) in New South Wales in 1885 to develop vast deposits of silver, lead, and zinc (Short, 2016; Selwyn, 2020). BHP subsequently diversified its business through global warfare, forming a coalition that led the settler colony's production of munitions and aircraft (Selwyn, 2020). Another London-based mega-mining corporation, Rio Tinto, is equally entangled with the destruction of Aboriginal land and life, including recent violations of the Juukan Gorge rock shelters, sacred sites for Puutu Kunti Kurrama and Pinikura nations (Wahlquist and Allen, 2020). However, the company's origins are more proximate to the metropole, where a consortium of Deutsche Bank and Matheson & Co—a British trading house smuggling opium, tea, and cotton from Hong Kong and Calcutta—purchased Andalusia's renowned mines in 1873 and became the world's premier copper producer (Selwyn, 2020). Rio Tinto was soon imbricated in Britain's military establishment, most notably by operating Rössing uranium mine in Namibia, which was 'as close as the UK would come to controlling its own uranium supply' (Hecht, 2012: 96–97; Selwyn, 2020). Despite international prohibitions on resource extraction due to Namibia's occupation by apartheid South Africa, Rio Tinto emerged at the forefront of a cartel developing Britain's nuclear arsenal (Hecht, 2012: 96–97). This extractivist settler colonial trend is reiterated by mining behemoth Anglo American, founded in Johannesburg in 1917 as the product of a race war that appropriated 87% of the colony's land for white occupation and bordered dispossessed Africans into reservations providing cheap labour for mines, plantations, factories, and homes (Grimm et al, 2018). By the time apartheid fell after a century in which at least 46,000 people died in gold mines, Anglo American controlled up to half of all South Africa's private industry (Mattera, 2013).

Evidently, genocide and ecocide have been structurally fundamental to the transformation of London's mining giants from settler colonial enterprises into multi-billion-dollar transnational corporations with a global network of mines. While every extractive zone is constituted by contingent socio-environmental relations, a totalising colonial gaze over five centuries—which reduces a multiplicity of lifeworlds into unproductive wastelands anticipating civilisation—emphasises that they are also deeply implicated with militarised practices, from land invasions and frontier massacres to mass displacements and bonded labour. Colonialism is a structure, and ecocide and genocide industrial projects, that unmistakably endure into the present (Wolfe, 2015).

Geographies of Racial Capitalism: British Martial Mining in Marikana and West Papua

The interdependent relationship between mineral extraction and arms production is a historical structure with contemporary articulations. While military hardware and strategies are essential to corporate control over land and people, the production of new military technologies is dependent on the ready supply of critical metals, from cobalt and copper to lithium and platinum. When data is not restricted by confidentiality for commercial or security purposes, mining companies conceal the role their minerals play in the arms trade, preferring to emphasise their contribution to socially useful products like schools, hospitals, and 'renewable energy'[4] (Selwyn, 2020). Downstream, weapons manufacturers admit only limited awareness of their material consumption and less about its origins. However, some arms industry policy documents are explicit about the raw and specialised materials essential to their business, going as far as praising their most important suppliers: Anglo American, Rio Tinto, and Glencore (Pavel and Tzimas, 2016).

[4] 'Renewable energy' discourses have been problematised by Dunlap (2021: 15), given their material demands for 'extreme hydrocarbon and mineral extraction, socio-ecologically abhorrent industrial facilities and labour conditions, as well as the production of large amounts of waste,' which disproportionately subordinate rural and Indigenous territories around the world. Dunlap (2021) offers 'fossil fuel + technologies' as a more appropriate framing.

Meanwhile, community resistance to mining operations is met with intimidation, surveillance, and harassment, as well as forced disappearances, invasions, and assassinations (Verweijen and Dunlap, 2021; Menton and Le Billon, 2021). On average, four land and environmental defenders have been killed every week since the Paris climate accords in 2015, with the mining sector linked to the highest number of murders (Global Witness, 2020). The operations or legacies of London's mining giants alone are responsible for at least 90 incidents of conflict, according to the *Environmental Justice Atlas*. This section connects seemingly disparate geographies of racial capitalism by tracing their routes through London, a global capital of organised violence from the City's stock exchange to the political halls of Westminster. As paradigmatic militarised extractive zones, the vast copper and gold Grasberg mine in West Papua and the Marikana mine in the Bushveld Complex of South Africa, where over 80% of global deposits of platinum are located, can be read generatively alongside each other (Selwyn, 2020). Without effacing incommensurable contexts of oppression and resistance, shared features are evident: the colonial constitution of each extractive zone; contemporary ecocides and plunder by London-based mining companies; and neo-colonial administrations militarised by British arms sales and trainings that, in turn, intensify martial mining operations (Images 5.1 and 5.2).

For several years, Marikana was the most productive platinum mine in the world, crystallising a profitable partnership between corporate descendants of the settler colonial London and Rhodesian Mining and Land Company and the IG Farben conglomerate, which supplied Zyklon B gas to the Nazi regime during the Holocaust (Lonmin and BASF respectively). Both companies supplied the apartheid South African regime with technology and weapons to circumvent international sanctions (van Vuuren, 2018: 375). Upholding vast global commodity chains, industrial production transforms Marikana's platinum into catalytic converters for engines, which purify carbon emissions in everything from public transport to armoured vehicles and combat aircraft (Selwyn, 2020). The extractive zone surrounding Grasberg in West Papua, one of the world's largest deposits of gold and copper, and a decades-long joint venture between Rio Tinto and American mining company Freeport McMoran, is equally imbricated in colonial violence. After West Papua's sovereignty was transferred from Dutch colonial rule to Indonesia in 1963, brutal anti-communist purges of up to a million people across the archipelago became known as the 'Jakarta method' of counterinsurgency worldwide

Image 5.1 Three Marikana miners stand in front of an armoured police vehicle while hundreds of their fellow workers gather behind them on Wonderkop hill (Source: Greg Marinovich)

Image 5.2 Two members of Indonesia's Mobile Brigade Corps—a special operations paramilitary unit of the Indonesian police—stand with their backs to Grasberg mine (Source: Muhammad Adimaja/ANTARA FOTO)

(Bevins, 2020). It also precipitated an ongoing military occupation saturated with torture, harassment, and forced evictions. The Indonesian military regularly engages in illegal economic activities and provides security to mining and plantation corporates in West Papua (TAPOL, 2020). Grasberg's copper and gold can be incorporated in a wide variety of military applications essential to the occupation, including navigation, radar and guidance systems, communications equipment, and armaments (Selwyn, 2020).

The extractive zones of Marikana and West Papua also experience routinised ecocide . Extracting an ounce of platinum results, on average, in sixteen tonnes of waste ore, while an ounce of gold can produce up to 79 tonnes of toxic sludge containing cyanide, mercury and arsenic, which are used to separate metals from ores (Earthworks and Oxfam, 2004; Selwyn, 2020). The world-leading platinum producer Anglo American records over 2 million ounces in annual production volumes, implying waste products of at least 32 million tonnes (Selwyn, 2020). An estimated 200,000 tonnes of waste are discharged from Grasberg every day, devastating landscapes, livelihoods, and sacred sites (McKenna, 2016), while in Marikana up to 8 tonnes of poisonous sulphur dioxide emissions are released daily, alongside illegal discharges which contaminate groundwaters (Bench Marks Foundation, 2013). Where water and soils are essential for sustenance and survival, the nexus between genocide and ecocide is abundantly clear. Equally apparent is how scales of destruction parallel those of theft. With an estimated $100 billion in reserves, Grasberg has become Indonesia's greatest taxpayer (Schulman, 2016). Lonmin executives, meanwhile, disbursed $607 million in dividends to shareholders—diverting another $160 million to a subsidiary in the tax haven Bermuda—while building only three homes for its workers in defiance of legal obligations to provide at least 5,500 (Forslund, 2015; Marinovich, 2017).

Polarising distributions of costs and benefits have escalated the struggles of oppressed peoples in both West Papua and Marikana. In 2012, platinum mineworkers across the Bushveld complex went on strike for a living wage. Insurgent forms of organised labour sustained five months of strike action at Marikana, despite rising tensions between unions and the mine's private security, including fatal episodes of violence (Sinwell and Mbatha, 2016). In response, Lonmin's acting chief executive—paid 236 times more than the wage demanded by workers—urged a 'massive police…and possible army presence' to break the strike, arguing that 'the state (must) bring its might to bear on this crucial sector of the economy' (Lonmin, 2014: 1–8). Another major shareholder on the board, now President of South Africa, Cyril Ramaphosa, called the workers 'dastardly criminals' and advised the police to 'act in a more pointed way' (ibid.: 55). Police commanders ordered four mortuary vans, assault rifles, and extra ammunition to the mining site (Marinovich, 2017: 176). Days later, on 16th August, the South African police committed two massacres of striking mineworkers, murdering 34 people, injuring 78, and arresting 272 more, dozens of whom remained incarcerated for years to come (Higginbottom, 2018). Militarisation of labour disputes and resource extraction is paralleled in Grasberg, where the mine's owners admitted to paying $20 million to the Indonesian military and police for security, with individual commanders receiving personal compensation up to $150,000 (Perlez and Bonner, 2005). Freeport spent a further $35 million on military infrastructure, including barracks, headquarters, and roads, and donated military vehicles to commanders (ibid). An estimated 160 West Papuans were killed by the military around the mine between 1975 and 1997, while another three deaths were reported in 2017, with villagers fleeing into the forests after calling for the mine's closure (Radio New Zealand, 2017).

While British mining companies make extortionate profits from the extractive zones in Marikana and West Papua, Britain's 'defence' industry also benefits from major arms deals with neo-colonial administrations under whose jurisdiction mining operations are carried out. Since 2010, British arms sales to Indonesia have totalled over £780 million, while military exports to South Africa exceeded £1 billion, including £369 million in the year of the massacres (Campaign Against the Arms Trade, 2021). Given that ammunition, assault rifles, and pistols form the bulk of export licences to South Africa, it is possible if not probable that UK-made weapons were used in Marikana. The nexus between militarism and mining was made even more apparent when, only three years after the massacre, officers from Britain's Ministry of Defence (MOD) visited South Africa and paid a direct visit to Lonmin's offices in Johannesburg, while devising a plan with 'concrete deliverables' to keep the ruling party in power (Miller, 2016). South Africa's police force murders people at three times the rate of their counterparts in the US (Tricontinental, 2020: 38) and frequently target social movements, such as the Amadiba Crisis Committee, whose successful resistance to a proposed titanium mine on the Eastern Cape came at the cost of assassinations and harassment of movement leaders (Higginbottom, 2016).

The geographies of Marikana and West Papua are further connected by the impunity with which British mining companies operate following their involvement in massacres and military occupations. The company executives and politicians who colluded to orchestrate the Marikana Massacre were exonerated by a government inquiry (Higginbottom, 2018). By 2019, Lonmin had merged its assets with South African mining giant Sibanye Stillwater and, in doing so, deferred the communities' long-standing demands for justice and reparations while holding onto shares worth 9% of Marikana's wealth (Noon, 2019). Similarly dismissive of calls for legal and financial restitution, Rio Tinto sold its interest in Grasberg for £2.77 billion in 2018, stating that its departure from West Papua included the 'selling on' of all liabilities and legacy issues (London Mining Network, 2020).

While West Papua and Marikana are separated by thousands of kilometres of land and ocean, they are connected by global circulations of natural resources, financial capital, and military equipment, with London situated at a crucial intersection of their shared exploitation and opposition. Without marginalising the specific historical and material contexts of each place-based struggle for justice and liberation, these extractive zones are intimately linked through the geographies of martial mining that underpin British colonial and imperial power.

Material Militaries: Greenwashing Imperial Warfare

At the height of the coronavirus pandemic, Prince Charles, heir to Britain's royal throne, called for the world to put itself on a 'warlike footing' to tackle the climate emergency: only by 'approaching our action from the perspective of a military-style campaign," he claimed, could planetary boundaries be sustained (Milman, 2020). The co-constitution of ecological crises and warfare extends further than martial metaphors. At the world's largest arms fair DSEI held biennially in London, the army's Chief of General Staff asserted that the military 'must exploit British industry's leadership in the clean environmental technology sector' (Sengupta, 2019). More than simply a 'commercial opportunity,' this intends to get the army 'on the right side of the environmental argument,' especially with younger recruits (ibid). Materialised by new solar-powered drones and hybrid technology for army vehicles, this pursuit aims to not only 'push the boundaries of military innovation' but 'incorporate sustainability into [the army's] operations' (MOD, 2020a).

With a larger carbon footprint than countries like Mozambique and Zambia, the decarbonisation of UK's military-industrial sector is part of a broader transition in energy systems that power empire (Parkinson, 2020). The 'green industrial revolution' is scaffolded by the state's commitment to reach 'net-zero' by 2050 and a £12 billion investment in 'gigafactories' for mass battery production, electric vehicle supply chains, 'low-carbon' maritime and aviation, carbon capture technologies, and quadrupled offshore wind capacity (Prime Minister's Office, 2020a; b). Demanding exhaustive mineral extraction, this energy transition not only multiplies 'green sacrifice zones' (Hitchcock Auciello, 2019; Zografos and Robbins, 2020; Whitmore, 2021) but occurs alongside the UK's largest military investment since the end of the Cold War (£16.5 billion), cementing its position as Europe's largest defence spender (MOD, 2020d). In response to climate breakdown, these investments emphasise an emerging eco-fascist consensus among transnational ruling classes: securing resources and energy for continued capital accumulation, while policing rising population displacements and social uprisings (Walia, 2021). Predictably, many of the arms traders that profit from war—including Airbus, Thales, and Leonardo—are also primary contractors for the drones, sensors, and technologies used to militarise and externalise borders, preventing safe routes of passage for people displaced by conflict and climate chaos (Akkerman, 2016). Nearly 300 people have died at Britain's borders since 1999, rising to tens of thousands in Europe (Galisson, 2020).

The MOD is the 'single largest customer' for UK industry, with a procurement budget of up to £350 billion scheduled to 'radically change the nature and environment of warfare' over the coming decades (MOD, 2022; Curtis, 2020). A 'technology-led modernisation' promises to enable 'enhanced lethality' across multiple domains of conflict: land, sea, air, space, and cyberspace (MOD, 2018). Being the world's second largest arms exporter (£14 billion a year) and fourth highest trader in security and surveillance equipment, Britain's military transformations have global implications (Defence & Security Organisation, 2019). Combat aircraft, armoured vehicles, missiles, and drones are not assembled simply to showcase power, but to execute wars, military occupations, and global policing. British armed forces are committed to over 30 operations in at least 25 countries, including covert wars in Afghanistan, Iraq, Libya, Mali, Pakistan, Somalia, Syria, Pakistan, and Yemen, with a network of 145 military bases in 42 territories around the world (Curtis and Kennard, 2019; Miller, 2020). These sites store weapons, toxic chemicals, and explosives for testing and training and reproduce social, spiritual, and ecological harm for dispossessed communities in Belize, Brunei, Cyprus, Kenya, and the Chagos Islands, among others (Miller, 2020) (Image 5.3).

The MOD declares that these military exports and engagements are vital to 'assuring the UK's access to secure and affordable resources,' and by extension its status as a global military power (MOD, 2015: 20). Industrial resource extraction thus parallels rampant military expenditures which, in turn, enable the accumulation of sufficient minerals for arms production and supply chains. The American National Mining Association (2017), which includes Rio Tinto among its members, estimates that the Pentagon consumes 750,000 tonnes of minerals a year. With over 800 bases around the world, upscaling the US empire's share of global military spending (38%) implies that war departments consume up to 2 million tonnes of natural resources annually; acid mine drainage, several orders of magnitude higher, is an inevitable by-product (Vine, 2015; Selwyn, 2020). Once assembled, offensive weapons are responsible for widescale habitat destruction, biodiversity loss, pollution of land, water, and air, and

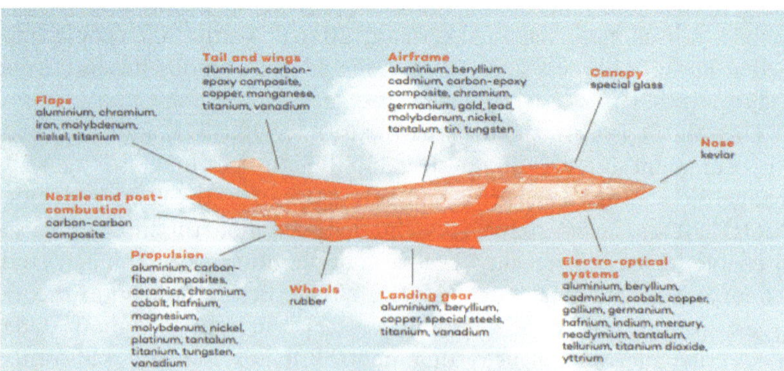

Image 5.3 Material Militaries. The natural resources required to assemble fighter aircraft (Selwyn, 2020). Design: Kay Stephens

other environmental catastrophes (Majeed, 2004). Beyond their contributions to droughts, heatwaves, tropical storms, and flooding, militarism and bordering also entrench a global system of apartheid tethered to racialised disposability (Besteman, 2020). While thousands of kilometres of fortified walls guard expropriated wealth, more human and non-human life is sacrificed to increasingly uninhabitable climates of insecurity (Walia, 2021). Anticipating polarisation, MOD-commissioned research forecasts contingency plans for over 3.5 °C warming in global temperatures by 2100 (RAND Corporation, 2020). This assumes 'increasing demand' to respond to extreme weather events, including disruptions to supply chains for 'minerals used in manufacturing defence equipment,' which may also be caused by 'violent conflict...in mineral-mining regions [due to] resource shortages' (ibid.: 8–12). In this context of militarised decarbonisation, 'Global Britain' represents an imperial (re)assertion to protect 'our people, our homeland and our democracy' in a 'competitive age' and digitalised era of racial capitalism (HM Government, 2021). Despite elevating climate change and biodiversity loss to policy precedence, Global Britain's continued prosperity is crucially dependent on martial mining in extractive zones across the Global South.

Martial Mining Frontiers: The Arctic, Abyssal Zones, and Asteroids

Since 1994, 28 trillion tonnes of ice have disappeared from the Earth's surface, with melting permafrost, glaciers, and ice sheets triggering sea level rises that displace a million people per centimetre (McKie, 2020). MOD research predicts an 'ice-free' Arctic Ocean by 2050, warming twice as fast as other parts of the world (RAND Corporation, 2020). The discovery of resources beneath the tundra and the emergence of new shipping routes has prompted an escalation of military and extractive interests seeking to capitalise on increasingly accessible polar regions (Buxton and Hayes, 2015). In September 2020, the UK led an expeditionary force into the Barents Sea for the first time in decades to improve Britain's capability to 'conduct warfare in one of the world's most challenging environments' (MOD, 2020b). Renewed emphasis on cold weather warfare builds on annual training exercises in northern Norway deploying over a thousand UK army personnel, alongside a military helicopter base 'Clockwork' deep in the Arctic circle (Miller, 2020). London-listed mining companies have been equally quick to exploit polar retreats. Rio Tinto signed an agreement with Bluejay Mining to develop ilmenite mineral sand ores in Greenland, which can be refined as titanium metal in airframes, jet engines, and turbines, or as titanium dioxide in electro-optical systems (Selwyn, 2020). Other companies circling Greenland include Red Rock Resources, exploring for iron ore, and Alba Mineral Resources excavating graphite and mineral sands. British martial mining coalesces with clarity in the Arctic biome as geopolitical competition intensifies alongside resource extraction.

Parallel climate wargames are notable in the Gulf state of Oman, where the British army is 'testing the ability of equipment to operate in different climates,' including extreme heatwaves (British Army, 2019). With an enduring history of British colonial occupation, Oman's average temperatures have increased 0.7 °C since 1950, often reaching 50 °C, while rising sea levels, desertification, and cyclones are only the most observable impacts of climate change (Choudri, 2012). By the military's logic, this invites a trebling of the size of the Royal Navy base in Duqm, which will host two new air force carriers, each longer than the Houses of Parliament, weighing 65,000 tonnes, and capable of embarking thirty-six F35 aircrafts (Miller, 2020; Selwyn, 2020). Another UK training facility and logistics centre operate around the port, while the Sultanate hosts three GCHQ surveillance bases (Curtis, 2019). Moreover, British and Omani armed forces engage in joint trainings involving thousands of armoured vehicles, with BAE Systems tanks and aircraft exchanged at such quantities that Oman reports the highest per capita military spending in the world (Dudley, 2020). Rooted in the 1908 discovery of oil in Iran and founding of the Anglo-Persian Oil Company (now BP), which enabled the navy's transition from coal to oil, Britain's extractive interests in the Gulf are equally evident in Oman: BP owns 40% of the world's largest onshore gas field in Khazzan, while over a third of Oman's six thousand oil-producing wells are owned by Shell (IHS Markit, 2021; Petroleum Development Oman, 2021).

Similar manoeuvres are underway in the ocean's abyssal zones, which provide 95% of all habitable space on Earth (Blue Marine Foundation, 2020). A new frontier of deep-sea mining for nickel, copper, cobalt, and rare earth elements[5] threatens the largest ecosystem on the planet

[5] Prized for their magnetic intensity, rare earth elements are critical to military production, including precision-guided missiles, hypersonic weapons and lasers, satellites, and drones (Gould, 2020; Scheyder, 2019). However, it is China's dominance in rare earth

with 'inevitable and irreparable harm' (Greenpeace, 2020). Over 1.5 million square kilometres of ocean floor has already been monopolised by a small number of corporations headquartered in Europe and North America, with links to fossil fuel, mining, and arms industries (ibid). In 2013, Prime Minister David Cameron pledged to put Britain at the forefront of the nascent industry; now, thanks to government sponsorship, UK Seabed Resources is the largest owner of the international seabed, holding exploration contracts for an area greater than England (ibid). The parent company of this novice prospector is Lockheed Martin, the world's largest military contractor. Meanwhile, the UK facilitates the death merchant's path through the International Seabed Authority, where it has elected the Secretary-General position since 2016, during which time Lockheed employees have attended 42 meetings with government ministers (Greenpeace, 2020; Sanderson, 2020). Urgency to complete a mining code and begin seabed extraction without public consultations obscure an asymmetrical arrangement: the public takes ultimate liability for damages, while profits accrue for private shareholders. Yet the removal of nodules by large machines, releasing stored carbon and contaminants that transform water properties, drastically compromises the ocean's ability to continue supporting multitudes of inhabiting life (Blue Marine Foundation, 2020). With impacts far beyond deep-sea mining footprints, species extinctions will result in profound changes to fisheries that feed billions of people (Greenpeace, 2020). Interacting with other ocean stressors, including climate change and waste pollution, the consequences of abyssal ecocides for Pacific Island and coastal communities are potentially genocidal.

As capitalism collapses under its socio-ecological contradictions—the exhaustive exploitation of workers and nature, to paraphrase Marx—frontiers of accumulation are being propelled beyond planetary boundaries. Lunar landscapes and asteroids have, like the ocean's abyssal zones, entered the scopes of extractive industries, always accompanied by martial actors. At the Defence Space Conference in 2020, Britain's Defence Secretary heralded a decade where 'Global Britain takes its place as a Space

production, rather than geological scarcity, that largely explains the rush for rare earth frontiers (Klinger, 2018). With Pentagon contracts lined up, British mining companies like Rainbow Rare Earths and Mkango Resources are developing deposits in Burundi and Malawi respectively (Selwyn, 2020).

Power in the new Space Age' (MOD, 2020c). While Britain's space presence Skynet, a £6 billion constellation of satellites securing intelligence and communications for British and NATO forces, was launched primarily from the European spaceport Kourou in French Guiana, a new 'Space Command' is scheduled to launch its first rocket from Scotland in 2022 and become 'fundamental to military operations' (Selwyn, 2020; Miller 2021). In addition to this, a £30 million partnership between the MOD and Virgin Orbit is developing 'military uses of small satellites,' which will enable the transformation of space 'into a warfighting domain' (MOD, 2019b). Significantly, the Five Eyes intelligence network, which binds the UK to its settler colonial prodigies—the US, Canada, Australia, and New Zealand—is already integral to global drone wars (Amnesty International, 2018). Through other alliances like the Combined Space Operations and US-led Operation Olympic Defender, which aims to 'strengthen deterrence against hostile actors in space,' military dominance in space becomes a prerequisite for correlates on land and sea, in the air and cyberspace, all increasingly reliant on digital imagery, communications, and navigation (Curtis, 2020).

Satellite technologies—composed of aluminium and titanium frames and electro-optical systems reliant on rare earth elements—also facilitate a 'new digital era of surveillance, data mining and the mapping of resource-rich territories,' which serve as visual gateways for multinational state investment in extractive industries (Gómez-Barris, 2017: 8). Expanding the Earth's natural resource base vertically, competition to extract minerals in outer space is also accelerating apace. The Artemis Accords spearheaded by the US classifies deep space exploration as a 'legally and physically unique domain of human activity,' with signatories including the UK, Australia, Canada, Japan, Luxembourg, Italy, and the United Arab Emirates (McFall-Johnsen, 2020; Jamasmie, 2021). This definition empowers private citizens and corporations to mine celestial bodies for commercial purposes, further facilitated by NASA's Gateway plans for a permanent moon-orbiting outpost. 'Cosmic colonialists' like Jeff Bezos and Elon Musk have already appropriated the opportunity to develop technologies for asteroid mining via holding companies Blue Origin and SpaceX (Boyle, 2017; Duffy, 2021) (Image 5.4).

Tracing the glacial, abyssal, and planetary frontiers of military activity and resource extraction affirms their interdependence and co-production. A warfare economy oriented towards nuclear annihilation and multi-domain dominance, alongside policing and incarceration, will always

Image 5.4 Martial mining cycle (Selwyn, 2020). Design: Kay Stephens

depend on a ravenous appetite for minerals to assemble these weapons of destruction and control. Moreover, irrespective of the sustainable pretensions in emerging extractive frontiers compared to terrestrial predecessors, racial capitalism ensures the burdens of martial mining will continue to be unequally distributed across axes of class, race, gender, (dis)ability, and nationality, and externalised to non-human animal species and future generations of life on Earth.

Conclusion

Since the Shinkolobwe mine closed in 2004, cobalt has become the DRC's most prized resource. The silvery-blue metal is central to the 'green industrial revolution,' storing energy in batteries for electric vehicles, as well as military applications (Selwyn, 2020). Yet home demolitions, evictions, and targeted violence against land and environmental defenders continue to saturate the DRC's mining industry (SOMO, 2016). At least 35,000 children as young as six currently work in criminally unsafe conditions in these mines (Kara, 2018). Cobalt extraction similarly results in water, air, and land pollution, which contributes to chronic illnesses in communities surrounding operations (Amnesty and Afrewatch, 2016). But the world's dominant cobalt producer Glencore is registered in Jersey, a British tax haven, and listed on the London Stock Exchange (Selwyn, 2020). In profit pursuit, Glencore's Congolese subsidiaries are known to have employed private security companies managed by former apartheid soldiers guilty of serious rights violations (Environmental Justice Atlas, 2019). Meanwhile, Glencore's lead is being followed by more London-listed mining companies, like Power Metal Resources and Red Rock Resources, developing cobalt licences in Katanga and Lubumbashi respectively (Selwyn, 2020). These activities are enabled by UN peacekeeping operations, which are mandated to establish trading centres for coltan and gold, train significant numbers of 'mining police,' and 'extend state authority into mining sites' (UNEP, 2012: 49–51). Before deploying to extractive warzones, battalions from Malawi and Zambia are trained by the British army (MOD, 2019a: 43).

Displacing and dispossessing communities, applying explosives and chemicals to remove industrial volumes of minerals from the earth, and securing extractive zones through military occupations and corporate counterinsurgency, extractivism is an inherently militarised process.

Equally, the production of technologies of war and policing is an extractive process, reliant on a periodic table of natural resources and global networks of mines, smelters, refineries, railways, shipping routes, and factories. This interdependent relationship between industries of war and extraction is the organising principle of martial mining, conceptually developed in this chapter through histories of colonial ecocide that enabled the emergence of London's mining giants; contemporary militarised extraction in West Papua and Marikana which buttress British capital accumulation and neo-colonial state power; and frontiers in the Arctic, deep-sea and outer space being forged by global military powers and extractive industries. Across these spatial and temporal contexts, martial mining excavates the co-production of extractivism and militarism: established and expanded through multi-dimensional warfare, extractive zones simultaneously materialise the arms trade and global military powers, with Global Britain a principal perpetrator and beneficiary.

Gómez-Barris (2017: 3) cautions that 'if we only track the purview of power's destruction and death force, we are forever analytically imprisoned to reproduce a totalising viewpoint that ignores (unbridled) life…and finds forms of resisting and living alternatively.' From insurgent unions in Marikana to liberation movements in West Papua, from Pacific Islanders defying deep-sea mining and climate genocide to Greenlandic Inuit opposing extractive and martial incursions, resistance to colonial ecocide and racial capitalism is global. Given the interconnected nature of oppressive systems—from warfare, bordering, and policing to industrial resource extraction—transformative resistance and solidarity must be equally impossible to isolate. During a global pandemic of devastating proportions but dizzying profits for a few, transnational social movements led by Black, Indigenous, and other racialised and exploited workers and youth, are coalescing around urgent demands to regenerate social and ecological relations beyond sacrificial extractive zones (Edwards, 2021). Primary among them is the abolition of martial ecologies that degrade and destroy life, including institutions of war and incarceration. Necessarily bold, abolition enables a wholesale transfer of resources from perpetrators of the climate emergency towards a multitude of liberated ecologies that sustain common needs for water, food, healthcare, education, and community—a 'reparations ecology' (Patel and Moore, 2018). The integrity and sanctity of our planetary habitat is at the forefront of this struggle and dismantling martial mining is a paramount imperative.

Bibliography

Agozino, B. (2014). 'The Africana paradigm in *Capital*: the debts of Karl Marx to people of African descent,' *Review of African Political Economy*, 41(140), 172–184.

Akkerman, M. (2016). 'Border wars: The arms dealers profiting from Europe's refugee tragedy,' *Transnational Institute*, 4 July 2016.

Amnesty International (2018). 'Deadly assistance: The role of European states in US drone strikes,' *Amnesty International*, April 2018.

Amnesty International and Afrewatch (2016). ''This is what we die for': Human rights abuses in the Democratic Republic of the Congo power the global trade in cobalt,' January 2016.

Arms Control Association (2020). *'The nuclear testing tally: Fact sheets & briefs.'* Arms Control Association report, July 2020, https://www.armscontrol.org/factsheets/nucleartesttally, accessed 11 June 2021.

Environmental Justice Atlas (2019). '*Glencore-Katanga Mining Ltd., Democratic Republic of Congo (DRC),*' Environmental Justice Atlas, 2 February, https://ejatlas.org/conflict/glencore-katanga-mining-ltd-in-the-democratic-republic-of-congo-drc, accessed 11 June 2021.

Beckles, H.M. (2018). 'Britain's Black Debt: Reparations owed for the crimes of Native genocide and chattel slavery in the Caribbean,' in Rhodes Must Fall Movement, Oxford (eds.) *Rhodes Must Fall: The Struggle to Decolonise the Racist Heart of Empire*. Zed Books.

Bench Marks Foundation (2013). 'Coping with unsustainability: Policy Gap 7, Lonmin 2003–2012,' *Business & Human Rights Resource Centre*, October 2013.

Besteman, C. (2020). *Militarized Global Apartheid*. Duke University Press.

Bevins, V. (2020). *The Jakarta Method: Washington's Anticommunist Crusade and the Mass Murder Program that Shaped Our World*. PublicAffairs.

Blue Marine Foundation (2020). 'In clear sight: Shining a light on the opaque deep-sea mining industry,' December 2020.

British Army (2019). 'Oman provides challenging environment for British Army exercise,' *British Army*, 12 March.

Brock, A. and A. Dunlap (2018). 'Normalising corporate counterinsurgency: Engineering consent, managing resistance and greening destruction around the Hambach coal mine and beyond,' *Political Geography*, 62: 33–47.

Boyle, A. (2017). 'Why asteroids loom as a future space frontier for mining and manufacturing,' *GeekWire*, 31 July.

Buxton, N. and Hayes, B. (2015). *The Secure and the Dispossessed: How the Military and Corporations are Shaping a Climate-Changed World*. Pluto Press.

Campaign Against Arms Trade (2021). '*UK Arms Export License Data*,' Campaign Against the Arms Trade, https://caat.org.uk/, accessed 12 June 2021.

Campaign for Nuclear Disarmament (2021). '*Hiroshima and Nagasaki*,' Campaign for Nuclear Disarmament, https://cnduk.org/resources/hiroshima-and-nagasaki/, accessed 2 July 2021.

Choudri, M.A. (2012). 'Climate change in Oman: current knowledge and way forward,' *Education, Business and Society: Contemporary Middle Eastern Issues*.

Churchill, W. (2003). 'A breach of trust: The radioactive colonization of Native North America,' In: Churchill W (ed). *The Ward Churchill Reader*. Routledge.

Connor, S. (2015) "The Anthropocene': The human epoch started with first atomic bomb test, scientists decide,' *The Independent*, 15 January.

Corera, G. (2013). 'MI6 and the death of Patrice Lumumba,' *BBC News*, 2 April.

Curtis, M. and Kennard, M. (2019). 'Britain's seven covert wars: An explainer,' *Declassified UK*, 17 September.

Curtis, M. (2019). 'Britain and Oman: Will their growing special relationship survive succession?,' *Middle East Eye*, 17 June 2019.

Curtis, M. (2020). 'The UK will spend over £350bn on extravagant military projects while failing to ensure national health security,' *Declassified UK*, 25 March.

Defence & Security Organisation. 'UK Defence & Security Export Statistics for 2018,' *Department for International Trade*, 30 July 2019.

Dovey, C (2017). 'The mapping of massacres,' *The New Yorker*, 7 December.

Dudley, D. (2020). 'Which country spends most per person on defense? It might not be the one you think,' *Forbes*, 19 February.

Duffy, J. (2021). 'Wannabe-Bond villain billionaires are launching us into a new era of cosmic colonialism,' *Novara Media*, 11 July 2021.

Dunlap, A. (2020). 'Wind, coal and copper: the politics of land grabbing, counterinsurgency, and the social engineering of extraction,' *Globalizations*, 17(4): 661–682.

Dunlap, A. (2021). 'Does renewable energy exist? Fossil fuel+ technologies and the search for renewable energy,' in Batel S. and D.P Rudolph (eds). *A Critical Approach to the Social Acceptance of Renewable Energy Infrastructures: Going Beyond Green Growth and Sustainability*. Palgrave.

Earthworks and Oxfam America (2004). 'Dirty Metals: Mining, Communities and the Environment.'

Edwards, Z. (2021). 'Racial capitalism and COVID-19,' *Monthly Review*, 1 March 2021.

Forslund, D. (2015). 'The Bermuda connection: Profit shifting, inequality and unaffordability at Lonmin 1999–2012,' *Alternative Information and Development Centre*.

Galisson, M. (2020). 'Deadly crossings and the militarisation of Britain's borders,' *Institute of Race Relations*, November 2020.

Global Witness (2020). 'Defending tomorrow: The climate crisis and threats against land and environmental defenders,' July 2020.

Gill, B. (2021a). 'Beyond the premise of conquest: Indigenous and Black earth-worlds in the Anthropocene debates,' *Globalizations*.

Gill, B. (2021b). 'A world in reverse: The political ecology of racial capitalism,' *Politics*.

Gómez-Barris, M. (2017). *The Extractive Zone: Social Ecologies and Decolonial Perspectives*. Duke University Press.

Gould, J. (2020). 'Pentagon legislation aims to end dependence on China for rare earth minerals,' *DefenseNews*, 18 May.

Greenpeace (2020). 'Deep trouble: The murky world of the deep sea mining industry,' December 2020.

Grimm, M., Krameritsch, J., and Becker, B. (2018). *Business as Usual After Marikana: Corporate Power and Human Rights*. Jacana Media (Pty).

Grove, J. (2019). *Savage Ecology: War and Geopolitics at the End of the World*. Duke University Press.

Hall, M. (2020). 'A new Cold War: mining geopolitics in the Arctic circle,' *Mining Technology*, 11 December.

Hecht, G. (2012). *Being Nuclear: Africans and the Global Uranium Trade*. MIT Press.

Higginbottom, A. (2016). 'Follow the money—From Pondoland to London Wall,' *London Mining Network*, 4 May 2016.

Higginbottom, A. (2018). 'The Marikana massacre in South Africa: the results of toxic collusion,' XVIII Congreso Colombiana de Historia.

Hitchcock Auciello, B. (2019). 'A just(ice) transition is a post-extractive transition: Centering the extractive frontier in climate justice,' *War on Want and London Mining Network*, September 2019.

HM Government (2021). 'Global Britain in a competitive age: The Integrated Review of Security, Defence, Development and Foreign Policy,' March 2021.

Hochschild, A. (1998). *King Leopold's Ghost: A Story of Greed, Terror and Heroism in Colonial Africa*. Mariner Books.

IHS Markit (2021). 'BP divests 20% of Oman's Khazzan gas field,' IHS Markit, 11 February https://ihsmarkit.com/research-analysis/bp-divests-20-percent-of-omans-khazzan-gas-field.html, accessed 11 June 2021.

Jamasmie, C. (2021). 'Experts warn of brewing space mining war among US, China and Russia,' *Mining.com*, 2 February. https://www.mining.com/experts-warn-of-brewing-space-mining-war-among-us-china-and-russia/, accessed 11 June 2021.

Kara, S. (2018). 'Is your phone tainted by the misery of the 35,000 children in Congo's mines?' *The Guardian*, 12 October.

King, T.L. (2019). *The Black Shoals: Offshore Formations of Black and Native Studies*. Duke University Press.

Klinger, J. (2018). *Rare Earth Frontiers: From Terrestrial Subsoils to Lunar Landscapes.* Cornell University Press.

London Mining Network. (2020). '*Cut and Run: How Britain's Top Two Mining Companies Have Wrecked Ecosystems Without Being Held to Account*,' February 2020.

Selwyn, D. (2020). 'Martial mining: Resisting war and extractivism together,' *London Mining Network*, November 2020.

Lonmin (2014). '*Exhibit VVVVI: Bundle of Emails in Evidence of Albert Jamieson*,' Centurion SA: Farlam Commission.

Losurdo, D. (2011). *Liberalism: A Counter-History*. Verso Books.

Marinovich, G. (2017). *Murder at Small Koppie: The Real Story of South Africa's Marikana Massacre*. Michigan State University Press.

Majeed, A (2004). 'The impact of militarism on the environment,' *Physicians for Global Survival.*

Manjapra, K. (2018). 'When will Britain face up to its crimes against humanity?,' *The Guardian*, 29 March.

Mattera, P (2013). '*Anglo American: Corporate Rap Sheet*,' Corporate Research Project, https://www.corpresearch.org/Anglo-American, accessed 11 June 2021.

McFall-Johnsen, M. (2020). 'China and Russia haven't signed on to NASA's new plan to unify how humanity explores space,' *Business Insider*, 13 October https://www.businessinsider.com/nasa-artemis-accords-deep-space-exploration-moon-mars-asteroids-comets-2020-10?r=US&IR=T, accessed 11 June 2021.

McKenna, K. (2016). *Corporate Social Responsibility and Natural Resource Conflict*. Routledge Research in Sustainability and Business.

McKie, R. (2020). 'Earth has lost 28 trillion tonnes of ice in less than 30 years,' *The Guardian*, 23 August.

McKittrick, K. (2015). *Sylvia Wynter: On Being Human as Praxis*. Duke University Press.

Menton, M. and P. Le Billon (2021). *Environmental Defenders: Deadly Struggles for Life and Territory*. London: Routledge.

Miller, P. (2016). 'Exclusive: The British army's secret plan to prop up South Africa's ruling party,' *Vice*, 16 May.

Miller, P. (2020). 'The UK military's overseas base network involves 145 sites in 42 countries,' *Declassified UK*, 24 November.

Miller, P. (2021). 'The final frontier: Could Britain's new spaceports embroil the UK in future star wars?,' *Declassified UK*, 24 March.

Milman, O. (2020). 'Prince Charles calls for 'Marshall-like plan' to combat climate crisis,' *The Guardian*, 21 September.

Ministry of Defence (2015). 'Joint Doctrine Note 1/15: Defence Engagement,' August 2015.

Ministry of Defence (2018). 'MOD Area of Research Interest,' Ministry of Defence and Defence and Security Accelerator, 15 January.
Ministry of Defence (2019a). 'Ministry of Defence Annual Report and Accounts, 2018–19,' Ministry of Defence, July 2019.
Ministry of Defence (2019b). 'RAF pilot seconded to Virgin Orbit space programme,' Ministry of Defence, 3 October.
Ministry of Defence (2020a). 'Armoured vehicles to test electric technology,' Ministry of Defence, 20 August.
Ministry of Defence (2020b). 'UK leads multi-national task group of warships above the Arctic Circle in demonstration of freedom of navigation,' *Ministry of Defence*, 10 September.
Ministry of Defence (2020c). 'Defence Secretary's speech at the Defence Space Conference 2020c,' *Ministry of Defence*, 18 November.
Ministry of Defence (2020d). 'Defence secures largest investment since the Cold War,' *Ministry of Defence*, 19 November.
Ministry of Defence (2022). '*Doing Business with Defence: Opportunities and Innovation in the Defence Supply Chain*.' April 2022.
Moore, J. (2015). *Capitalism in the Web of Life: Ecology and the Accumulation of Capital*. Verso Books.
Moreton-Robinson, A. (2015). *The White Possessive: Property, Power and Indigenous Sovereignty*. University of Minnesota Press.
Mumford, L. (2010). *Technics and Civilisation*. University of Chicago Press.
National Mining Association (2017). 'Minerals and metals help improve our national security, 3 February https://nma.org/2017/02/03/minerals-and-metals-help-improveour-national-security/, accessed 11 June 2021.
Noon, L. (2019). 'Lonmin's profits rise as Marikana community continues to suffer,' *London Mining Network*, 26 March 2019.
Parkinson, S. (2020). 'The environmental impacts of the UK military sector,' *Scientists for Global Responsibility and Declassified UK*, May 2020.
Patel, R. and Moore, J. (2018). *A History of the World in Seven Cheap Things: A Guide to Capitalism, Nature and the Future of the Planet*. Verso Books.
Pavel, C. and Tzimas, E. (2016). 'Raw materials in the European defence industry,' *JRC Science for Policy Report*, European Commission.
Perlez, J. and R. Bonner. (2005). 'Below a mountain of wealth, a river of waste,' *The New York Times*, 27 December.
Petroleum Development Oman (2021). '*Petroleum Development Oman (PDO)*,' Shell Development Oman https://www.shell.com.om/en_om/about-us/what-we-do/qalhat-lng-saoc.html, accessed 10 June 2021.
Prime Minister's Office (2020a). '*A New Deal for Britain*,' Prime Minister's Office, 30 June.

Prime Minister's Office (2020b). '*New plans to make UK world leader in green energy*,' Prime Minister's Office and Department for Business, Energy & Industrial Strategy, 6 October.

Radio New Zealand (2017). 'West Papuans call for closure of Freeport gold mine,' Radio New Zealand, 24 November https://www.rnz.co.nz/international/pacific-news/344637/west-papuans-call-for-closure-of-freeport-gold-mine.

RAND Europe and University of Exeter (2020). 'A changing climate: Exploring the implications of climate change for UK Defence and Security,' *RAND Corporation*.

Reif, K. and S. Bugos (2021). 'UK to increase cap on nuclear warhead stockpile,' *Arms Control Association*, April 2021.

Reyntjens, F. (2004). 'Rwanda, ten years on: From genocide to dictatorship,' *African Affairs*, 103(411), pp. 177–210.

Robinson, C. (2000). *Black Marxism: The Making of the Black Radical Tradition*. University of North Carolina Press.

Sanderson, H. (2020). 'Lockheed subsidiary calls for global agreement on deep sea mining code,' *Financial Times*, 10 September.

Satia, P. (2018). *Empire of Guns: The Violent Making of the Industrial Revolution*. Penguin Press.

Sengupta, K. (2019). 'British army says it must get green to protect environment and maintain recruitment,' *The Independent*, 12 September.

Scheyder, E. (2019). 'U.S. Army will fund rare earths plant for weapons development,' *Reuters*, 11 December.

Schulman, S. (2016). 'The $100bn gold mine and the West Papuans who say they are counting the cost,' *The Guardian*, 2 November.

Short, D. (2016). *Redefining Genocide: Settler Colonialism, Social Death and Ecocide*. Zed Books.

Sinwell, L. and Mbatha, S. (2016). *The Spirit of Marikana: The Rise of Insurgent Trade Unionism in South Africa*. Pluto Press.

SOMO: Centre for Research on Multinational Corporations (2016). 'Cobalt blues: Environmental pollution and human rights violations in Katanga's copper and cobalt mines,' April 2016.

TAPOL (2020). 'Entrenched militarism, the 'prosperity' approach and governance by exception in West Papua,' *TAPOL*, 8 December.

Tricontinental (2020). ''The politic of blood': Political repression in South Africa,' Dossier 31, 4 August 2020.

United Nations Environment Programme (2012). 'Greening the Blue Helmets: Environment, Natural Resources and UN peacekeeping operations,' May 2012.

Van Vuuren, H. (2018). *Apartheid, Guns and Money: A Tale of Profit*. Hurst Publishers.

Vergès, F. (2017). 'Racial Capitalocene,' In: Johnson GT and Lubin A (eds.) *Futures of Black Radicalism*. Verso Books.

Verweijen, J. and A. Dunlap (2021). 'The evolving techniques of social engineering, land control and managing protest against extractivism: Introducing political (re)actions 'from above,'' *Political Geography*, 83, pp. 1–9

Vine, D. (2015). *Base Nation: How US Military Bases Abroad Harm America and the World*. Metropolitan Books.

Wahlquist, C. and Allen, L. (2020). 'Juukan Gorge inquiry: Rio Tinto's decision to blow up Indigenous rock shelters 'inexcusable,'' *The Guardian*, 9 December.

Walia, H. (2021). *Border and Rule: Global Migration, Capitalism, and the Rise of Racist Nationalism*. Haymarket Books.

Whitmore, A. (2021). 'A material transition: Exploring supply and demand solutions for renewable energy minerals,' *War on Want*, March 2021.

Wolfe, P. (2015). *Traces of History: Elementary Structures of Race*. Verso Books.

Yussof, K. (2018). *A Billion Black Anthropocenes or None*. University of Minnesota Press.

Zeilig, L. (2008). *Lumumba: Africa's Lost Leader*. Haus Publishing Ltd.

Zografos, C. and P. Robbins (2020). 'Green sacrifice zones, or why a Green New Deal cannot ignore the cost shifts of just transitions,' *One Earth*, 3(5), pp. 543–546

CHAPTER 6

The Self-Reinforcing Cycle of Ecological Degradation and Repression: Revealing the Ecological Cost of Policing and Militarization

Alexander Dunlap

Introduction

It is a rather obvious, but often forgotten, fact that political violence requires an immense amount of resources. The police, military, private security and, overall, efforts of governmental and corporate control require the accumulation and organization of resources (Tilly, 1985; Rodgers, 2006; Simpson, 2019), making the rise of exchange and monetary systems crucial mechanisms for political and land control (Polanyi, 2001; Graeber, 2012). This chapter, however, is concerned with the ecological requirements of policing and political repression. This means

A. Dunlap (✉)
Centre for Development and the Environment, University of Oslo, Oslo, Norway
e-mail: alexander.dunlap@sum.uio.no

A. Dunlap and A. Brock (eds.), *Enforcing Ecocide*,
https://doi.org/10.1007/978-3-030-99646-8_6

recognizing an obvious, yet largely under-estimated reality: The military, but also the police, private security and mercenary forces, *inherently retain significant extractive costs and requirements.* Warfare and policing exist because of entire material and financial supply webs based on mining, smelting, manufacturing, transportation logistics, decommissioning processes and, of course, market logics. With the exception of Green New Deal (GND) proposals targeting military budgets (e.g. Sanders, 2019), the issue of ecological requirements for repressive forces, most notably the police, has been missing or publicly unacknowledged within the environmental and climate debate.

The framework of climate change, and its *distancing effect*, allows the ecological cost of policing to get lost within climate debates. Redirecting and distancing issues by enlarging the scale of ecological concern to the entire climate of the planet naturally minimizes, or erases in some cases, localized or everyday processes of socio-ecological injustice and degradation (Dunlap and Jakobsen, 2020). This is known as "climate reductionism." Climate change policy debates tend toward subordinating everyday practices of ecological degradation—specific factories, mines, production processes, or products—to the "big," "planetary," and "earth system" problem. Thinking of Cara Daggett (2018), Elizabeth Bell and colleagues (2020), we might call this an expression of "climate masculinities," emphasizing grand scales, reductionary data, rooted in approaches of scientific domination of the planet, which marginalize or ignore other scientific approaches (e.g. ontologies and epistemologies; Belfer et al., 2019). This logic of scaling-up, which in the case of climate change, depends on reducing all issues—habitat loss, ecological toxification, political violence, relational practices with lands—to carbon emission reduction and, consequently, the economic logic and arithmetic of carbon accounting (Gills and Morgan, 2020). "Big" problems, then require "big" actors—governments, corporations, transnational NGOs, and international committees. This is complemented by "big" solutions, such as geoengineering (Dalby, 2015), green economy practices related to large-scale conservation enclosures (Büscher and Fletcher, 2020; Dunlap and Sullivan, 2020; Huff, 2021), low-carbon infrastructures (Dunlap, 2021a, 2021b; Sovacool, 2021) "green" mining and so-called "transition metals" (Whitmore, 2021; Bolger et al., 2021). While climate change has undoubtedly "raised awareness" and genuine concern about ecological issues, international meetings (e.g. COP26), socio-ecological degradation, and business profiteering practices continue. This scaling-up and distancing of local issues allow climate change to smoother issues of Indigenous self-determination (and decolonization), extractive industries and, the focus of this chapter, the ecological cost of repressive forces.

The issue of climate change, but more specifically ecological catastrophes, has an uncanny affinity with military and police violence. While climate change results from mines, factories, energy demand, and, overall, capitalistproduction, repressive forces serve to enforce these economic and extractive practices. When people protect their land, livelihoods, and agricultural practices by refusing pipelines (Gedicks, 1994; Estes, 2019), mines (Brock, 2020; Verweijen and Dunlap, 2021), and wind turbines (Dunlap, 2018a, 2021b)—the police, military, and mercenary forces are there to intimidate locals, break road blockades, and subdue recalcitrant land defenders (Menton and Le Billon, 2021). This, moreover, includes the systemic process of land grabbing, opening up mines, and allowing the building of toxic industries (Shapiro and McNeish, 2021). The vehicles, equipment, and weapons of repressive forces could not exist without the destruction and sacrifice of ecologies, socio-cultural relationships, and often various Indigenous peoples themselves across the Global South and North (Rodney, 2009/1972; Galeano, 1997/1973; Gedicks, 1994; Short, 2016). The common issues associated with the police, such as, racial discrimination, enforcing class structures, white supremacy, and police brutality *are in realityinherently ecological and extractive issues.* Said differently, all the harassment, fines, arrests, beatings, and murders, underlining the George Floyd rebellions across the United States, which resonated across the world, are fundamentally also ecological issues. Anti-police outrage and rioting, then, can—and should be read—as outrage and, even, ecological riots against the extractivism and ecological destruction that supports police power and makes police brutality possible in the first place. The political ecology of policing and repression still has enormous room for exploration. The police could not have access to the technology, logistics, and equipment to carry out their repression without the supply chains and logistics predicated on destroying watersheds, forests, habitats, and the peoples—human and nonhuman—who live there.

This chapter employs a heuristic approach of deduction from available information, which inherently challenges the "black boxing" of models, as well as their assumptions and data points (Bolger et al., 2021, Annex; Dunlap, 2022). An accurate assessment of the ecological cost of repressive forces requires a mixed-methods approach, triangulating quantitative research with qualitative research. This chapter, however, is rather limited. Raising, instead, the general issue and highlighting further the ecological problem of repressive forces. The next section acknowledges that military and police budgets are among the largest in the world, before shifting to examine the *material requirements* and *ecological impacts* of the military and police. This exploration serves to indicate the severity of ecological issues associated with organizing "security," policing and warfare.

The Ecological Cost of Policing

The military, police forces, and private contractors are all symptoms and guardians of the state and capitalism. These repressive institutions of scientific violence are born from and continue to uphold colonial-capitalist infrastructures and political arrangements. Repressive forces receive enormous amounts of funding, while enjoying a heightened political status. The US military's budget was of USD 47.35 billion in 1960 and has incrementally increased to USD 731.75 billion in 2019. The same year, Saudi Arabia spent USD 61.87 billion, France USD 50.12 billion, Germany USD 49.28 Billion, the United Kingdom USD 48.65 Billion, and Japan USD 47.61 Billion (Macrotrends, 2021). This does not account for other state agencies designed to enforce political order within those respective countries. In the United States, for example, these include the National Security Agency (NSA), the Central Intelligence Agency (CIA), the Drug Enforcement Agency (DEA), the Federal Bureau of Investigation (FBI), the Alcohol Tobacco and Firearms (ATF), the Department of Home Land Security (DHS) as well as various police departments: highway patrols, Sheriffs, the municipal police, Special Weapons and Tactics (SWAT) teams and unarmed policing units.[1] Every country has—to various degrees—a similar arrangement of different counter-intelligence, espionage, and political policing agencies. This includes a proliferation of private security guards, as Provost (2017) reveals how they now outnumber the police in the United Kingdom and India.

The proliferation of agencies, institutions, training, and weaponry—or "violence work" (Huggins et al., 2002; Seigel, 2018)—needs to be discussed within the framework of ecological and climate crises.

[1] The list is extensive, see: https://www.thetruthaboutguns.com/full-list-of-armed-federal-agencies/.

Policing, as mentioned above, is hugely resource-intensive, ecologically destructive, and largely neglected in discussions on climate change and ecological crises. Recently, there has been greater attention given to the US military (among others) as a leading contributor to climate change and exempt from national carbon emissions accounting (Ambrose, 2021). This general acknowledgment, however, only scratches the surface in recognizing how the profound and multi-dimensional impact the police and military forces have on the climate and different ecosystems. Moreover, Other armed and policing forces are largely ignored in this conversation, whether the different police departments, paramilitaries, and counter-intelligence divisions. From the energy-intensive data and fusion centers, to the counter-intelligence and paramilitaries operations, the fleets of idling police cars, cyber warfare, and other aspects of police logistics.

The numbers and descriptions offered in this chapter are necessarily incomplete and many of the impacts of policing technologies and war making are not quantifiable—in fact, these numbers might hide as much as they reveal. What constitutes data, how it is gathered and how models are used—with all of their underlying assumptions—remain a serious scientific (e.g. epistemological) and public policy issue (Bolger et al., 2021, Annex; Dunlap, 2022). The natural sciences and much of policy-oriented research might omit qualitative socio-ecological features, due to their reductionist quantitative epistemologies. Data ignores the political violence necessary to establish mines. For instance, it relies uncritically on the contested Environmental Impact Statements (EIAs) that are criticized for lacking local participation, de-valuing local knowledge and, all the while, offering conservative estimations (Kirsch, 2014; Lawrence and Larsen, 2017). These estimations downplay or omit air, land, and water toxification as well as accidents (Bolger et al., 2021), which arise from pipeline leaks or tailing dam failures.[2] Quantitative studies can only characterize mining environments, given their basic attributes. Much of the research is conducted by institutions that are dependent on and are organized to support state and corporate endeavors, the institutions of which, in turn, are organized to preserve extractive patterns and political control (see also by Chatterjee and Maira, 2014). Research, implicitly and explicitly, is frequently rigged in favor of resource extractivism (Liboiron,

[2] Note there has been, at least, 130 major tailing dam failures between 1961 and 2021. See the list at: http://www.wise-uranium.org/mdaf.html.

2021), expansionist interests, and, by default, policing institutions and operations. Ideological, institutional, and epistemological bias deserve serious critical self-reflection and immediate change.

In the next subsection, the chapter discusses some key ecological impacts of the military and police. This does not include approximations of irregular forces (e.g. paramilitaries, mercenaries, and criminal enterprises) or aspects of digital logistics related to the military and police. Paramilitary groups, mercenaries, and criminal organizations are increasingly important policing and militarized forces, but tracking and calculating the ecological cost for operating these organizations is outside the scope of this chapter. This chapter attests to the challenge policing forces pose to genuine socio-ecological sustainability and climate change mitigation. Within each of these subsections, publicly available research data is reviewed to offer a general overview of the ecological impact of the military and police.

The Military

Militaries, Joni Seager (1992: 198) argued two decades ago, are "the most destructive environmental institutions in the modern world; they have the technology and the global reach to destroy and poison entire regions and vast ecosystems." War continues to kill humans and ecosystems long after shooting ends (see also Nixon, 2011; Smith, 2017). In addition to creating wastelands, deserts, and dead zones, militaries have huge and underreported ecological footprints; emitting vast amounts of Green House Gases (GHGs); using large amounts of resources (that require extraction and processing); transforming large tracts of land; and causing air, soil, and water pollution as well as radioactive waste (Churchill, 2003; Smith, 2017). The US military alone "consumes more liquid fuels and emits more CO_2e (carbon-dioxide equivalent) than many medium-sized countries," explained Oliver Belcher and colleagues (2020: 72). In 2017, it bought about 269,230 barrels of oil every day and "emitted 25,375.8 kt-$CO2$ by burning those fuels," which excludes emissions from electricity, food, and land use changes (Belcher et al., 2020: 72). Since the US invasion of Afghanistan, the US military has emitted 1.2 billion metric tons of greenhouse gases, according to the Brown University Cost of War project (CoW, 2019b).The US Air Force is by far the largest $CO2e$ emitter (Belcher et al., 2020: 73). Meanwhile, the "US government is conducting counterterror activities in 85 countries, vastly expanding this war across the globe" (CoW, 2019a). Government' foreign policy interests, across the world, are always prioritized over ecological and climate catastrophes (see de Vries, this volume).

In 2019, the EU military expenditure, according to a conservative estimate (due to lack of quality data), emitted at least 24.8 million tCO2e: the equivalent of the annual emissions of approximately 14 million cars (Parkinson and Cottrell, 2021). France, the EU country with the highest military expenditure, is estimated to emit over 8 million tCO2e/year. EU military emissions exclude the UK's. Stuart Parkinson (2020: 2) from Scientists for Global Responsibility calculated the total GHG emissions of the UK's military spending in 2018 to be approximately 11 million tCO2e, including all lifecycle emissions, such as those from raw material extraction abroad. This number "is more than 3.5 times larger than the total direct GHG emissions of the MOD, and more than 11 times larger than the GHG figures quoted in the main text of MOD" (Parkinson 2020: 2) and exceeds the emissions of more than 60 individual countries like Madagascar and Zambia (Selwyn 2020: 2). This shows again that military emissions are chronically underreported.

While these carbon statistics are useful to get a sense of the climate impacts of militaries, as mentioned above, this is only a small part of their impact on ecosystems. The next section outlines some key areas to conceptualize the extent of the social and ecological problem posed by the military. This entails assessing the military's—or other such institution's—*material requirements* and *ecological impacts*.

Material requirements include the resources that are needed for operating or "running" a military. Here we include the extractive, processing, manufacturing, and transportation costs of military's equipment and infrastructures. Military personnel are ignored in this study. The material requirement data is also difficult to obtain. Not only does it relate to national security interests (de Vries, this volume), but raw material and manufacturing supply webs are notoriously complex and lack transparency. Material requirements can be broken down into several categories: personnel equipment (e.g. bulletproof vests, night vision, knives, etc.), arms (e.g. small, medium, and large), vehicles (e.g. armored and non-armored), aircraft (transportation, fighter jets, and helicopters), naval vessels (e.g. aircraft carriers, cruisers, submarines) and infrastructure (e.g. buildings, roads, power lines, power generation, bridges, tents, etc.). This categorizing is provisional, and can be developed and subcategorized further. Examining the material requirements and supply web processes of each category will begin to reveal the true socio-ecological cost of military and policing technologies.

War making requires enormous amounts of resources and materials—from steel to plastics, from food to medical equipment, from water to rare earth minerals, all of which require extraction and have catastrophic ecological impacts. Take, for example, the importance of germanium and other minerals for the scientific developments that have advanced military and police equipment. David S. Abraham (2015: 164) locates the "new electronics and Rare Metal Age" alongside the Manhattan Project at Purdue University with their experiments on ***germanium*** in an effort to improve radar performance. "Purdue researchers," Abraham (2015: 165) explains, by introducing "slight impurities such as arsenic and phosphorus into high-purity germanium, they could create a diode that would allow" the transformation of "radio signals into audible sounds." While "seemingly insignificant" Abraham (2015: 165) notes, this not only allowed a radical increase in telecommunication possibilities, but also hardware durability. Laying the foundation for transistors, integrated circuits, and "semiconductors critical to the electronics we use now." Thermal-imaging systems, such as in night vision goggles, navel radar, tanks, and riflescopes could not be possible without germanium. Since the 2001 invasion of Iraq and Afghanistan, according to Abraham (2015: 164), the "overall US demand for germanium in thermal optics used in defense applications jumped from 5,000 tons in 2003 to about 30,000 tons four years later." Institutions of military science, such as Los Alamos in New Mexico, Livermore Labs and Sandia National Laboratory in California, Beijing's Tsinghua University, and the United Kingdom's BAE Systems Lab, among many others, remain energy and material intensive institutions producing technologies of warfare, surveillance, and mass killing. These technologies have enormous extractive costs and, later, proliferate further with consumer market applications.

The London Mining Networks report, *Martial Mining: Resisting Extractivism and War Together* (Selwyn, 2020; see also Selwyn this issue), a foundational expose into the material requirements of the British military, illustrates this point further. In addition to being responsible for enormous carbon emissions, the British military requires huge amounts of raw materials for the production of new weapons and equipment, in addition to maintaining old ones. The report also reveals that "6% of the earth's land mass is used for military training," which combines with the land use for the extraction, smelting, and manufacturing of weapons of war. The UK's Ministry of Defense's (MOD) "next generation of military hardware needs over half a million tons of raw materials (Selwyn, 2020: 3). Given the UK's important role as arms exporter, not least to countries committing war crimes (including Saudi Arabia and Israel) arms manufacturing demands a minimum of 20.6 million tons of minerals to "upgrade" militaries across the world (Selwyn, 2020: 3). The extractive demands of militaries are extensive, but this remains a potent research frontier for people, and if possible governments, to begin addressing in order to make real efforts toward a socio-ecological transition.

Some of the key materials required by the military are aluminum, cobalt, and copper. ***Aluminum*** is instrumental to the automobile, aerospace, and optic industries (Selwyn, 2020). Its production tends to be based on coal-powered electricity for primary smelting. The production of one ton of aluminum generates four tons of bauxite ore (Selwyn, 2020: 14). ***Cobalt*** is used in battery components and propulsion systems in aircraft, naval vessels, and missiles. Cobalt mining is not only highly polluting and degrading, it is directly related to "biodiversity loss and deforestation through mines and disposal sites, as well as air pollution through emissions and discharges," which often involves the exploitation of children (Sovacool and Del Rio, 2020: 10). ***Copper*** is instrumental to military electronics, vehicles, and infrastructures. The mining and smelting of one ton of copper produces 110 tons of waste ore (Selwyn, 2020: 17). This waste frequently contains rhenium, "used in military explosive and super alloys for jet engines. 1,000 tons of refined copper ore can also yield 1 kilogram of tellurium, which is critical for solar and infrared materials" (Selwyn, 2020: 17). The making of ***steel*** is not only energy-intensive but also relies on coal burning; steel alloy production requires 770 kilograms of metallurgical coal to make 1 ton of steel, which includes splicing iron ore with manganese, nickel, and zinc to enhance steel durability. Moreover,

> 1 ton iron ore = 3 tonnes waste;
> 1 tonne manganese = up to 2.5 tonnes ore;
> 1 tonne zinc = up to 33 tonnes ore;
> 1 tonne nickel = up to 100 tones ore. (Selwyn, 2020: 25)

There are 39 minerals instrumental to the defense industry (EC JRC, 2020).[3] In addition to other metals and minerals, ***rare earth elements*** are crucial for the military sector. "Rare earths are raw materials of strategic significance for the economic and military security of the West," as a Working Paper published by the German Federal Academy for Security Policy states, they are: "essential to numerous civilian and military technologies" (Kullik, 2019: 1). According to the US Government, every nuclear-powered submarine uses 9,200 pounds of rare earth materials—and the US alone maintain almost fifty of them (Abraham, 2015: 168). Warships and fighter jets have similar extractive footprints. Each of the US military's "seventy-seven DDG 51 Aegis destroyers uses 5,200 pounds, and each of the forthcoming F-35 Lightning II aircraft requires approximately 920 pounds" (Abraham, 2015: 168).[4] Fighter jets, likewise, require numerous tons of rare earth minerals for "everything from starting the engines to controlling wings flaps, for landing to assisting the electrical interface between the plane's controls and its components" (Abraham, 2015: 168). Rare earth elements are also necessary for aerospace engineering, drones, surveillance systems, and lasers (Kullik, 2019; EC JRC, 2020)—in addition to lower-carbon and e-mobility technologies.

The military's reliance on rare earth minerals—and the radical expansion of military and consumer markets—also has geopolitical implications, necessitating reliable supply chains that are themselves in need of securing through military and political economic intervention. "NATO is almost 100% dependent on rare earths imports from China" (Kullik, 2019: 2; de Vries, this volume). The European Commission (EC JRC, 2020: 70) recognizes that the EU is fully dependent on imports of 13 of the 39 raw materials (i.e. boron (as borates), dysprosium, gold, magnesium, molybdenum, neodymium, niobium, praseodymium, samarium, tantalum, titanium, yttrium and other REEs [Rare Earth Element]). Overall, for more than two thirds of those raw materials, the share of imports exceeds 50%. This indicates an extremely high level of foreign

[3] For detailed statistics (including for platinum, gold, solver, titanium, and uranium) see the Martial Mining report (Selwyn, 2020).

[4] This document has EU stats. EC. 2020. "Critical Raw Materials for Strategic Technologies and Sectors in the EU—A Foresight Study." European Commission: Brussels, Belgium. European Commission. Accessed October 15, 2020. https://ec.europa.eu/docsroom/documents/42881. Not for military though?

dependence not only for weapons of war, but also for electronic vehicles, wind, solar, and digitalized technologies, which also implies the militaries' need to secure strong geopolitical positions. "Every tonne of rare earth produced generates approximately one tonne of radioactive wastewater," explains Julie Michelle Klinger (2017: 55), as well as seventy-five cubic meters of acid wastewater: 9,600 to 12,000 cubic meters of waste gas containing hydrofluoric acid, sulfur dioxide, and sulfuric acid as well as approximately 8.5 kilograms of fluorine. In another estimate, according to the London Mining Network (Selwyn, 2020: 23), every ton of rare earth minerals generates "1.4 tonnes radioactive waste, 60 million litres waste gas with hydrochloric acid, and 200,000 litres acid-waste water." A lot of toxic and radioactive waste is produced in order to obtain a small amount of rare earth minerals. Figure 6.1 illustrates where and how these elements are used on a fighter aircraft. These rare earth statistics, we must acknowledge, naturally normalize and decontextualize the violent processes of extraction necessary to create these war machines. Mining and militarization are thus two sides of the capitalist coin, which we see increasingly re-branded through equipment, vehicles, and infrastructure as "green," "environmentally friendly," and "sustainable."

Institutions of military science, such as Sandia National Laboratory in California, Beijing's Tsinghua University, or the United Kingdom's BAE Systems Lab, that develop and produce technologies of warfare, surveillance, and mass killing, generate their own extractive and energy-intensive ecological impacts. They include not only emissions and waste, but can involve accidents and spillages, as the Hanford uranium production site showed. Since it was established in 1943, the site has leaked over 900,000 gallons of highly radioactive fluids into the Colombia River (Churchill, 2003: 120). Hanford "Tank B-109, which is at least 75 years old, is estimated to be leaking 3.5 gallons a day, or nearly 1,300 gallons per year," explains the Washington state Department of Ecology (WSDoE, 2021) This Tank B-109 leakage, they explain, is "leaking into an area where other tanks have already leaked 200,000 gallons into the soil." While the Hanford facility was shut down in 1990, it periodically still makes news headlines for radioactive waste leakage today.

Having offered some figures illustrating the resource intensity of military production and war making, we now turn to the *ecological impacts* of warfare and the use of these technologies. These are by no means accidental, but often the intended consequence of warfare to eliminate resistance, destroy livelihoods, and subjugate people to political and

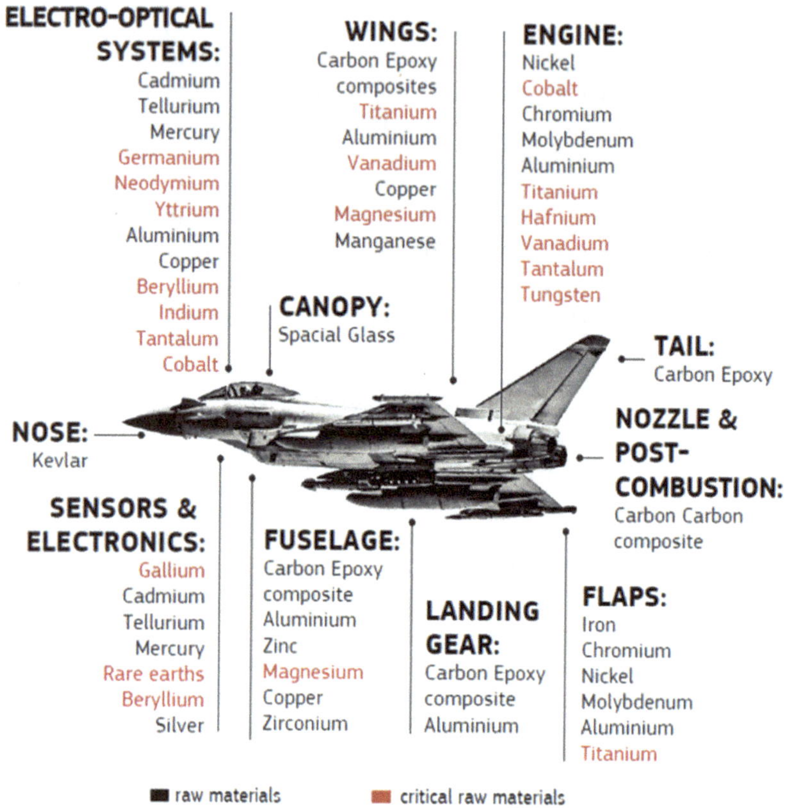

Fig. 6.1 Aircraft material requirements (European Commission, 2020: 73)

economic regimes. Global environmental disruption and destruction has been a strategy of war from time immemorial (Dunlap, 2018b; Kaur, 2021), but this has only intensified over time with new and more noxious chemical and biological agents that have long lasting and far reaching effects (Levy and Vaillancourt, 2011). These impacts start before war making begins—with weapon testing and training, for instance. "[E]ntire islands have been obliterated or rendered completely uninhabitable by humans for indefinite time periods, not to mention the inestimable effects on other species," Levy and Vaillancourt (2011: 219) state, citing British testing of anthrax on the Scottish island of Gruinard (World War II) and United States, British, and French nuclear testing in the Pacific (1950s and 1960s) as important examples. "Most nuclear weapons in the United States and Europe have been tested on indigenous peoples' land with dramatic health consequences," which are especially severe for women (e.g. birth defects) and children (Churchill, 2003; see also LaDuke and Cruz, 2012). "[O]nce real ventilation of the [uranium] mines began to occur during the mid-60s, the vents were often situated right in the middle of residential areas," explains Ward Churchill (2003: 122). This allowed "yellow cake" dust, which retains 85% of its radioactivity, to spew into [Native's] reservations, forcing the inhabitants "to breathe the same potent mixtures of radon, thoron, and other toxic substances which were plaguing their husbands, fathers, and neighbors below," working the mines (Churchill, 2003: 124, 122). This is a testament to the genocidal relationship between nuclear development and Native Americans, which extended, as Jensen (2006) reminds us, to paving roads mixed with uranium tailings around and through US Native reservations. This, we must remember, is focusing on the production of nuclear weapons, not their ground and atmospheric testing nor their actual use on Japan with the most atrocious and exterminating human and nonhuman consequences.

As the Vietnam and Central American wars showed most infamously, "scorched earth" approaches to war making entail the elimination of everything in the "enemy" territory, which has severe—if not ecocidal—consequences for lands and ecosystems. Even if unsuccessful at total elimination (as most are), scorched earth approaches still resulted in mass human and nonhuman death, long-term positioning, birth defects, and alterations to the genetic quality of all species (Churchill, 2003; Nixon, 2011; Smith, 2017). Scorched earth strategies are common in conventional wars (e.g. the Vietnam War, the US invasion of Panama in 1990 and the Gulf War in 1991–1992) and "Dirty Wars" in Latin America (1960–1990) (Menjívar and Rodríguez, 2005). The Gulf War witnessed the weaponizing of depleted Uranium-238 and "5.5 million [oil] barrel total—equal to 25 Exxon Valdez spills'—scattered across 'more than 460 miles of Arabian Gulf coastline' contaminated with heavy oil" (Thomas, 1994: 117, 115). More than 200 species were inundated and were dying from oil spills, refineries set on fire and attacks against critical infrastructure by US and Iraqi forces (Thomas, 1994), which entails an unknown death toll of 100,000s of people. This does not take into account the "slow violence" from oil spills as well as napalm and depleted uranium contamination. According to the UK Atomic Energy Authority, "there is enough depleted uranium-238 in Kuwait and Iraq to cause "tens of thousands of deaths," explains Thomas (1994: 123; Picard and Beigi, *this volume*). Children and developing fetuses are particularly vulnerable to uranium poisoning.[5]

This in no way adequately describes the socio-ecological cost of war, but offers an indication to its ecocidal reality, which has only intensified overtime. Indeed, the very concept of *ecocide* arose in reaction to the Vietnam War (Short, 2016), where US military forces "sprayed 79 million liters of herbicides and defoliants over about one-seventh of the land of southern Vietnam" (DeWeerdt, 2008). The "jungle" was regarded as a safe haven for the insurgents and, in essence, had to be eliminated to stifle and eliminate the Vietcong. Forests retain a historically important relationship to rebellions, with great efforts being undertaken to destroy the means of subsistence and illegible spaces that they represent (Scott, 1998; Peluso and Vandergeest, 2011; Boot, 2013; Vidalou, 2017). The US application of chemical herbicides (such as Agent Orange) to defoliate and mass murder nonhuman peoples was not a strategy without precursors. It was inspired by British efforts to counter anti-colonial insurgent fighters in Malay by spraying chemicals to eliminate "jungle crops planted by rebels" (DeWeerdt, 2008). Mesopotamian marshes which once covered over 20,000km^2 of lakes, reed marshes, mudflats, seasonal lagoons, and salt-tolerant scrubs were drained by Saddam Hussein's regime in the beginning of the 1990s to control a Shia uprising after the Gulf War (DeWeerdt, 2008). Less than 10% of the marshes remained, while much of the area became a desert covered by salt crusts (DeWeerdt, 2008; Priestley, 2021). Local temperatures in the region started to rise, the frequency of dust storms began increasing, meanwhile extinctions and species decline began to intensify as efforts to control and repress the Kurdish population continued (see also Picard and Beigi, *this volume;* Priestley, 2021). The impacts of war and its implicit resource control are

[5] Maybe we should add these reports from the recent war in Iraq: https://www.al-monitor.com/originals/2019/06/iraq-health-basra-cancer.html, https://www.theguardian.com/commentisfree/2012/oct/25/fallujah-iraq-health-crisis-silence, https://www.thedailybeast.com/cheats/2012/10/15/iraq-sees-surge-in-birth-defects. And this update on depleted uranium: https://foreignpolicy.com/2017/02/14/the-united-states-used-depleted-uranium-in-syria/. Deforestation & ISIS? https://foreignpolicy.com/2019/07/15/afghanistans-forests-are-turning-a-profit-for-the-islamic-state/.

significant and are certainly a major root of the ecological and climate crisis.

Warfare is ecological destruction: poisoned trees, rivers, seas, animals, and birds; mountains destroyed by bombs (e.g. Afghanistan); fires set; and the consequences of using chemical and biological weapons cannot be overstated. The earth has entered a stage of generalized ecological catastrophe. John Clark (2020), argues that we are entering the era of the *Necrocene*, the era of mass death and extinction. The military, warfare and the production of war machines are certainly emblems of the *Necrocene*, yet they are not the only forces involved in warfare and ecocide. To illustrate this, we now turn to the socio-ecological impacts of the police, introducing another research frontier that is in urgent need of investigation.

The Police

Police forces use some of the same types of weaponry and equipment as the military, which require some of the same raw materials, manufacturing processes, and transportation logistics. Yet, despite these parallels and the increasing militarization of police forces, the police operate within a different context, or "battlefield," and retain distinctly different procedures, equipment, and consumption patterns. While militaries use tanks, aircrafts, and battleships, police forces employ large numbers of automobiles and vans; militaries use long distance rifles and guns, whereas police forces rely predominantly on small arms, vehicles, drones, bulletproof vests, and other—overlapping—(telecommunication) technologies. Yet, structurally speaking, the police and military are comparable in their proliferation across space and time, as well as their increasing use of similar equipment, which demand intensive material and energy consumption.

In 2017, Aaron Karp (2018) from the Small Arms Survey estimated that there were at least 22.7 millions of firearms in use by police forces, and likely many more. This excludes customs officers, wildlife management, and prison authorities. With the widespread militarization of police forces, police personnel are becoming increasingly armed with high-caliber weapons, "more alike to military armament" (Karp, 2018: 3). Police across the world have typical equipment requirements: Handcuffs, nightsticks, handguns, pepper spray, tasers, radio communication, bullet proof vests and, with more "modernized" police forces, body cameras, drones, digitalized communication, computerized data bases (e.g. in each car), and so on. More militarized police forces like in the United States increasingly have grenades and grenade launchers, fully automatic rifles, Mine-Resistant Ambush Protected (MRAP) vehicles, tanks, and drones.[6]

Assessing the ecological cost of the police gets rather specific, especially in terms of the types of metals used in weapons; minerals used in particular cameras, microchips in computers, and batteries in drones. This equipment requires special alloys, rare earth elements, copper, and the whole gamut of the periodic table. Guns, for example, include metals with nickel, chromium, and tungsten as well as non-metallic elements like molybdenum, sulfur, and silicon (Sweeney, 2011). There are specific metallurgy and molding processes for guns (that might have different heat/energy requirements), where the numbers assigned to the metal (and guns), such as 1060, 4140, or 5150 all designate the mineral blends within them (Sweeney, 2011). "4140," Patrick Sweeney (2011) explains, "has about 1 percent chromium, 0.25 percent molybdenum, 0.4 percent carbon, 1 percent manganese, around 0.2 percent silicon and no more than 0.035 percent phosphorus and no more than 0.04 percent sulphur. That leaves most of it, 94.25 percent, iron." Alternatively, "4150 has 0.5 percent carbon" added to it, which makes the alloy much harder—and difficult to work with—but is more desirable for rugged conditions (Sweeney, 2011). Considering most police officers across the world have a handgun, let alone the US' modernized and highly armed police forces with their shotguns, submachine guns, assault, and sniper rifles, there is an enormous production of weapons requiring specialized minerals and energy-intensive smelting and manufacturing processes.

The production of firearms and other policing equipment is highly resource-intensive. Tasers, for instance, need copper, compressed nitrogen (to launch the darts), and lithium (for batteries), among others. Tear gas, likewise, requires charcoal, potassium nitrate, silicon, sucrose, potassium chlorate, magnesium carbonate, O-Chlorobenzalmalononitrile, and nitrocellulose (Wired, 2011). Tear gas is encased in aluminum canisters. Tens of thousands of tear gas canisters are shot regularly; at Anti-Globalization protesters (1998–2003), Plaza occupations (e.g. Spain, Greece, the Occupy movement, Arab Spring, between 2010 and 2012), and anti-police riots in the United States (2014–2020). Tear gas is a weapon of war that is highly toxic, leading to illness, death, and injury

[6] https://www.copblock.org/147358/war-equipment-police-receive-from-federal-1033-program/.

(Feigenbaum, 2017). The tear gas industry, dominated by the United States, Israel, and Brazil, is profiting from social discontent and producing more and more chemical weapons that end up polluting the environment (Feigenbaum, 2017). Every piece of police equipment and armament is based on complex mineral compositions, mining, smelting, manufacturing, and logistical dynamics, which represent important research frontiers for exploring the ecological realities of police repression.

In addition, the police use vast amounts of fossil fuels and mineral resources through motorized vehicles. Between 2014 and 2019, US police departments bought 208,461 police cars, sports utility vehicles, and prisoner transport trucks from Ford motor company alone (Ford Authority, n.d.). This vehicle purchase does not include police pickup trucks, motorcycles, and other makes and models of police cars. Police vehicles require vast amounts of metals, copper, cobalt, rare earth elements and other minerals (in need of further research), which—of course—demand energy-intensive mining, smelting, and manufacturing processes. The fuel requirements and usage of these vehicles are equally shocking. The earlier Ford Crown Victoria Police Interceptor, last produced in 2011, was rated poorly by the US Environmental Protection Agency (EPA) for burning 16 miles per gallon when driven in cities (Welty, 2015). According to the Ford calculator (Ford, 2021), gas tanks are 21.4 gallons (approx. 97 liters) and the 2018 Police Interceptor Utility gets an EPA estimated 17mpg and consumes 1,176 gallons of fuel a year. These figures represent an ideal and do not take into account engine ware and the various factor related to operating vehicles over time. More still, these figures are presented on the Ford website in comparison to the new hybrid 2021 Police Interceptor Utility vehicles that accordingly only consume 833 Gallons of fuel a year (Ford, 2021), marketing attempts at reducing fuel consumption through electric vehicle technology. Police vehicles, moreover, are constantly "idling" (parked with the engine running), which consumes greater quantities of fuel use, thereby producing more emissions. The police is an ecologically costly organization.

This resource consumption only intensifies with police militarization. Between 2006 and 2016, Adam Andrzejewski (2016) reveals, US police forces have obtained:

> 7,091 trucks ($400.9 million); 625 mine-resistant vehicles (421.1 million); 471 helicopters ($158.3 million); 56 airplanes ($271.5 million); and 329 armored trucks and cars ($21.3 million);
>
> 83,122 M16/M14 rifles (5.56mm and 7.62mm) ($31.2 million); 8,198 pistols (.38 and .45 caliber) ($491,769); and 1,385 riot 12-guage shotguns ($137,265);
>
> 18,299 night-vision sights, sniper scopes, binoculars, goggles, infrared and image magnifiers ($98.5 million); 5,518 infrared, articulated, panoramic and laser telescopes ($5.5 million);
>
> 866 mine detecting sets, marking kits, and probes ($3.3 million); 57 grenade launchers ($41,040); 5,638 bayonets ($307,769) and 36 swords and scabbards.

This also includes the distribution of military armored vehicles to US police departments: 865 mine-resistant ambush protected (MRAP) vehicles and 335 armored trucks (Margulies, 2019). Morgan Margulies (2019: 32) calculates that the MRAP vehicles average about 3 miles per gallon (MPG) while armored trucks average 5mpg. In addition, military vehicles such as MRAP vehicles are exempt from diesel engine emission standard requirements meant to protect local air quality; illustrating their special status.

The purpose here is to point to the severity of the operational needs and ecological impacts incurred by police forces, as well as the spiraling cycle of political repression and ecological degradation, which has been set in motion by ideologies of progress, (internal and external) colonization and industrialization. Police and the industries producing their equipment drive mining and profit from maintaining political order and disorder. While more research, specific qualitative and compiled quantitative data as well as mixed-method approaches can strengthen this argument, it is safe to approximate that the police—especially given their enforcement of extractive activities—are central contributors to the ongoing ecological and climate catastrophe.

Conclusion

The resource consumption, violence, and ecological destruction presented here is overwhelming. This chapter seeks to widen the door to further consider how the military, police, and mercenary forces are contributing to ecological and climate catastrophe. The role of the police and military in the enforcement of ecologically destructive mining and infrastructure projects are largely acknowledged, yet little has emerged in political ecology literature, and the social sciences in general, discussing the inherent material requirements and production processes necessary to orchestrate political repression. This chapter, while stressing concerns with the reductive qualities of quantitative data that conceal the intensive ecological impacts of repressive forces, seeks to heuristically reveal the enormous amounts of fossil fuels, minerals, and energy that are necessary for the weapons, vehicles, aircrafts, and drones employed by the military and police. Moreover, this chapter serves to remind readers that warfare not only has horrible human and nonhuman costs, but also that the production of these technologies implies severe ecological, even ecocidal, consequences that implicitly support anti-Indigenous and white supremacist objectives of political control and resource acquisition.

The purpose here is to initiate a process of thinking deeply and realistically about extractive supply chains necessary for repressive forces. Consistent with questioning reductivist methods, researchers need to begin adequately accounting for the socio-ecological damages that militaries and police forces incur. What does it require to produce germanium and then apply it as a component in military/police equipment? This question indicates the importance of accounting for the infrastructure, energy, and personnel costs related to research and development of repressive technologies. The socio-ecological costs of cybernetic and computational technologies are enormously high, demanding greater consideration into the complex financial and material supply webs that underline the equipment used to execute violence and enforce political order. This chapter, more than anything, applies the basic lessons of political ecology; reminding us that classism, racism, patriarchy, violence, and brutality associated with the police and military are fundamentally an ecological issue. Everyone, arguably, has a common interest in aiming for total liberation, while focusing on the general qualitative improvement of lives, habitats, and our general relationships that will seek to eliminate coercion and domination against all life.

Bibliography

Abraham DS. (2015) *The Elements of Power: Gadgets, Guns, and the Struggle for a Sustainable Future in the Rare Metal Age*. Yale University Press.

Ambrose T. (2021) *World's Militaries Avoiding Scrutiny Over Emissions, Scientists Say*. Available at: https://www.theguardian.com/environment/2021/nov/11/worlds-militaries-avoiding-scrutiny-over-emissions.

Andrzejewski A. (2016) *War Weapons For America's Police Departments: New Data Shows Feds Transfer $2.2B in Military Gear*. Available at: https://www.forbes.com/sites/adamandrzejewski/2016/05/10/war-weapons-for-americas-local-police-departments/.

Authority F. (n.d.) *Ford Police Vehicle Sales Numbers*. Available at: https://fordauthority.com/fmc/ford-motor-company-sales-numbers/ford-sales-numbers/ford-police-vehicle-sales-numbers/.

Belcher O, Bigger P, Neimark B, et al. (2020) Hidden carbon costs of the "everywhere war": Logistics, geopolitical ecology, and the carbon boot-print of the US military. *Transactions of the Institute of British Geographers* 45: 65–80.

Belfer E, Ford JD, Maillet M, et al. (2019) Pursuing an indigenous platform: exploring opportunities and constraints for indigenous participation in the UNFCCC. *Global Environmental Politics* 19: 12–33.

Bell SE, Daggett C and Labuski C. (2020). Toward feminist energy systems: Why adding women and solar panels is not enough☆. *Energy research & social science* 68 (1): 1–13.

Bolger M, Marin D, Tofighi-Niaki A, et al. (2021) *'Green Mining' Is a Myth: The Case for Cutting EU Resource Consumption*. Available at: https://friendsoftheearth.eu/wp-content/uploads/2021/09/Methodology-considerations-Annex-to-green-mining-is-a-myth.pdf.

Boot M. (2013) *Invisible Armies: An Epic History of Guerrilla Warfare from Ancient Times to the Present*. New York: Liveright Publishing Corporation.

Boukherroub T, Bouchery Y, Corbett CJ, et al. (2017) Carbon footprinting in supply chains. In: Bouchery Y, Corbett CJ, Fransoo JC, et al. (eds) *Sustainable Supply Chains*. Switzerland: Springer, 43–64.

Brock A. (2020) Securing accumulation by restoration—Exploring spectacular corporate conservation, coal mining and biodiversity compensation in the German Rhineland. *Environment and Planning E: Nature and Space*: 2514848620924597.

Büscher B and Fletcher R. (2020) *The Conservation Revolution: Radical Ideas for Saving Nature Beyond the Anthropocene*. New York: Verso

Cavanagh CJ and Benjaminsen TA. (2017) Political ecology, variegated green economies, and the foreclosure of alternative sustainabilities. *Journal of Political Ecology* 24: 200–341.

Chatterjee P and Maira S. (2014) *The Imperial University: Academic Repression and Scholarly Dissent*. Minneapolis: University of Minnesota Press.

Churchill W. (2003a) A breach of trust: The radioactive colonization of native North America. In: *Acts of Rebellion: The Ward Churchill Reader*. New York: Routledge, 111–140.

Clark J. (2020) What is eco-anarchism? *The Ecological Citizen* 3: 9–14.
CoW. (2019a) *Costs of War: Summary Findings*. Available at: https://watson.brown.edu/costsofwar/papers/summary.
CoW. (2019b) *Summary: Pentagon Fuel Use, Climate Change, and the Costs of WarJune 12, 2019*. Available at: https://watson.brown.edu/costsofwar/files/cow/imce/papers/2019/Summary_Pentagon%20Fuel%20Use%2C%20Climate%20Change%2C%20and%20the%20Costs%20of%20War%20%281%29.pdf.
Daggett C. (2018) Petro-masculinity: fossil fuels and authoritarian desire. *Millennium* 47: 25–44.
Dalby S. (2015) Geoengineering: The next era of geopolitics? *Geography Compass* 9: 190–201.
DeWeerdt, S. (2008). War and the environment: War can wreck landscapes and ecosystems as well as people. *World Watch Magazine*, 21(1) [Online]. Available from: http://www.worldwatch.org/node/5520. Accessed 3 Jan 2015.
Dunlap A. (2018a) *End the "Green" Delusions: Industrial-Scale Renewable Energy Is Fossil Fuel+*. Available at: https://www.versobooks.com/blogs/3797-end-the-green-delusions-industrial-scale-renewable-energy-is-fossil-fuel.
Dunlap A. (2018b) The 'solution' is now the 'problem:' Wind energy, colonization and the 'genocide-ecocide nexus' in the Isthmus of Tehuantepec, Oaxaca. *The International Journal of Human Rights* 42(4): 550–573.
Dunlap A and Sullivan S. (2020) A faultline in neoliberal environmental governance scholarship? Or, why accumulation-by-alienation matters. *Environment and Planning E: Nature and Space* 3(2): 552–579.
Dunlap A. (2021a) Does renewable energy exist? Fossil fuel+ technologies and the search for renewable energy. In: Batel S and Rudolph DP (eds) *A Critical Approach to the Social Acceptance of Renewable Energy Infrastructures—Going Beyond Green Growth and Sustainability*. London: Palgrave, 83–102.
Dunlap A. (2021b) Spreading 'green' infrastructural harm: Mapping conflicts and socioecological disruptions within the European Union's transnational energy grid. *Globalizations*: 1–25. https://doi.org/10.1080/14747731.14742021b.11996518.
Dunlap A. (2022) Conclusion: A call to action, towards an insurrection in energy research. In: Majia H, Nadesan MJP, and Keahey J (eds) *Energy Democracies for Sustainable Futures*. Amsterdam Academic Press, 285–301.
Dunlap A and Jakobsen J. (2020) *The Violent Technologies of Extraction: Political Ecology, Critical Agrarian Studies and The Capitalist Worldeater*. London: Palgrave.
EC. (2020) *Critical Raw Materials for Strategic Technologies and Sectors in the EU—A Foresight Study*. Available at: https://ec.europa.eu/docsroom/documents/42881.

Estes N. (2019) *Our History Is the Future: Standing Rock Versus the Dakota Access Pipeline, and the Long Tradition of Indigenous Resistance*. New York: Verso.

Feigenbaum A. (2017) *Tear Gas: From the Battlefields of World War I to the Streets of Today*. New York: Verso Books.

Ford. (2021) *Responsive to Your Budget*. Available at: https://www.ford.com/police-vehicles/police-interceptor/hybrid-utility/calculator/.

Galeano E. (1997 [1973]) *Open Veins of Latin America: Five Centuries of the Pillage of a Continent*. London: Monthly Review Press.

Gedicks A. (1994) *The New Resource Wars: Native and Environmental Struggles Against Multinational Corporations*. Montreal: Black Rose Books Ltd.

Gills B and Morgan J. (2020) Economics and climate emergency. *Globalizations*: 1–16.

Graeber D. (2012) *Debt: The First 5,000 Years*. London: Melville House.

Huff A. (2021) Frictitious commodities: Virtuality, virtue and value in the carbon economy of repair. *Environment and Planning E: Nature and Space*: 1–21.

Huggins MK, Haritos-Fatouros M and Zimbardo PG. (2002) *Violence Workers*, Berkelely: University of California Press.

Jensen D. (2006) *Endgame, Vol. 1: The Problem of Civilisation*. New York: Seveb Stories Press.

Karp A. (2018) *Estimating Global Civilian-held Fire Arms Numbers*. Available at: https://www.smallarmssurvey.org/sites/default/files/resources/SAS-BP-Civilian-Firearms-Numbers.pdf

Kaur R. (2021) Nuclear necropower: The engineering of death conditions around a nuclear power plant in south India. *Political Geography* 85: 1–15

Kirsch S. (2014) Mining capitalism: The relationship between corporations and their critics, Berkeley: University of California Press.

Klinger JM. (2017) *Rare Earth Frontiers: From Terrestrial Subsoils to Lunar Landscapes*. Ithaca: Cornell University Press.

Kullik VJ. (2019) *Unter dem RadarDie strategische Bedeutung Seltener Erden für die wirtschaftliche und militärische Sicherheit des Westens*. Available at: https://www.baks.bund.de/sites/baks010/files/arbeitspapier_sicherheitspolitik_2019_13.pdf.

LaDuke W and Cruz SA. (2012) *The Militarization of Indian Country*. East Lansing: Michigan State University Press.

Lawrence R and Larsen RK. (2017) The politics of planning: Assessing the impacts of mining on Sami lands. *Third World Quarterly* 38: 1164–1180.

Levy A and Vaillancourt J-G. (2011) War on earth? Junctures between peace and the environment. In: Matyók T, Senehi J, Byrne S (eds) *Critical Issues in Peace and Conflict Studies: Theory, Practice, and Pedagogy*. Lanham: Lexington Books, 217–244.

Liboiron M. (2021) *Pollution Is Colonialism*. Durham: Duke University Press.

Macrotrends. (2021) *U.S. Military Spending/Defense Budget 1960–2021.* Available at: https://www.macrotrends.net/countries/USA/united-states/military-spending-defense-budget.

Margulies M. (2019) Decarbonize to Demilitarize Local Law Enforcement Agencies. *The Roosevelt Review* 3: 32–33.

Menjívar C and Rodríguez N. (2005) *When States Kill: Latin America, the U.S., and Technologies of Terror.* Austin: University of Texas Press.

Menton M and Le Billon P. (2021) *Environmental Defenders: Deadly Struggles for Life and Territory.* London: Routledge.

Nixon R. (2011) *Slow Violence and the Environmentalism of the Poor.* Cambridge: Harvard University Press.

Parkinson S. (2020) The environmental impacts of the UK military sector. In: *Scientists for Global Responsibility-Declassified UK.* Halton: Lancaster.

Parkinson S and Cottrell L. (2021) *Under the Radar: The Carbon Footprint of Europe's Military Sectors.* Available at: https://www.sgr.org.uk/publications/under-radar-carbon-footprint-europe-s-military-sectors.

Peluso NL and Vandergeest P. (2011) Political ecologies of war and forests: Counterinsurgencies and the making of national natures. *Annals of the Association of American Geographers* 101: 587–608.

Polanyi K. (2001 [1944]) *The Great Transformation: The Political and Economic Origins of Our Time.* Boston: Beacon Press.

Priestley C. (2021) "We won't survive in a city. The marshes are our life": An analysis of ecologically induced genocide in the Iraqi Marshes. *Journal of Genocide Research* 23: 279–301.

Provost C. (2017) *The Industry of Inequality: Why the World is Obsessed With Private Security.* Available at: https://www.theguardian.com/inequality/2017/may/12/industry-of-inequality-why-world-is-obsessed-with-private-security

Rodgers D. (2006) The state as a gang conceptualizing the governmentality of violence in contemporary Nicaragua. *Critique of Anthropology* 26: 315–330.

Rodney W. (1972 [2009]) *How Europe Underdeveloped Africa.* Washington, DC: Howard University Press.

Sanders B. (2019) *The Green New Deal.* Available at: https://berniesanders.com/en/issues/green-new-deal/.

Scott JC. (1998) *Seeing Like a State: How Certain Schemes to Improve the Human Condition Have Failed.* New Haven: Yale University Press.

Seager J. (1992) Operation desert disaster. In: Peters C (ed) *Collateral Damage: The New World Order at Home and Abroad.* Boston: South End Press, 197–216.

Seigel M. (2018) *Violence Work: State Power and the Limits of Police.* Duke University Press.

Selwyn D. (2020) *Martial Mining: Resisting Extractivism and War Together.* Available at: https://londonminingnetwork.org/wp-content/uploads/2020/04/Martial-Mining.pdf.

Shapiro J and McNeish J-A. (2021) *Our Extractive Age: Expressions of Violence and Resistance.* Oxon: Routledge.

Short D. (2016) *Redefining Genocide: Settler Colonialism, Social Death and Ecocide.* London: Zed Books.

Simpson M. (2019) Resource desiring machines: The production of settler colonial space, violence, and the making of a resource in the Athabasca tar sands. *Political Geography* 74: 1–12.

Smith G. (2017) *The War and Environment Reader.* Washington, DC: Just World Books.

Sovacool BK. (2021) Who are the victims of low-carbon transitions? Towards a political ecology of climate change mitigation. *Energy Research & Social Science* 73(1): 1–16.

Sovacool BK and Del Rio DDF. (2020) Smart home technologies in Europe: A critical review of concepts, benefits, risks and policies. *Renewable and Sustainable Energy Reviews* 120: 1–20.

Sweeney P. (2011) *Guide to Gunmetal.* Available at: https://www.rifleshootermag.com/editorial/guide-to-gun-metal/83987.

Thomas W. (1994) *Scorched Earth: Military's Assault on the Environment.* Philadelphia: New Society Publishers

Tilly C. (1985) War making and state making as organized crime. In: Evans P, Rueschemeyer D, Skocpol T (eds) *Bringing the State Back In.* Cambridge: Cambridge University Press, 169–191.

Verweijen J and Dunlap A. (2021) The evolving techniques of social engineering, land control and managing protest against extractivism: Introducing political (re)actions 'from above'. *Political Geography* 83: 1–9.

Vidalou J-B. (2017) *Être forêts: habiter des territoires en lutte.* Paris: Zones.

Welty J. (2015) *The Fuel Efficiency of Law Enforcement Vehicles.* Available at: https://nccriminallaw.sog.unc.edu/the-fuel-efficiency-of-law-enforcement-vehicles/.

Whitmore A. (2021) *A Material Transition: Exploring Supply and Demand Solutions for Renewable Energy Minerals.* Available at: https://www.waronwant.org/sites/default/files/2021-03/A%20Material%20Transition_report_War%20on%20Want.pdf.

Wiired. (2011) *What's Inside: Tear Gas.* Available at: https://www.wired.com/2011/05/st-whatinside-teargas/.

WSDoE. (2021) *Ecology Tracking Hanford Waste Tank Leak.* Available at: https://ecology.wa.gov/About-us/Who-we-are/News/2021/Ecology-tracking-Hanford-waste-tank-leak.

CHAPTER 7

Oil, Arms and Emissions: The Role of the Military in a Changing Climate

Wendela de Vries

Introduction

In a world where essential living conditions, including arable land and fresh water, are becoming increasingly scarce, the role of western militaries in supporting an economic system based on unlimited exploitation of natural resources is becoming increasingly problematic. Peace and environmental activists and scholars are starting to recognize that the military is playing a central role in climate destruction and injustice. From the beginning of the global expansion of western economies, control over resources has been at the core of military strategies. At present, this finds its expression in NATO, the North Atlantic Treaty Organization. NATO is not only designed to defend homelands but also to control countries with abundant raw materials and critical earth minerals, and to dominate essential shipping routes. This has dire consequences for people from the Global South as well as for the climate.

W. de Vries (✉)
Stop Wapenhandel, Amsterdam, The Netherlands
e-mail: w.de.vries@stopwapenhandel.org

A. Dunlap and A. Brock (eds.), *Enforcing Ecocide*,
https://doi.org/10.1007/978-3-030-99646-8_7

While activists demand *climate justice* by challenging power inequalities and a change in production and consumption patterns, military strategists are calling for *climate security*, which means protecting the status quo where the wealthy are sheltered from the damaging consequences of climate change (if that is possible). At the same time, armed forces contribute to climate change with unrestricted greenhouse gas emissions. This article explores how western military strategies and values contribute to climate change and how armed forces respond to this looming disaster. What role do militaries play in the extraction economy? How is the involvement of militaries legitimized, for instance in the NATO Strategic Concept, the basic document setting out NATO policies? And what is the role of the global arms trade in facilitating control over resources?

Armed forces have no reduction obligation under international climate agreements, specifically the Paris Treaty. Military planners recognize climate change as a problem and are implementing adaptation measures for their infrastructure, energy consumption and military strategy. Yet, I argue that these measures are ineffective in reducing the military carbon boot print, by exploring the role of fossil fuel and alternatives in NATO strategies and the feasibility of military emission reduction. To understand the role of militaries in climate crisis, this article starts out investigating the relationship between Western values, resource control and climate catastrophe by pointing to the ways these are politically mobilized. I then examine how *energy security* is a necessary prerequisite to war making and informs the militarized strife for control over see lines of communication and resources. Resource control and arms trade, I then argue, override humanitarian and environmental concerns, as deliberately vague arms treaties and the increase in arms trade illustrate. Arms trade figures illustrate the money flows between arms exporting and oil exporting countries. The safety, livelihoods and human rights of people from the Global South are subordinated to the interest of the extraction economy. The violent repression of local resistance, arms trade with repressive regimes, and the militarization of borders to stop people from finding refuge and a better life all illustrate this point. Western military activity not only harms humans, however, but has disastrous planetary consequences. In the next section, I briefly examine some of the military's key contributions to climate change and their possibilities to decarbonize. This leads me to finish by exploring critiques of greenwashing and possibilities for future change.

Western Values and Control Over Resources

When analysing the role of arms and armed forces in climate change it makes sense to focus on the role of western armed forces. NATO brings together the countries with the biggest greenhouse gas emissions per capita and the largest global arms exports, responsible for more than half of all military expenditure in the world. NATO estimates its 2020 military spending at US$ 1,092,482 million (NATO 2020a). Global military spending in 2020 is calculated at US$ 1981 billion by the Stockholm International Peace Research Institute (SIPRI 2021). NATO describes itself as "a unique community of values, committed to the principles of individual liberty, democracy, human rights and the rule of law" (NATO 2010). NATO not only helps to defend the territory of its members, but also "engages where possible and when necessary to project its values further afield" (NATO 2020b). This NATO strategy translates into the military policies, strategies and structures of all individual NATO member states.

The type of freedom and values that NATO protects include unlimited access to raw materials for ever-growing production and consumption. Access to fossil fuels and, increasingly, to rare earth minerals are the most important priorities (Stavridis 2021). According to the NATO Strategic Concept, "[a]s a larger share of world consumption is transported across the globe, energy supplies are increasingly exposed to disruption." NATO heads of state and government thus declare that they "will ensure that NATO has the full range of capabilities necessary to deter and defend against any threat to the safety and security of our populations… Therefore, we will develop the capacity to contribute to energy security, including protection of critical energy infrastructure and transit areas and lines, cooperation with partners, and consultations among Allies on the basis of strategic assessments and contingency planning" (NATO 2010). The protection and security of energy infrastructures and climate security are thus clearly important priorities.

NATO armed forces are equipped and deployed to defend the economic interests and ways of life of NATO member countries. The fact that these ways of life are unsustainable is well-known to the military planners and anticipated for in strategies: "Key environmental and resource constraints, including health risks, climate change, water scarcity and increasing energy needs will further shape the future security environment in areas of concern to NATO and have the potential to significantly affect NATO planning and operations" (NATO 2010). The interests of NATO countries include domination over other economies and/or values. NATO armies have frequently "protected values and liberties" with military interventions and acts of forced regime change, most notably in the Gulf region. Among the most notorious is the 1953 US- and British-sponsored overthrow of the democratically elected government of Mossadeq of Iran after it nationalized the Anglo-Iranian Oil Company. More recent are the first and second Gulf Wars against Iraq in 1991 and 2003, eventually killing Iraqi dictator Saddam Hussein, who used to be armed and military supported as a Western ally right up to 1990, notwithstanding his already well-known atrocities (Alkadiri and Mohamedi 2003). This illustrates the changing and strategic instrumentalization of arguments based on alleged security concerns and values, including democracy or human rights, for (geo)political purposes.

Western governments consider their national interests as legitimate reasons for *power projection*, a military term for the implementation of policy by means of force, or the threat thereof, outside one's own territory. The rhetoric about "values" is helping to cover up self-interested military interference and dominance. The "unique community of values" has destabilized many countries and sometimes entire regions. The Western public's acceptance of violent interference by armed forces in other parts of the world is supported by the racists sentiment that non-western people need "protection" (e.g. Rodney 2009 [1972]). An inheritance from colonial times, non-western people are easily depicted as violent and backward, not able to install stable governments themselves. NATO countries claim the right to impose their values as if these countries had no values of their own. In reality, the violence and instability in many countries is caused by the political and military interventions of exactly the countries that are supposed to "protect" (Mundy 2020). This is very visible in the international rights concept of Responsibility to Protect (R2P), first officially used in the UN-authorized attack on Libya in 2011. From a military point of view, the attack could be called a success. Looking at the continuing violent instability in Libya and several other affected countries in the region, one can wonder for what purpose the attack took place and whose interests were served.

Energy Security

Wars tend to have multiple causes, but the access to and price of oil is a major driver of many recent wars and military interventions (Aarts 1992; Faulconer 2020). Oil is essential for the mass consumption-based economies of NATO countries. As formulated in the Strategic Concept: "All countries are increasingly reliant on the vital communication, transport and transit routes on which international trade, energy security and prosperity depend" (NATO 2010). And not only is it essential for economies, it is also a precondition for military action, as essential as bombs and ammunition. Uninterrupted access to energy, *energy security*, is a precondition to win a war. "Energy security should be a constant item to be monitored, assessed, and consulted among Allies" writes the high-level NATO Reflection Group in a forward-looking paper toward a new NATO strategy (NATO 2020c).

To provide armies, navies and air forces with fuel in case of war, countries need control over energy sources and freedom of transport of this fuel and other supplies. "A stable and reliable energy supply, by diversification of routes, suppliers and energy resources, and the inter-connectivity of energy networks are of critical importance and increase resilience. Assuring energy supplies to military operations is important for NATO and the Allies" writes the Reflection Group (NATO 2020c) At the same time, control over transport routes makes it possible to deny an opponent access to energy resources, which gives a huge strategic advantage, notably over chief competitor China (Morcos 2021). Although the United States is on the brink of losing its position as the world's strongest economy to China, the United States is still by far the world's strongest military power, responsible for more than one-third of all global military spending (Marksteiner et al. 2021) and most of the global arms exports (Wezeman et al. 2021). For the protection of their own resources and denial of access to resources to adversaries, NATO armies strive to control over "sea lines of communication" (SLOCs), essential shipping routes for fuel and supplies. International trade also has an interest in these sea lines, as the volume of shipping trade continues to grow with the ever-growing hunger for fast and cheap consumer goods (Research Department 2021). The desire to control SLOCs—similarly to the control over resources—is motivating military action, such as NATO naval presence in the Persian Gulf. The continued strive to dominate SLOCs is most visible in the South Chinese Sea and Indian Ocean Region where tensions rise over maritime control (Ghiasy et al. 2018; Euraktiv with Reuters 2021). Control over SLOCs and other (fossil fuel) resources is thus integral to NATO's conceptualization of energy security, and energy security central to understandings of military security.

For the United States, the role of imported oil has diminished due to booming domestic production of shale oil. The United States has turned from an oil importing country into the world's largest oil producer: By the last week of 2019, US oil exports had reached nearly 4.5 million barrels a day and were shipped to more than 50 countries (Tobben and Merrill 2021). The largest buyer of US oil is China, with two million barrels a day. It is remarkable that this trade seems not to have suffered under the Trump's anti-China rhetoric. To avoid too much dependency, China is strategically spreading its fuel imports, Saudi Arabia and Russia also remaining major suppliers. Increased US energy self-sufficiency does not mean that the United States has lost interest in controlling sea routes and fuel resources, as well as, more recently, sources of rare earth minerals. According to James Stravridis, former supreme commander of NATO, the United States must design a policy for "rare-earth independence" including ensuring supply chains, mandating defence contractors to wean themselves off Chinese rare earths and sponsoring research and development to find alternative materials. Critical raw materials are essential for most advanced electronics including defence electronics as well as lower-carbon energy technologies—leading Stravridis to call on US climate envoy Kerry to "advocate for a sensible strategy on securing rare earths" in the National Security Council (EC JRC, 2020; Stavridis 2021). The supply and political economy of fossil fuels and strategic minerals are deeply interconnected, both instrumental to military and consumer technologies.

Resource Control, Arms and Human Rights

Control of resources and sea lines of communication can be outsourced to allies by supporting (often authoritarian) governments with military equipment (including weapons) and training. This leads to extensive and uninterrupted arms transfers to strategically located countries such as Egypt, Saudi Arabia and Israel, despite their serious human rights violations. When the Biden government announced its major arms sales to the Al-Sisi military regime in Egypt—criticized for rampant disregard of human rights, democratic process and freedom of the press

(Farooq 2021)—it argued that the sale would "support the foreign policy and national security of the United States by helping to improve the security of a Major Non-NATO Ally country that continues to be an important strategic partner in the Middle East... The proposed sale will support the Egyptian Navy's Fast Missile Craft ships and provide significantly enhanced area defense capabilities over Egypt's coastal areas and approaches to the Suez Canal" (Defense Security Cooperation Agency 2021). These US sales come on top of an already gigantic Egypt acquisition program including German MEKO-frigates with Dutch equipment, a helicopter landing ship from France and a number of submarines from Germany. It will make the Egyptian navy capable to control and police the Suez channel and Eastern Mediterranean (Broek 2020).

Treaties and policies to control arms trade, such as the UN Arms Trade Treaty (UN 2014) and the EU Common Position on arms export (EU 2008) are formulated in deliberately vague terms to leave space for arms exporting countries to let military and strategic interests overrule peace and human rights (Vries 2013). Arms imports into the Middle East increased by 61% between 2010 and 2019 and accounted for 35% of total global arms imports between 2015 and 2019 (SIPRI 2019). The majority of these arms are provided by the USA and European states, also major buyers of fossil fuels from the Middle East. Middle Eastern fossil fuel revenues are thus "recycled" back into arms exporting countries (Wearing 2018). After the Middle East, East Asia is the second largest arms importer in the world. Competition with China, with sometimes open hostilities, has made India the world's second importing country of major weapons (SIPRI 2018). For arms supplying countries, military export is a way to exert political influence over the importers. Being a dominant arms supplier creates a long-time dependency relation from buyer to seller for spare parts, training and upgrades.

Resource extraction is often met by resistance of local communities and/or environmental activists (Selwyn 2020; Verweijen and Dunlap 2021). Armed forces and/or militias, often equipped with Western arms, are involved in violent repression of such resistance. Nomadic and indigenous populations trying to protect their land and livelihoods are particularly vulnerable, their legal position often not strong or simply ignored. Local activists are at risk of being detained, mistreated or even killed. In 2019, Global Witness recorded 212 murdered land and environmental defenders (Global Witness 2020). In many cases, like in West Papua, repression involves armed forces equipped with western weapons

(International Coalition for Papua 2020). While Western countries are arming repressive regimes and fuelling wars to protect their unsustainable economic system, migrants and refugees are fleeing from poverty, repression and war (Centre Delàs, n.d.). Their number will increase dramatically when even more people from the global South will be driven from their homes by extreme weather, sea level rise, desertification and food and water shortage, as many do not have the means to cope with the consequences of climate change. According to UNHCR, already around 23 million people have to leave their homes each year due to extreme weather conditions (UNHCR 2021). Most become internally displaced, while some are trying to cross borders in search of safety and protection.

Although the distinction between war refugees, labour migrants and climate refugees is difficult or even impossible, as migration motives are multiple and intertwined, military planners prepare for more instability and mass displacements of people due to climate change (IMCC 2021). They introduce climate indicators into warning systems and are finding legitimation for the use of force in a concept of Responsibility to Protect and Prepare (R2PP). Instabilities that arise due to water and food shortages and the shrinking of habitable land followed by increase of migration are included in military scenarios. Climate change is called a *threat multiplier* (EEAS 2016). But instability and migration are not caused by climate change itself, they are caused by insufficient means to cope and as such, are the result of unfair distribution of resources and power. These conflicts are not nature-made but human-made. By labelling conflicts as "climate conflicts" the root cause of the problem is covered up and the use of force is legitimized (Stop Wapenhandel 2019).

In response to the increasing number of refugees and migrants, and legitimized through the depiction of migrants and refugees as security threats, Western countries—and the European Union as a political bloc—are militarizing their borders. The global market for border technology is estimated to be worth approximately €17.5 billion in 2018, with annual growth of at least 8% expected in coming years (Akkerman 2019). Big arms companies like Airbus, Thales, Leonardo, Lockheed Martin, General Dynamics, Northrop Grumman and L3 Technologies are among the biggest profiteers of this market (Ruiz Benedicto et al. 2020). The dramatic consequences can be seen at the EU borders and the US-Mexico border. NATO is supporting military action of EU border agency Frontex in the Mediterranean, forcing desperate people to take even more dangerous migration routes. (Akkerman 2017) Increasingly and with support of the arms industry, Western countries are closing off for the rest of the world with military means.

Power Projection and Military Emissions

NATO weaponry not only harms humans but the entire planet. Western armed forces are huge climate polluters. Accurate data on military emissions are often kept secret, using national security as the reason. It is obvious that the amount of fossil fuels needed to endure in war is relevant military information. But without this data it is difficult to discuss military emissions, set emission reduction targets and monitor results. One can question this military secrecy, considering the democratic need for policies based on accurate figures. Military emissions concern direct emissions from systems and operations (scope 1 emissions) and infrastructure (scope 2 emissions) as well as indirect emissions from the military supply chain (scope 3). Estimates deduced from the US Department of Defence energy consumption data show that in the period from 2001, when the United States invaded Afghanistan, to 2018, military forces emitted 1.3 billion tonnes of carbon dioxide equivalent (tCO2e) (Crawford 2019). The war-related portion of those emissions—including for the major war zones of Afghanistan, Pakistan, Iraq and Syria—is estimated to be more than 440 tCO2 (Crawford 2019).

In a conservative estimate of the EU's military carbon footprint, including both scope 1 and 2 and "upstream" scope 3 emissions, researchers calculated that in 2019, military emissions of the six largest EU countries amounted to approximately 25 billion tCO2e—equivalent to the annual CO2 emissions of about 14 million cars (Parkinson and Cottrell 2021). Based on these figures, one can conclude that the total global military carbon footprint (scope 1, 2 and 3) equals that of a large European nation. The US military alone creates more planet-warming greenhouse gas emissions through its defence operations than industrialized countries such as Sweden and Portugal (Crawford 2019). In other words, militaries not only protect and enforce ecologically disastrous operations including fossil fuel extraction and transport, but they cause climate damage doing so.

It is the armament for *power projection*, for expeditionary warfare far from national territory, which is contributing most to military emissions.

Power projection needs heavy transport capacity to far-off places and military equipment able to go fast and far, demanding a huge share of Western defence budgets and providing huge profits for the military industry sector (Mazarr 2020).As then-NATO Secretary General Robertson said in 2002 at a Defence Industry Conference: "Alliance military capabilities will be developing further towards long-range power projection… We need forces that are slimmer, tougher, and faster; forces that reach further, and can stay in the field longer" (Robertson 2002). If Western militaries were equipped only to defend their own territory, they could do with lighter, less fuel-consuming weaponry. *Power projection* capacity makes the US military the largest institutional energy consumer in the world; the US Navy uses around 5 billion litre of fuel per year, the Air Force 9 billion (Crawford 2019). A military strategy that was not grounded in *power projection* could substantially contribute to military emission reduction.

The Paris Climate Treaty does not include targets for the military sector, but NATO agreed at its 2014 summit that members should work "towards significantly improving the energy efficiency of our military forces" (NATO 2014a). The dangerous transport of fossil fuels, especially in conflict zones, brings together ecological and social costs and illustrates the link. The road through rough Pakistan's mountains—often thousands of kilometres long—that fuel convoys had to take to supply military compounds in Afghanistan are one notorious example. Between 2008 and 2014, these convoys were attacked en route 485 times, causing 167 deaths and 450 injuries (Rosenthal 2010).

Enhancing energy efficiency is on the agenda for Western militaries. Solar panels and bio-waste installations increasingly contribute to the energy supply of military installations and are widely presented in the media as the military contribution to emission reduction. Yet, these emissions are small when compared with emissions from ships, aircraft and combat vehicles, leading to critiques of greenwashing and articulations of "sustainable violence" (Bigger and Neimark, 2017; Dunlap, 2017, 2021). Emissions from transport fuels, on the other hand, are difficult to decarbonize. Like in the civil sector, propulsion fuel is a bottleneck for a transition to sustainability, especially in case of heavy and/or fast military systems for global power projection.

Greenwashing

There are similarities in the ways both the civilian and the military sectors are ignoring that for the foreseeable future there is no sustainable propulsion fuel available to reduce the global transport carbon footprint. Fighter jet emissions illustrate the point. Aviation is using roughly two-thirds of all fuel in defence (Crawford 2019). Fighter jets are able to fly deep into enemy territory, especially when taking off from aircraft carriers, in essence floating military air bases. Fighter jets are central in *power projection* strategies and high on the shopping list of many countries.

Fighter jet fuel is basically high-quality kerosene, the same as used in civil aviation. Experiments with plant-based biofuels are rejected by environmentalists and others as being unsustainable; it will use too much arable land that is badly needed for food production (Bigger and Neimark 2017; Hargreaves 2021). Experiments with F-16 fuel blended with 5% biofuel from cooking oil and household waste, as experimented by the Royal Netherlands Air Force, are quickly meeting the limit of available waste products (McCue 2019). The replacement of fossil-based kerosene with sustainable air fuel (SAF) still has a long way to go. Production of synthetic kerosene is also still in a very experimental phase, as yet with a high energy inefficiency rate (Stay Grounded, n.d.). Nor is electricity close to providing an alternative. Electric planes are at the very beginning of development, and so far only suitable for short distance light planes. And although hydrogen has many supporters, it is an energy carrier, not an energy source, and as with electric planes, hydrogen is only sustainable when it comes from renewable sources. There are no serious calculations yet of the availability of sufficient renewable energy for a whole military air fleet, but it is highly unlikely that these needs can easily be met. In sum, despite optimistic stories about zero emission planes, it will take many years to develop these for civilian transport, let alone for the superfast and/or heavy military air fleet (NLR/SEO 2021).

Yet, civil and military aviation industries are presenting future to-be-developed technologies as solutions for urgently needed emission reduction (Clean Sky 2018). This is distracting political efforts and investments from realistic and feasible steps toward sustainability. Considering the long time that weapon development takes—from first design to final product—the end of fossil fuel-based military systems is far away. Armed forces are locked into fossil fuel technologies. Like in the civil sector, the only possible way to seriously limit emissions in the military sector is to change from *more* and *bigger* to *less* and *smaller*. With no sustainable fuel available yet, less emission could be reached by lighter weapon systems. But alternatives such as using drones instead of fighter jets only make sense when fighter jets will be *replaced* by drones, not by *adding* drones, as is now in the planning with the Future Combat Air System, for instance (Airbus Defence 2020). Moreover, the use of drones comes with ethical problems of its own (Cole 2014). New systems might include more energy-efficient technologies but still not reduce emissions or mining. In a recent Dutch replacement of a series of naval support ships, the ship engines were made more efficient, but as the weight of the ships had increased, no energy efficiency was achieved (Ministerie van Defensie 2020). The current efforts at decarbonizing the armed services appear half-hearted at best.

Curiously, the UK Ministry of Defence announced plans for up to 50% sustainable aviation fuel for RAF aircraft in the future, including for F-35 and Typhoon fighter jets. "The UK is leading the way in sustainability and by refining our aviation fuel standards we are taking simple yet effective steps to reduce the environmental footprint of defence… As we strive to meet this government's Net Zero carbon emissions target by 2050, it is right that we step up to spearhead these positive changes across both military and civilian sectors" according to the UK ministry (Mönch 2020). Other national armed forces are setting targets as well. The Royal Netherlands Air Force intends to operate all aircraft on sustainable air fuel (SAF) and reduce dependency on fossil fuels by 20% in 2030, and by 70% in 2050. The technologies to meet these targets still have to be developed, which have an unknown extractive and manufacturing cost.

The goal is to "raising the green profile of NATO" as proposed in the Green Defence framework (NATO 2014b). The public is starting to demand action so some ambition must be demonstrated. "Increased energy efficiency responds to environmental concerns in Allies' public opinion and demonstrates that NATO is responsive to them. Saving energy and demonstrating environmental awareness are enduring benefits" writes NATO in the Green Defence framework (ibid). The NATO Summit 2021 communiqué shows the limits of low-emission warfare. NATO has agreed "to significantly reduce greenhouse gas emissions from military activities and installations without impairing personnel safety, operational effectiveness and our deterrence and defence posture. We invite the Secretary General to formulate a realistic, ambitious and concrete target for the reduction of greenhouse gas emissions by the

NATO political and military structures and facilities and assess the feasibility of reaching net zero emissions by 2050" (NATO 2021). So instead of setting targets, as some individual countries have done, NATO is asking for a feasibility assessment. But military effectiveness, deterrence and defence come first. And more importantly: the military goal of protecting an unsustainable economic system is not questioned at all.

Will the Military Change?

The military is not ignoring or denying climate change. It is taking it seriously as "a threat multiplier that impacts Allied security, both in the Euro-Atlantic area and in the Alliance's broader neighbourhood. Climate change makes it harder for militaries to carry out their tasks" (NATO 2021). In 2020, the Reflection Group for a new NATO strategy says: "Climate change is a driver of NATO's security environment. Its effects can be seen in, inter alia, the intensity of geopolitical competition, freedom of navigation in the High North, and migration streams from the south, all of which involve vital Allied interests" (NATO 2020c).

Armed forces have to deal with extreme weather threatening military bases and influencing readiness of forces. Crucial US military strongholds such as the Pacific island of Guam, home of nuclear submarines and B-52 bombers, face increasingly serious freshwater problems due to salinization (Kodack 2019). US warships had to flee harbour to open sea when hurricane Dorian hit the US coast in 2019, bringing extreme flood and rain (US Naval Institute 2019). UK troops in Iraq desert bases suffered from extreme heat while equipment was faltering. During the huge Californian fires of 2020, commanders expressed worry about troop readiness for major conflict when soldiers were engaged in fighting climate disasters like wildfires. In 2018, the American Ministry of Defence released a "Climate-Related Risk to DoD Infrastructure: Initial Vulnerability Assessment Survey," reporting that thousands of US American military bases and installations are exposed to climate-related impacts. Long-time resource war researcher Michael Klare even suggests that senior military commanders might be the ones to bridge the divide between climate believers and deniers (Klare 2020). The climatic threats to military operations appear more than what planners have anticipated.

Institutes such as the International Military Council on Climate and Security (IMCCS) also continue to put climate change on the military agenda. IMCCS defines itself as "non-partisan." The institute of senior military leaders, security experts and security institutions is busy "anticipating, analysing and addressing the security risks of a changing climate" (2021). IMCCS is pushing the issue on military agendas, e.g. the Munich Security Conference, an annual military forum with participation of heads of state, ministers and senior military officials. The first World Climate and Security Report of IMCCS identifies climate change as a "significant driver of instability" calling security institutions and militaries to prepare "for water and food security and their associated effects on stability, conflict and displacement" (IMCCS 2020), not only outside but also inside allied territory (EPRS 2019). In its second edition, published in 2021 ahead of the G7 and NATO summit, emphasis lies on including "climate security" in military planning. The report includes recommendations for developing new global governance tools "to cope with the emerging climate security implications for a range of issues in the international domain to include human rights, international maritime law, and geo-engineering."

This military attention for climate change raises the question of how far armed forces can be allies for activists against climate change. General Tom Middendorp, chair of IMCCS and former commander of the Dutch armed forces, was invited to contribute to a Dutch Extinction Rebellion handbook, featuring besides ecologists such as Vananda Shiva (Baars et al. 2020). He presented a twenty-first-century adaptation of the White Man's Burden and propagated involvement of western militaries in climate conflicts in vulnerable areas because, according to the retired general, they "bring an impartial voice to the table, a voice that is not political, a voice that is not domestic, and a voice that has a certain authority" (General Middendorp 2020). When military actors start depicting their involvement in conflicts as "impartial" and "non-political," one should be wary. It is ignoring the basic rule formulated by Von Clausewitz, that war is the continuation of politics by other means. At best, Western military forces will add one more warring party to a conflict. In most cases, they will start supporting the local party that best serves their country's interests, with operations, arms and training, adding more violence to an already bad situation. Afghanistan is the most recent dramatic example.

To limit military emissions there should be more seriousness about non-military ways of addressing conflict, including revaluation of disarmament treaties, tension-reduction and trust-building policies. This would be beneficial in more than one way. In a world with global problems like pandemics and climate change, we need to address conflicts with diplomacy not arms, and find common solutions for what is really threatening us. Unfortunately, we see the opposite, with strong pressure on countries to raise defence budgets, extensive technology modernization programmes and NATO/EU deployments continuing outside Europe (Klare 2021). Boosting NATO budgets up to 2% of BNP with 20% to be spent on new armaments will increase military emissions even further.

For solely defending national territory one does not need extensive air fleets of fighter bombers. While military training can sometimes be made less carbon intensive by using simulation technologies, the production of arms, the use of arms, the damage done by arms (including the burning of oil wells and refineries) and the reconstruction of damage after war are contributing to climate pollution and ecocide . War is not only deadly for people, it is also killing the climate. If the new NATO Strategic Concept, expected by the end of 2021, does not change its direction, military emissions will increase and NATO will continue to protect the interest of the extractive economy.

Conclusion

This chapter demonstrates that the armed forces are huge climate polluters and constitute an important pillar to the global extraction economy, a system of fossil fuel-based overconsumption and unfair distribution of resources, at the expense of the global South. This is legitimized by the racist assumption that western values are superior. Despite high military carbon emissions, NATO concerns about climate change do not primarily deal with environmental interests but first and for all with military security interests. Plans to reduce military emissions seem to have primarily strategic and financial reasons. NATO worries about deployability of forces and how to adapt military strategy to new kinds of resource conflicts. The expected increase of migration, partly due to climate change, is defined as a security problem; military strategies include anti-migration strategies, to the benefit of the military industry. For military organizations, climate change is dealt with in a military context and with military methods, where the use of force is central. This dominant approach must be reconsidered.

Contrary to claims of the arms industry and military organizations, armed forces are not making much progress on becoming sustainable. While emission reductions in military infrastructure are possible, the technology to lower the emissions of military transport, large military platforms and weapon systems simply does not exist. Only by reducing military equipment, exercises and deployment is it possible to reduce military emissions substantially. Unfortunately, the contrary is happening, and military emissions continue to grow. The problems of climate change are clear, yet how to make these changes—and peoples determination to make them—is what remains to be seen.

Bibliography

Aarts P. (1992) Democracy, Oil and the Gulf War. *Third World Quarterly* 13(3): 525–538.

Airbus Defence (2020, June 17) Future Combat Air System: Owning the Sky with the Next Generation Weapons System. Retrieved from https://www.airbus.com/newsroom/stories/Future-Combat-Air-System-Owning-the-sky-with-the-Next-Generation-Weapons-System.html.

Akkerman M. (2017, May). *NATO and EU Border Security in the Mediterranean*. Stop Wapenhandel.

Akkerman M. (2019, Nov.). *The Business of Building Walls*. Stop Wapenhandel, Centre Delas, TNI

Alkadiri R & Mohamedi F. (2003) World Oil Markets and the Invasion of Iraq. *Middle East Report* 227 (Summer 2003).

Baars, B, Geuzebroek, C and Esajas, M. (2020, May 18) *Een groen leger bestaat niet*. OneWorld.

Bigger P and Neimark BD. (2017) Weaponizing Nature: The Geopolitical Ecology of the US Navy's Biofuel Program. *Political Geography* 60: 13–22.

Broek M. (2020) Mega Egyptian order for Dutch branch of French Thales. Retrieved from https://stopwapenhandel.org/node/2419.

Centre Delàs d'Estudis per la Pau. (n.d.) Interactive Map "European weapons and Refugees". http://www.centredelas.org/mapa-interactiu-armes-europees-i-refugiats/?lang=en.

Clean Sky (2018, Oct. 8) EU Synergies with EU Funding for Defence and Dual-Use. Retrieved from https://www.cleansky.eu/news/synergies-with-eu-funding-for-defence-and-dual-use.

Cole, C. (2014) *What's Wrong with Drones?* Drone Wars UK

Crawford NC. (2019 revised and updated). *Pentagon Fuel Use, Climate Change,and the Costs of War*. Watson Institute, Brown University.

Defense Security Cooperation Agency (2021, Feb. 16) Egypt—Rolling Airframe Missiles (RAM) Block 2 Tactical Missiles (Press Release). Retrieved from https://www.dsca.mil/press-media/major-arms-sales/egypt-rolling-airframe-missiles-ram-block-2-tactical-missiles.

Dunlap A. (2017) Wind Energy: Toward a "Sustainable Violence" in Oaxaca, Mexico. *NACLA* 49: 483–488.

Dunlap A. (2021) Does Renewable Energy Exist? Fossil Fuel+ Technologies and the Search for Renewable Energy. In: Batel S and Rudolph DP (eds) *A Critical Approach to the Social Acceptance of Renewable Energy Infrastructures—Going Beyond Green Growth and Sustainability.* London: Palgrave, 83–102.

EC JRC (European Commission Joint Research Centre) (2020) *Critical Raw Materials for Strategic Technologies and Sectors in the EU—A Foresight Study.* Available at: https://ec.europa.eu/docsroom/documents/42881.

Euraktiv.com with Reuters (2021, Jan. 25) South China Sea tension flares again as Biden Takes Charge. Retrieved from https://www.euractiv.com/section/global-europe/news/south-china-sea-tension-flares-again-as-biden-takes-charge/.

EEAS (2016, June) Shared Vision, Common Action: A Stronger Europe—A Global Strategy for the European Union's Foreign and Security Policy. European External Action Service.

EPRS (2019, Jan. 29) European Parliamentary Research Service. The Concept of 'Climate Refugee' Towards a Possible Definition.

European Union (2008, Dec. 8) Council Common Position 2008/944/CFSP Defining Common Rules Governing Control of Exports of Military Technology and Equipment.

Farooq UA. (2021, Feb. 17) Rights Groups Slam US Arms Sale to Egypt as Activist's Relatives Arrested. Middle East Eye. Retrieved from https://www.middleeasteye.net/news/egypt-weapons-sales-denounced-mohamed-soltan-relatives-arrested.

Faulconer M. (2020) The Mosaddeq Coup: Cold War Strategy, Oil, and American Ideals. *The Macksey Journal* 1, Article 214.

General Middendorp T. (2020) Munich Security Conference. In Climate Change More Prominent Than Ever at Munich Security Conference with "World Climate Security Report 2020" Release. Council on Strategic Risks. https://councilonstrategicrisks.org/2020/02/19/climate-change-more-prominent-than-ever-at-munich-security-conference-with-world-climate-security-report-2020-release/.

Ghiasy R, Su F, and Saalman L. (2018, Sept.) *The 21st Century Maritime Silk Road: Security Implications and Ways Forward for the European Union.* SIPRI.

Global Witness (2020, July). Defending Tomorrow.

Hargreaves S. (2021, Jan. 21) Green Jet Fuel Plans Risk History Repeating Itself. Transport & Environment. Retrieved from https://www.transportenvironment.org/news/green-jet-fuel-plans-risk-history-repeating-itself.

IMCCS (2020) International Military Council on Climate and Security World Climate and Security Report 2020.

IMCCS (2021) Expert Group of the International Military Council on Climate and Security World Climate and Security Report 2021.
International Coalition for Papua (2020, Jan.) Human Rights and Conflict Escalation in West Papua, sixth report.
Klare M. (2020, Feb. 18) A Military Perspective on Climate Change Could Bridge the Gap Between Believers and Doubters. *The Conversation.*
Klare M. (2021, Feb. 25) *Heating the Planet Through a New Cold War.* Tomdispatch
Kodack M. (2019, Dec. 2) Future Climate Scenario Projects Water Woes for U.S. Military Bases in Guam. Centre for Climate & Security. Retrieved from https://climateandsecurity.org/2019/12/future-climate-scenario-projects-water-woes-for-u-s-military-bases-in-guam/.
Lopes da Silva D, Tian N, and Marksteiner A. (2021, April). *Trends in World Military Expenditure, 2020.* SIPRI
Mazarr MJ. (2020, April 15) Toward a New Theory of Power Projection. Toward a New Theory of Power Projection. Commentary. Retrieved from https://warontherocks.com/2020/04/toward-a-new-theory-of-power-projection/.
McCue D. (2019, Jan 16) Royal Netherlands Air Force Operating F-16 Fighting Falcons on Sustainable Fuel. Renewable Energy Magazine. Retrieved from https://www.renewableenergymagazine.com/biofuels/royal-netherlands-air-force-operates-f16-fighting-20190116.
Ministerie van Defensie (2020, June 22) Antwoorden op feitelijke vragen inzake de A-brief project 'Vervanging hulpvaartuigen CZSK'. Kamerstuk 27830, nr.305.
Morcos P. (2021 June 8) NATO's Pivot to China: A Challenging Path. Commentary. Centre for Strategic and International Studies CSIS. Retrieved from https://www.csis.org/analysis/natos-pivot-china-challenging-path.
Mönch (2020, Dec. 14) RAF Aircraft to Run on Alcohol and Household Waste. Mönch Publishing Group retrieved from https://www.monch.com/mpg/news/air/7884-raf-aircraft-to-run-on-alcohol-and-household-waste.html.
Mundy J. (2020, June 9) The Oil for Security Myth and Middle East Insecurity. Middle East Report *Online.*
NATO (2010) Active Engagement, Modern Defence. Strategic Concept for the Defence and Security of the Members of the North Atlantic Treaty Organisation. Adopted by Heads of State and Government in Lisbon 2010.
NATO (2014a, Sept. 5) Wales Summit Declaration. Issued by the Heads of State and Government participating in the meeting of the North Atlantic Council in Wales.
NATO (2014b, Feb.) Green Defence Framework.

NATO (2020a) PR/CP (2020a)104 Defence Expenditure of NATO Countries (2013–2020a) Retrieved from: https://www.nato.int/nato_static_fl2014/assets/pdf/2020/10/pdf/pr-2020-104-en.pdf.

NATO (2020b last updated Sept. 24) NATO's Purpose. Retrieved from: https://www.nato.int/cps/en/natohq/topics_68144.htm.

NATO (2020c, Nov. 25) NATO 2030: United for a New Era. Analysis and Recommendations of the Reflection Group Appointed by the NATO Secretary General.

NATO (2021, July 14) NATO Climate Change and Security Action Plan.

NLR/SEO. (2021, Feb. 11) Destination 2050—A Route to Net Zero European Aviation. NLR—Royal Netherlands Aerospace Centre & Amsterdam Economics.

Parkinson S and Cottrell L. (2021, Feb.) Under the Radar. The Carbon Footprint of Europe's Military Sectors. CEOBS, SGR, GUE/NGL.

Research Department (2021, April 1) Container Shipping—Statistics & Facts. Retrieved from https://www.statista.com/topics/1367/container-shipping/.

Robertson L. (2002, Oct.14) The World in 2015—Predicting the Unpredictable. Keynote Speech at Defence Industry Conference. Available at https://www.nato.int/cps/en/natohq/opinions_19705.htm?selectedLocale=en.

Rodney W. (1972 [2009]) *How Europe Underdeveloped Africa*. Washington, DC: Howard University Press.

Rosenthal E. (2010, Oct. 4) U.S. Military Orders Less Dependence on Fossil Fuels. *The New York Times*.

Ruiz Benedicto A, Akkerman M, and Brunet P. (2020) A Walled World, towards a Global Apartheid. Centre Delas, Stop Wapenhandel, TNI.

Selwyn D. (2020, Nov.) Martial Mining. Resisting Extracivism and War Together. London Mining Network.

SIPRI (2018, March 12) Asia and the Middle East lead rising trend in arms imports, US exports grow significantly, says SIPRI. Press Release.

SIPRI (2019, March 11) Global Arms Trade: USA Increases Dominance; Arms Flows to the Middle East Surge, says SIPRI. Press Release.

SIPRI (2021, April 26) World Military Spending Rises to Almost $2 Trillion in 2020. Press Release.

Stavridis J. (2021, March 5) US Needs a Strong Defense Against China's Rare-Earth Weapon. Bloomberg Opinion. Retrieved from https://www.bloomberg.com/opinion/articles/2021-03-04/u-s-needs-a-strong-defense-against-china-s-rare-earth-weapon

Stay Grounded (n.d.) Get Information! Available at: https://stay-grounded.org/get-information/#greenwashing.

Stop Wapenhandel (2019) Report on Expert Meeting on Climate Change and the Military. Unpublished Paper.

Tobben S and Merrill D. (2021, Jan. 7) Exporting the U.S. Shale Boom Has Changed Oil Markets Forever. Bloomberg. Retrieved from: www.bloomberg.com/news/articles/2021-01-07/exporting-the-u-s-shale-boom-has-changed-oil-markets-forever.
United Nations. Arms Trade Treaty (2014).
UNHCR (2021) Strategic Framework for Climate Action https://www.unhcr.org/604a26d84/strategic-framework-for-climate-action.
US Naval Institute (2019, Sept. 3) Coast Guard Prep for Hurricane Dorian. Retrieved from https://news.usni.org/2019/09/03/navy-coast-guard-prep-for-hurricane-dorian.
Verweijen J and Dunlap A. (2021) The Evolving Techniques of Social Engineering, Land Control and Managing Protest Against Extractivism: Introducing Political (Re)Actions 'from Above'. *Political Geography* 83: 1–9.
Vries, W de. (2013, April 23) We Have an Arms Trade Treaty. What Difference Does It Make? *War Profiteers News* 38.
Wearing D. (2018) *AngloArabia. Why Gulf Wealth Matters to Britain*. Polity.
Wezeman PD, Kuimova A, and Wezeman ST. (2021, March 15) Trends in International Arms Transfers, 2020. SIPRI.

PART III

Policing Ecosystems

CHAPTER 8

If the Army Cuts Trees, Why Can't We? Resource Extraction, Hunting and the Impacts of Militaries on Biodiversity Conservation

Anwesha Dutta and Trishant Simlai

Introduction

The connections between environmental destruction and the military within mainstream academic literature date to the work of Joni Seager, who, back in 1993, wrote that "anywhere in the world, a military presence is virtually the single most reliable predictor of environmental damage" (Seager 1992, 201). This environmental damage is also evident in the context of militarized biodiversity conservation, which replicates a fortress conservation model (Brockington 2002). It is also intricately linked to

A. Dutta (✉)
Chr. Michelsen Institute, Bergen, Norway
e-mail: anwesha.dutta@cmi.no

T. Simlai
Department of Geography and Selwyn College, Cambridge, UK

A. Dunlap and A. Brock (eds.), *Enforcing Ecocide*,
https://doi.org/10.1007/978-3-030-99646-8_8

the lives and livelihoods of communities that reside within and around conservation spaces. There are two dominant narratives on the role of armed forces in environmental conservation. The first suggests that environmental issues and national security stand opposed to one another, and that environmental needs must be subordinated to national security issues (O'Brien and Barnett 2013, among others). The second is the deployment of armed forces for environmental safeguarding and protection, which is now gaining ground as part of the militaries' strategies in most of the world (Duffy 2014; Duffy et al. 2019). With these contrasting narratives in mind, in this chapter we (re)visit and present the body of work on military–environment relations. Our aim here is to unravel empirically the complexities associated with military presence in areas of high biodiversity that are also simultaneously inhabited by indigenous populations and have emerged as sites for counterinsurgency operations. We argue that military entanglements with both the bio-physical environment as well as communities inhabiting those spaces are not always fraught with spectacular forms of violence or overt destruction of nature but is expressed through quotidian ways of resource extraction (in connivance with local syndicates or contractors), occupying territory leading to evictions and displacement or blockades of essential animal corridors and so on. We also point out that the actual impact these military operations have mainly on the natural environment and endangered species is hard to document due to a lack of transparency and the level of immunity towards non-disclosure of information that is inherent to these military operations. Finally, we strive to do the above with empirical material gathered across conservation sites in north and northeast India. Both authors have carried out primarily qualitative fieldwork in the Manas Biosphere Reserve and the Kaziranga National Park in Assam as well as the Corbett Tiger Reserve in Uttarakhand since 2014. Methodologically the authors engaged in focused group discussions, participant observations, life histories, transect walks and semi structured interviews in and with communities residing in the fringes of the park. Authors further engaged with forestry officials, forest guards, village institution leaders, local Non-Governmental Organization (NGO) staff as well as high-level forest bureaucrats in the capital city of Assam, Guwahati. Given the nature of research and the authors' long-term association in these regions an exact estimation of the number of interlocutors is hard to arrive at.

Clearly, the military's environmental considerations, in a broader sense, encompass anything related to the environment that either affects the planning and execution of military operations, positively or negatively, or people affected by those operations. These include (but are not limited to) environmental conditions affecting soldiers, the health of personnel, access to clean water, sanitation, and other environment-related infrastructures; compliance with regional and international environmental laws; pollution prevention and environmental management; protection of historical and cultural sites; sustainability; and management of agricultural and natural resources. However existing research on the militarization of nature documents that local communities that reside in spaces also occupied by the military invariably become entrenched in processes of violence and environmental injustices due to also the very nature of the militaries' expansionist agendas that render the lived environment of the communities insecure (Dunlap and Fairhead 2014; Münster and Münster 2012). Yet, as we shall go on to demonstrate in this chapter, the military tries to greenwash its environmental interactions. Take for example, the United Kingdom (UK) has recorded progress by jointly managing Ministry of Defense (MoD) lands with local civilian communities such as the Army welfare trusts and the Wildlife trusts. The MoD has a Defense Environmental unit to coordinate these activities and its army produces the periodical *Sanctuary*, which records the environmental activities of the British Army in the UK and overseas (Ministry of Defense United Kingdom 2017).

Similarly, in Venezuela, one of the important roles of the Venezuelan National Guards is the protection of natural resources and the Brazilian military have replicated a piece of the Amazon jungle complete with Amazonian wildlife at the jungle warfare training school in Manaus (Mendel 1999). Moreover, the United States military permits scientists to carry out scientific environmental studies on their vast military bases and land owned by the ministry of defense (Mehta 2021). Closer to our fieldwork sites, the Nepalese army helps protect wildlife in the Chitwan National Park (Mahatara et al. 2018) while the Vietnamese army hand-plants trees in the areas degraded by Agent Orange during the Vietnam war (Scarpelli 2018). In India, the Ecological Task Force comprising of retired armed forces personnel was constituted for land restoration, afforestation, and aid in the cleaning of the Ganges River (Dutta 2020). Additionally, in June 2009 the UN Department of Peacekeeping Operations (UNDPKO) promulgated its environmental protection policy for UN field missions. Consequently, the UNDPKO has also drafted environmental protection guidelines for UN field missions. The guidelines are designed to assist the staff at United Nations peacekeeping field missions, including military, police and civilian components, in addressing environmental issues that are likely to arise from their operations (Asiedu 2010).

The above-mentioned environmental military initiatives around the world do not compensate for the adverse ecological consequences of the armed forces. Illustratively, the ministry of Defense of the United Kingdom is also the largest single contributor to greenhouse-gas emissions within the United Kingdom, responsible for half of the total national emissions (Ben et al. 2021). Moreover, globally military training bases are designed to store military equipment and personnel as well as facilitating tactical operations (Owen 1990; Goldsmith 2010). These bases can have broad ranging anthropogenic impacts on the local ecosystem like contamination of water bodies, alteration of landscape ecologies and occasional accidental oil spills (Lawrence et al. 2015). In the South Asian context, the continuing militarization of the Siachen glacier has serious social ecological implications for the South Asian countries of India, Pakistan, Nepal and Bangladesh which are already one of the most vulnerable regions to climate change (IPCC 2018).

It is important to note that globally military training areas (MTAs) are estimated to constitute approximately 6% of the Earth's surface, encompassing a multitude of environments and ecosystems (Lawrence et al. 2015). More precisely, the size of the MTA estate is as a minimum 50 million hectares, although the actual figure may be closer to 200–250 million hectares (Zentelis and Lindenmayer 2015). Zentelis and Lindenmayer (2015) further suggest that MTAs are likely to be present across all major global ecosystems and, if appropriately managed, with involvement of local communities, could have the potential to contribute significantly to biodiversity conservation. Yet No MTAs are explicitly managed for their environmental values (e.g., biodiversity conservation, scenic values, cultural heritage sites), rather these are managed to ensure military training is not concerned with environmental issues (Fiott 2014; Zentelis and Lindenmayer 2015; Havlick 2011), instead allowing security issues to take precedent over conservation. Moreover, the environmental values associated with MTAs could be designated as all those aspects of the environment that are valued by society, in general and these occur on nearly all MTAs globally (Zentelis et al. 2017). Zentiles and colleagues highlight the causes behind some of the important environmental values that are found at MTAs. Several authors in the past have assigned these to military training disturbance in actuality creating new habitats (e.g. Jentsch et al. 2009; Cizek et al. 2013). Additionally, several MTAs contain either remnant vegetation and/or disturbance-dependent communities, that have been extinct in the surrounding environment (e.g. Gazenbeek 2005). Illustratively, the intensification of agricultural practices in Europe led to the loss of many heathlands that now occur only in MTAs due to military training-related disturbance (Natura 2000; Gazenbeek 2005). In fact, the vestiges of coastal heathland at the Shoalwater Bay MTA in Australia is the largest surviving area of coastal heathland on the Australian east coast. This is attributed to the area being used solely for military training (Zentelis et al. 2017; Keith et al. 2019).

Nevertheless, apart from some key literature on adverse ecological impacts of militaries on bio-physical environments (Thomas, Thomas and Seager 1994., Smith 2017), and the green militarization literature (Duffy 2014; Hübschle 2017; Lunstrum 2014; Massé et al. 2018) focusing on the negative consequences of military styled conservation on local populations, there is little that focuses on the ways in which MTA´s are vital for conservation and the ways in which they can be managed. Only more recently Zentelis et al. (2017) have suggested a set of management principles that integrate the management of both military training objectives and environmental values. They arrived at these principles following a desk review of Australian and German MTA management documentation which revealed that neither regime had a systematic management principle for MTAs although they are among countries at the forefront of MTA management globally. Therefore, it is a counterintuitive conjecture when it comes to conservation importance of military lands and operations particularly in the more developing nations and how these are detrimental to both people and species. Zentelis et al. (2017) also note that in order to achieve this military-conservation integration it is imperative to understand the intersection of the impacts of military training on the environment, alongside the known, or potential, environmental values of a particular training area. Clearly, there appears to be a dearth of systematic studies of the direct impacts of militaries on ecosystems, endangered species and climate change, but there is still little in the literature that examines the ecological impacts of conservation militarization itself. We continue to unearth this empirically in the section below in the context of military operations within fragile political environments in India.

This chapter demonstrates how military operations and training areas in India are in contravention with nature and people, further complicating community–military interactions particularly in overlapping zones of political conflict and biodiversity conservation, either leading to dispossession or blockades of important animal corridors, among other consequences. Towards this end, in this chapter we shall strive to both problematize and critically analyse the links between military and environment, with a special focus on biodiversity conservation in the global south and case studies from India. More importantly, we also show how the military through its functions of disaster management, efficient contributions to civil infrastructure construction for development projects like building of roads and bridges paints a positive public image while fostering civil–military relations. This in turn offsets the untransparent and mainly negative consequences the military has on the ecologies and local populations. The next section begins by first drawing on the general literature around green militarization and green violence and situates the role of armed forces there in, we then move on to an empirical section highlighting the role of security forces in India in resource extraction and hunting in conservation areas. Thereafter, we describe the impacts of militarization and securitization of forests in inhibiting the implementation of plural and progressive legislations such as the Scheduled Tribes and Other Traditional Forest Dwellers (Recognition of Forest Rights) Act, 2006 in India. We also demonstrate how militaristic installations block potential forest corridors and impede movement of dispersing large mammals outside protected areas. We then conclude by providing a short analysis of the ways in which the nationalistic glorification of armed forces and notions of patriotism make invisible the environmental and ecological costs of their presence in fragile ecosystems among vulnerable populations.

Situating Armed Forces in Green Militarization and Green Securitization

Historically, national armed forces have performed important roles in establishing conservation measures, albeit often forcibly, across developing countries in South America, Asia and Africa. This has been observed in Nepal (Ethirajan 2013), Colombia (Ojeda 2012), Indonesia (Peluso 1993; Dorr et al. 2013), Guatemala (Ybarra 2016), Congo (Verweijen and Marijnen 2018), Cameroon (Pennaz et al. 2018) South Africa

(Lunstrum 2014; Piombo 2013), among many others. In the case of Botswana, one of its Defense Force's primary responsibilities is the protection of its national parks (Henk 2006). In the case of India, drones were recently introduced in the Kaziranga National Park in Assam (Simlai 2015). Moreover, around twenty-four persons (allegedly poachers) were killed in and around the core area of Kaziranga National Park in 2014 (Barbora 2017). The death toll has risen since then, with the arming of forest guards with sophisticated arms and ammunitions and shoot at sight operations (Simlai and Kazmi 2017). According to Lunstrum (2014), militarized activities, during and post periods of conflict, cause alarming harm to the bio-physical environment, and that the two are indeed antithetical. Similarly, Ybarra (2016) has found in the case of Guatemala how the forest is a site of guerrilla refuge that has led to the military's perceived need to control the jungle. Duffy (2014) furthers this discourse by re-emphasizing that this whole notion of "militarized forms of anti-poaching are not new: for example, early game wardens in British colonial administrations were often ex-military personnel." This has also been systematically analysed in the context of the formation of political forests particularly in South East Asia where counterinsurgency techniques were used by the colonial state to bring both people and forests under imperial jurisdictions (Duffy 2014, 821; Peluso and Vandergeest 2011). In fact, the renewed war for conservation, wherein counterinsurgency and security approaches are being increasingly used in conservation practice mirrors the language of interventionism (Duffy 2014). Here the onus of wildlife and forest preservation, especially endangered and/or charismatic species lies with the international community and those military forms of intervention may be brought into effect to save them (Duffy 2014). Biodiversity conservation, then, is a perceived issue of national security (Duffy 2016; Büscher and Whande 2007), which is having disastrous results within ecosystems and communities, especially since integrating militaries' interest with that of environmental protection remains distant on the horizon.

Not surprisingly, therefore, states in the Global North along with big NGOs increasingly connote rural protected areas in economically poorer countries as sites of (in)security. This creates opportunities to categorize deforestation and biodiversity loss as threats, while promoting ecosystem services and presenting opportunities to control natural resources and state borders (Kelly and Ybarra 2016). This is how conservation carves out the path towards securitization, the process by which spaces and human subjectivities become targets of regulation and surveillance in the name of conservation and security. Since it is usually the policing and military agencies alongside war veterans from the global north, that are incorporated into protected area surveillance and enforcement strategies (Peluso and Vandergeest 2011; Ybarra 2016; Lunstrum 2014; Dwyer et al. 2016), both remain detached to interests of the local and forest dependent communities, rendering it easier to exercise forms of environmental injustice through displacements, territorial occupation and in some cases even acts violence particularly during counterinsurgency operations. Author A witnessed this during fieldwork in reserved forests along the India–Bhutan borderlands which was also a counterinsurgency zone, instances of villagers being harassed (verbally and physically) by military personnel with accusations of supporting the local militia or being involved in illegal timber trade.

This also links to the broader discourse relating to the moral force of conservation, with the heightened awareness of ecological crisis. Additionally, as has also been highlighted in the literature on political economy of "lootable" resources and crisis conservation (Le Billon 2004; Duffy 2014; Lunstrum 2016), conservation hotspots are politically fraught and fragile areas. The rainforests of the Indian northeast, marine national parks as well as the biodiversity regions of the Indian Himalaya are good illustrations of this (Dutta 2020; Murlidharan and Rai 2020). These areas have witnessed long-term political violence through local insurgency movements and the deployment of Indian armed forces to counter them. The counterinsurgency operations take place also in these evergreen forest spaces habitat to both megafauna and indigenous communities. This also links to the growing body of work linking conservation and counterinsurgency. Verweijen and Marijnen (2018) show how these two phenomena intersect in the Virunga National Park, located in the war-ridden east of the Democratic Republic of the Congo. Their work highlights that strict law enforcement along with joint operations of the Congolese army and park guards fuel, rather than mitigate, the dynamics of conflict and violence, thereby feeding into armed mobilization.

The above discussion leads us to demonstrate the implications of growing linkages between counterinsurgency, political violence and conflict in conservation hotspots in India, particularly showcasing the adverse effects that conservation militarization has on the ecosystem. Our empirics specifically demonstrate that military impact on local populations and the ecological environment is not necessarily showcased through spectacular modes of violence, torture, shootings of either people or species but can manifest in more mundane everyday forms through intricate relationships with local elites.

Empirical Findings Using Case Studies from Assam and Beyond

The first author carried out fieldwork (intermittently) in the Manas biosphere reserve on the intersection between conservation and counterinsurgency, forest rights and dispossessions from 2009 to 2017 and subsequently in Kaziranga national park on issues of poaching, park–people relationship and forest rangers from November to January 2019–2020. The second author carried out fieldwork in the Kaziranga National Park and the Manas biosphere reserve on the socio-political dimensions of interactions between forest staff and security forces from 2016 to 2017 and subsequently in the Corbett Tiger Reserve between 2018 and 2020. Now relegating back to our field site in Kaziranga, When the first author visited the park in 2019–2020 and spoke to a few of the squad members, it was found that a Special Rhino Protection Force (SRPF) was set up in 2019 with the sole aim of protecting the one horned Rhinoceros considered to be the state animal of Assam in India. Equipped with AK 47s and skilled in commando training, this squad is organized in collaboration with the Indian federal and state governments and is an initiative of the National Tiger Conservation Authority (NTCA). Interviews revealed that the squad members do not receive any education in forestry or biodiversity conservation or working with local communities and are primarily given arms training to track down and kill poachers in the park.[1] Although the purpose of this chapter is not to delve into the complexities of poaching and poachers which has beget a rich literature (Lunstrum and Givá 2020; Hübschle 2016; Massé 2018), it is worth mentioning that our research from the region reveals that extra judicial killings of residents in the name of conservation are not uncommon. Driven by rewards and incentives provided by the park authority to target poachers, forest guards in Kaziranga use lethal force to target anyone found inside the park boundary. There is evidence of staged encounters, use of torture and the regular framing of forest produce collectors as poachers.

[1] Interview with sqad member 2, 3, 6 of the SRPF 6, 7 and 9 December 2019.

One guard recounted that the rule is if they find a poacher inside the forest then there is a direct encounter, meaning they will not give a verbal warning. Guards do not arrest inside the forest, they kill.[2] A ranger colleague continued by saying that once a poacher is arrested, they take him to the rangers' office to a special room. He is then beaten up with cricket bats, thick rod wrapped with rhino skin. He emphasized that Rhino skin is so thick that the poachers skin comes off. He added, "All of us at the ranger's office hit the poacher and take turns. Usually, we beat the poacher through the night and then we give them injections to relieve them of pain and feed them pork and chicken curry and then we beat them again. This goes on for a couple of days."[3] It is worthwhile to mention here that a poacher is referred to anyone that engages in illegal harvesting of wildlife including those that provide logistical support.

These acts were rooted in structure of the forest department that incentivized such behaviour. Another forest guard said, "I shot a poacher in 1996 and again in 2005. It was a casual contract forest guard but after I shot a poacher my job was made permanent."[4]

While the larger conservation fraternity encompassing the forest department and conservation NGOs reel under a conservation success story (Balmford 2012), instances like these increase mistrust of local communities towards forest authorities which in turn can cause forest degradation due to conflicts between conservation goals of communities and the state. Several residents felt that even if they did not enter the forest, they were being killed, so it did not matter then if they grazed cattle inside or outside the forest."[5] Everyday forms of violence and fortress conservation methods of Kaziranga authorities seemed to have created a great deal of anger and mistrust among the local communities.[6] Growing mistrust also adds to instances of resistance in form of local residents letting their livestock inside the park which in turn aggravates concerns about foot and mouth disease spread in wild species populations. In fact, there are several examples of killing of wildlife or poaching or illicit usage of protected land (read green enclosures) by local

[2] Interview with forest guard 2 on 20 December 2019.

[3] Interview with forest guard 6 inside Kaziranga National Park on 2 January 2020.

[4] Interview with forest guard 11 inside Kaziranga National Park, 5 January 2020.

[5] interview with local resident in fringe village on 5 December 2016.

[6] Interview with local social activist in Kaziranga area on 8 December 2016.

communities dwelling in and around protected areas both in the global north, like Sweden and Greece (Larsson 2012), as well as across conservation hotspots in Africa (Neumann 1998; Weladji and Tchamba 2003; Cavanagh and Benjaminsen 2015; Shafer 1999).

State presence has always been minimal in the reserved forests within the Indo-Bhutanese borderlands in Assam in India. Three decades of militia violence left forest officials of Assam further weakened, offering little security and very limited funds to function under the Forest Conservation Act of 1980 and to penalize forest crime. A weakened state does not mean a weakening of state legitimacy per se and that the withdrawal of the forest department led to the dotting of the forest landscape with army and paramilitary camps for counterinsurgency operations (Dutta 2020). Forest department's camps were converted to Army bases. Therefore, the forest emerged as a twofold space of resource extraction and counterinsurgency operation. Additionally, the deployment of the 135 Infantry Battalion of the Ecological Task Force of Territorial Army of the Indian Army with the aim of afforestation through plantation activities in degraded forests around Manas National Park (Dutta 2020). The Ecological Task Force (ETF) was first commissioned in India in 1982 with the notion of Indian Army undertaking environmental conservation on "war footing" with the level of discipline and preparedness required to fight a war (Dutta 2020). This led to the ETF fencing off areas for plantation purposes which were hitherto albeit illicitly used by surrounding communities for grazing, subsistence food or crops like chilies and mustard, thereby leading to the emergence of what has been conceptualized in the broader critical conservation social sciences literature as "green" enclosures that habitually displaces populations or disrupt livelihood practices in an attempt to mitigate the deleterious environmental impacts of the former (Cavanagh and Benjaminsen 2015). This shows that the environment-military integration often tends to alienate people. In the case of the ETF, although the intention is forest restoration through plantation of native but noncommercial value timber species, this invariably did lead to barricading of grazing land for communities, highlighting also the importance of including communities in the equation.

The Ecological value of militarized conservation:

Decades of political violence between two opposing ethnic groups led to periodic violent clashes resulting in internal displacement of forest dwelling communities who were temporarily re-settled inside the forest in makeshift relief camps and second, paramilitary presence was established to protect these communities, guard the borders and fight insurgents.

Paramilitary presence inside these forests was meant for counterinsurgency operations. This meant regular patrolling of sensitive areas and setting ambushes to surprise insurgents that would use the forest as a refuge or as a means to pass through to villages located on the Indian side from Bhutan. Setting an ambush in the middle of a forest requires major habitat alteration. Ambushes require large-scale environmental alterations in the Manas Tiger reserve on the Indo-Bhutanese borderlands. Tall growing grass from grasslands in this landscape was chopped down to benefit vision during patrols. Furthermore, the construction of makeshift bases inside forests was done using locally available timber. Fuelwood for cooking and heating was also obtained by harvesting nearby short trees and other vegetation. The researcher also noted the presence of hunting dogs with soldiers, and interviews with local forest guards revealed that they were often used to bring down small herbivores such as barking deer, wild boar and hog deer, all of which are protected species, for a "feast" inside the forest. Moreover, the army's ecological presence goes beyond just extraction of forest resources, for counterinsurgency operations require the operation of vehicles and convoys including big trucks, tanks and road infrastructure inside the forest, which frequently requires clearing large swathes of land. Counterinsurgency operation inside dense forest has also led to clearing of trees which led to one respondent recounting that, "look at that area, it used to be dense forest till five years ago and then the army came and cleared it to catch militants. If the army can cut trees, so can we."[7] Further interviews with local NGOs and forest guards also showed that habitat loss was being caused due to the presence of security forces and their operations.

A conservation NGO staff expressed that grassland is an important habitat for endangered species such as Bengal florican, rare partridges and even the hispid hare and that the presence of the security forces was leading to the loss of this important habitat as they use open areas to set up makeshift ambush camps.[8] Additionally, a local village leader recounted that "Most amount of hunting is done by these soldiers, it all goes unnoticed because they are inside the forests for weeks together fighting militants and our presence is not appreciated by them."[9]

[7] Interview with local resident in a fringe village on 5 May 2016.

[8] Interview with local NGO staff on 10 December 2016.

[9] Interview with local forest guard on 6 December 2016.

While the army continued to extract forest resources without repercussions, the internally displaced persons and forest dwelling communities' access was severely curtailed. A forest department staff recounted to the first author, "since we do not really patrol inside the forest anymore, the paramilitary from time to time inform us of new forest land encroachments or hunting operations by villagers."[10]

Security forces in India are deployed in many ecologically fragile areas of conservation importance. Mishra et al. (2006) have discussed the role of army bases as a main form of disturbance in the ecologically sensitive and biodiversity rich areas of Arunachal Pradesh, India. This region of Kashmir is known as the most militarized site in the world and has immense adverse impacts on the local environment. Many security establishments are on forest land that overtime has caused large-scale forest degradation. Athar Parvaiz (2020) has argued that forests in Kashmir suffered huge damages from timber smuggling operated by a vigilante non state counterinsurgency militia called the "Ikhwanis" that were given patronage by security forces. It is to be noted that the Ikhwanis were an Indian state backed militia that played a prominent role in fighting the Kashmir militancy at its peak in the 1990s.

Forests of the resource rich central Indian state of Chhattisgarh are also heavily militarized and large-scale forest loss has been recorded due to road construction work for military vehicles to pass through. The forest lands of central India harbour many different indigenous groups with rich traditional and cultural belief systems associated with forests. Militarization of such landscapes not only causes degradation of these landscapes from an ecological sense but also is detrimental to Adivasi cultures that are intertwined with forests and the natural environment. In 2021, an Adivasi man participating in a local hunting festival called "*Sarhul*" was shot dead by security forces on suspicion of being a Maoist insurgent as he was carrying a locally made gun (Sharma 2021). Many activists in the region have raised concerns of rapidly deteriorating cultural values associated with forests due to the constant and omnipresent militarization of their forests. The ecological impacts of India's security forces remain a highly under researched topic and there is an urgent need for site-based detailed ethnographies that explore such impacts in different ecologies. In the next section we explore such impacts in a little more detail.

[10] Interview with forest range officer on 8 May 2015.

Hunting by Security Forces

Hunting of wildlife for trade in animal parts is seen as a major threat to wildlife across the tropics (Wright 2010; Milner-Gulland et al. 2003; Dobson et al. 2019). Many indigenous people and local communities living alongside forests and other protected areas depend on native wildlife for subsistence, trade, cultural and spiritual purposes (Robinson and Redford 1991; Fa et al. 1995). It is often argued that a growing human population with increased accessibility to remote regions, coupled with modern hunting methods are driving species to local extinctions across the globe (Hilaluddin et al. 2011; Karanth et al. 2010). Conservation science and related disciplines have mostly focused on hunting related to subsistence, often also blaming communities for over exploitation and conniving with poaching syndicates or on the illegal wildlife trade being driven by organized criminal cartels. However, the role played by standing armies and militaries in driving hunting and poaching in remote biodiversity regions including areas of political and civil unrest has not received much critical attention. The diverse geographic, cultural and socio ecological context of India has shaped hunting practices by resident indigenous communities over many centuries (Aiyadurai et al. 2010). However, it has been argued that colonial rule created a fervour for wildlife extermination through a system of rewards and bounties rapidly changing the intensity of hunting (Rangarajan 2001). It has also been argued that rapidly changing land use systems and personal aspirations of communities are altering the context in which subsistence hunting is done making it more intensive in nature (Velho et al. 2012). Furthermore, bushmeat hunting by local communities and indigenous peoples has been termed as a major conservation problem, that is driving species extinctions across India (Karanth et al. 2010; Velho et al. 2012).

The discourse on hunting in India remains focused on local communities and indigenous peoples. This makes invisible the role played by other actors such as deployed armies and paramilitary forces in the forests of India. Security forces in India are deployed in some of the most remote parts of the country including in forests that are of extreme conservation importance. Given the sensitive nature of these operations, most of these areas are then not present for mapping land use change via satellites and neither does military release data on the extent of territory used for such operations. Conservation scientists in India have noted that local residents often engaged in hunting of wildlife in these remote regions in order to barter them as trophies for fuel, groceries and alcohol from deployed soldiers (Mishra et al. 2006; Aiyadurai et al. 2010). Security forces often conduct major drills that occur in landscapes inhabited by protected and endangered species. During one such drill, soldiers from the Indian security forces engaged in the hunting of chinkaras (Gazella bennetii) or Indian Gazelle which is a protected species under the Wildlife Protection Act of India (Mishra 2011). Similarly, in 2015 Indian paramilitary soldiers hunted Indian Peafowl, a bird of national importance from a protected area near their deployment site (Express News Service 2015). There are also recorded examples where Indian security forces have shot at and killed endangered species in apparent acts of self-defence. However, forest departments and wildlife activists have noted that these are avoidable situations and that security forces often violate protocols when a situation of human–wildlife conflict arises. For example, in 2019 cases were registered against a unit of the Indian armed forces as per local wildlife laws for the killing of a leopard (Panthera pardus). Managing incidents of human–wildlife conflict are replete with challenging circumstances for forest staff and the presence of military forces in the equation problematizes the issue further.

Wider Implications in Conservation Practice

Deforestation and degradation of forest ecosystems have been directly linked to climate change and species extinctions across the globe. Some research has highlighted large-scale deforestation in conflict-affected and post-conflict countries. For example, recent research has revealed that across all conflict areas around the world, forest loss increased by 10% in 2020 amounting to 1.1 mega tonnes of CO_2 which is more than the total emissions of most western countries in a given year (Darbyshire 2021). Forest corridors that connect protected areas over large landscapes maintain species persistence and augment genetic flows especially in wide-ranging species such as Asiatic elephants (Elephas maximus) and tigers (Panthera tigris) (Sharma et al. 2013; Wikramanayake et al. 2004). Many of these corridors in countries like India are located in regions that have a strong military presence deployed for counterinsurgency operations such as Maoist insurgency in central India and ethnic self-determination groups in northeast India. Stationed security forces in such forest corridors often require infrastructure development such as widened roads for troop movement, garrisons and forward bases. Invariably these structures are set up in areas that end up impeding or blocking movement of wide-ranging species such as the Asiatic elephant. A well-researched example for this is a military ammunition warehouse in the north Indian state of Uttarakhand that is located in a forest corridor connecting two protected areas and has blocked the movement of Asiatic elephants between the two forest blocks (Johnsingh and Williams 1999). Military infrastructural development contradicts the supposed environmental imperatives of the Indian government.

Furthermore, the securitization by militaries of forested landscapes has also been linked to the curtailment of rights of indigenous communities. Many indigenous communities across the world have managed forest ecosystems through local institutions and traditional practices. In India, a landmark legislation called the Forest Rights Act (FRA) was introduced in 2006 as a means to ensure land tenure, food security and livelihoods of forest dwelling scheduled tribes and other traditional forest dwellers. More than a decade later, the FRA remains poorly implemented throughout India and especially in the central Indian tribal belt (Sarin and Springate-Baginsky 2010). This region has seen very intensive militarization in the last decade due to one of the longest running armed conflicts against Maoists, which relates to controlling large deposits of mineral resources (Sundar 2016). Scholars have argued that the securitization of these forest blocks by security forces has resulted in a hurdle for the implementation of the FRA. Regular counterinsurgency operations by security forces render these areas inaccessible for tribal communities who cannot initiate the process of the FRA as national security takes precedence over tribal rights and forest conservation. Furthermore, militarization in the state of Chhattisgarh alone has resulted in thousands of displaced indigenous groups from remote forested villages making the implementation of the FRA a near impossibility in the region.

National Heroes or Environmental Annihilators? Critically Assessing the Janus-Faced Armed Forces and Its Multiple Roles

As reported by the United Nations' Centre for Disarmament, armed forces globally have used a steadily increasing amount of land for bases, other installations and training exercises over the last century. Clearly military operations, whether domestic, during peacekeeping operations or wartime, the environmental boot print remains considerable, contributing to increases in both scale and intensity of anthropogenic carbon dioxide emissions leading to ecological degradation. For example, the US military is the nation's largest consumer of fossil fuels (Bigger et al. 2021). We mention the US military as a means of measurement to gauge the impact of military operations in most of the developed world on conservation. We have already discussed above the large-scale environmental implications of both military equipment and personnel in contributing to increases in the scale and intensity of anthropogenic carbon dioxide emissions (Bigger et al. 2021; Belcher et al. 2020). Running these bases requisites vast massive amounts of natural resources—particularly fossil fuels—to staff, operate and transport equipment and personnel between and within destinations (Jorgenson et al. 2010). It is therefore counter intuitive to arrive at the conjecture that several of these bases would be set up in ecologically sensitive and biologically diverse areas with buffer forest lands surrounding these to ensure both, easier access to natural resources as well as to reduce compromise on military assets, training and operations. The conversion of huge tracks of jungles to which local populations had free access to historically into demarcated forest areas relates to Peluso and Vendergeest's (2011) work on the influence that counterinsurgency operations have in shaping areas of conservation as well as the placement of towns. Moreover, the presence of military in highly eco-sensitive zones like the Siachen glacier with over 30,000 soldiers are stationed along the 70 kms long glacier, has along with global warming exacerbated glacial melt resulting in cascading negative environmental consequences like catastrophic floods, droughts and food shortages for millions of South Asians. Although, the Indian Army came out with a joint doctrine in 2017 that acknowledges the natural environment as a critical area of the security paradigm, and also warns that if environmental degradation and related issues increase security risks, the military will need to respond. Yet this view appears to be extremely myopic since the military and security policy of the Indian state continues to view glacial melt in the Himalayas only within the context of endangering water supplies for military installations.

Yet simultaneously within the Indian context except for in politically sensitive areas of northeast, and the states of Chhattisgarh and Jammu & Kashmir, the military has been revered mainly for its services in maintenance of essential services either in times of natural disasters or other calamities, such as earthquakes, floods, riots, famine and fires. This also includes providing assistance in development projects like building of bridges, dams as well as in rescue and rehabilitation operations during adverse events of building collapse and so on. Beginning in the 1980s, as already mentioned the Ecological Task Force (ETF) in India has actively undertaken environment-related activities particularly in ecologically and politically sensitive landscapes comprising of ecological restoration of dilapidated areas due to limestone mining in Shivalik Hills, afforestation through plantation drives across several Indian states. More recently through the curation of a Ganga Task Force from within the ETF works towards spreading awareness among the public on ways to keep the river clean. Together with the Indian military's role in protecting territorial borders and auxiliary role in disaster, environment and a history of stable civil–military relationship in the nation, the military continues to hold a glorified position within the popular discourse relegating its adverse impact on the bio-physical environment and local forest and indigenous communities to the backseat. The militaries involvement in social development projects as mentioned through engagement in disaster operations, speedy construction of bridges and roads, rescue operations during landslides and flash floods, although are important and notable for the society, it however absolves the institution from the more negative counterinsurgency and environmental consequences of their actions, which to begin with are yet to be systematically and accurately documented.

Conclusion

In this chapter we have tried to unearth the complex entanglements between the military and the bio-physical environment. This entailed focusing on the Indian military's adverse consequences on local and tribal populations and biodiversity conservation. We have strived to bring together some of the distinct narratives pertaining to, first, the role military operations like counterinsurgency, running of huge military bases has on the bio-physical and human environments, while also showcasing everyday forms of environmental securitization which takes precedence over biodiversity conservation and environmental justice of indigenous forest populations. Secondly, drawing on the green militarization literature, which so far has been focused on spectacular forms of violence, we have demonstrated some of the material costs of the military observed across conservation sites in north and northeastern India. This relates to more mundane and everyday ways in which the Indian armed forces stationed in and around conservation spaces, including animal corridors engage in environmental destruction by either conniving with local timber and illegal wildlife syndicates or through pleasure activities (not for subsistence) like hunting of small game and avian species.

This contribution is particularly important since the militarization literature above is fragmented and often does not explicitly mention the role played by state militaries in the degradation of environment and contributing to anthropogenic forms of climate change. We have also situated our work within the emerging body of work in the space of green violence and militarization thereby further accentuating the current debates underpinning exploitation and alienation of people and species through the intersection between counterinsurgency and crisis conservation. We have also provided a review of existing work on militaries impact on the environment including a historical overview of such involvement. Further, using empirical material from India and with examples wherever possible we have shown the often-mundane ways in which the military engages with local populations and how some enjoy forms of impunity when it comes to both natural resource use, extraction and degradation. This area of inquiry calls for more empirical and fine-grained research to unearth the relationships between militarization and the environment further, particularly in the context of rising temperatures and glacial melts and ever increasing geopolitical tensions globally.

Bibliography

Aiyadurai, A., Singh, N. J., and Milner-Gulland, E. J. (2010) 'Wildlife hunting by indigenous tribes: a case study from Arunachal Pradesh, north-east India', *Oryx*, 44(4), pp. 564–572. https://doi.org/10.1017/S0030605309990937.

Asiedu, C. (2010) *Environmental Review of United Nations Peacekeeping Operations for Sustainability*. Kivu, DR Congo.

Aycrigg, J. L., Belote, R. T., Dietz, M. S., Aplet, G. H., and Fischer, R. A. (2015) 'Bombing for biodiversity in the United States: response to Zentelis & Lindenmayer 2015'. *Conservation Letters*, 8(4), pp. 306–307.

Balmford, A. (2012) *Wild hope: On the front lines of conservation success*. University of Chicago Press.

Barbora, S. (2017) 'Riding the Rhino: conservation, conflicts, and militarisation of Kaziranga National Park in Assam', *Antipode*, 49(5), pp. 1145–1163. https://doi.org/10.1111/anti.12329.

Ben, B., Barrie, B., and Chids, N (2021) 'Dealing with hot air: UK defence and climate change', *IISS Military Balance Blog*. Available at: https://www.iiss.org/blogs/military-balance/2021/04/uk-defence-climate-change.

Baumann, M., and Kuemmerle, T. (2016) 'The impacts of warfare and armed conflict on land systems', *Journal of Land Use Science*, 11(6), pp. 672–688. https://doi.org/10.1080/1747423X.2016.1241317.

Belcher, O., Bigger, P., Neimark, B., and Kennelly, C. (2020) 'Hidden carbon costs of the "everywhere war": Logistics, geopolitical ecology, and the carbon boot-print of the US military', *Transactions of the Institute of British Geographers*, 45(1), pp. 65–80.

Bennett, E. L., et al. (2002) 'Hunting the world's wildlife to extinction', *Oryx*, 36(4), pp. 328–329. https://doi.org/10.1017/S0030605302000637.

Bigger, P., Kennelly, C., Belcher, O., and Neimark, B. (2021) 5. 'The Carbon Bootprint of the US Military and Prospects for a Safer Climate', *Negotiating Climate Change in Crisis*, 53.

Billon, P. L. (2004) 'The geopolitical economy of "resource wars"', *Geopolitics*, 9(1), pp. 1–28. https://doi.org/10.1080/14650040412331307812.

Brockington, D. (2002) *Fortress conservation: The preservation of the Mkomazi game reserve*. Tanzania: Indiana University Press.

Büscher, B., and Whande, W. (2007) 'Whims of the winds of time? Emerging trends in biodiversity conservation and protected area management', *Conservation and Society*, 5(1), pp. 22–43.

Burns, E., Lowe, A., Lindenmayer, D., and Thurgate, N. (Eds.). (2014) *Biodiversity and environmental change: monitoring, challenges and direction*. CSIRO Publishing.

Cavanagh, C. J., and Benjaminsen, T. A. (2015) 'Guerrilla agriculture? A biopolitical guide to illicit cultivation within an IUCN Category II protected area', *The Journal of Peasant Studies*, 42(3–4), pp. 725–745. https://doi.org/10.1080/03066150.2014.993623.

Cizek, O., Vrba, P., Benes, J., Hrazsky, Z., Koptik, J., Kucera, T., ... Konvicka, M. (2013) 'Conservation potential of abandoned military areas matches that of established reserves: plants and butterflies in the Czech Republic', *PloS One*, 8(1), e53124.

'Civil-military relations transform in Nepal's Chitwan Park' (2020) *The Third Pole*, 15 January. Available at: https://www.thethirdpole.net/en/nature/civil-military-relations-transform-in-nepals-chitwan-park/ (Accessed: 17 September 2021).

Darbyshire, E. (2021) 'Deforestation in conflict areas in 2020', *Conflict and Environment Laboratory*. https://ceobs.org/assessment-of-recent-forest-loss-in-conflictareas/.

Deepak. (2019) 'Siachen glacier is turning into a high-altitude dumping site', *The Third Pole*, 22 August. Available at: https://www.thethirdpole.net/en/regional-cooperation/siachen-glacier/ (Accessed: 4 June 2021).

Dobson, A. D., Milner-Gulland, E. J., Ingram, D. J., and Keane, A. (2019) 'A framework for assessing impacts of wild meat hunting practices in the tropics', *Human Ecology*, 47(3), pp. 449–464.

Dorr, T. F., Heskamp, A. B., Madison, I. B., and Reichel, K. D. (2013) *Missing the poorest for the trees? Redd+ and the links between forestry, resilience and peacebuilding*. London School of Economics (LSE) (Alert International).

Duffy, R. (2014) 'Waging a war to save biodiversity: the rise of militarized conservation', *International Affairs*, 90(4), pp. 819–834. https://doi.org/10.1111/1468-2346.12142.

Duffy, R. (2016) 'War, by conservation', *Geoforum*, 69, pp. 238–248.

Duffy, R., Massé, F., Smidt, E., Marijnen, E., Büscher, B., Verweijen, J., ... Lunstrum, E. (2019) 'Why we must question the militarisation of conservation', *Biological Conservation*, 232, pp. 66–73.

Dunlap, A., and Fairhead, J. (2014) 'The militarisation and marketisation of nature: an alternative lens to "climate-conflict"', *Geopolitics*, 19(4), pp. 937–961. https://doi.org/10.1080/14650045.2014.964864.

Dutta, A. (2020) 'Forest becomes frontline: conservation and counter-insurgency in a space of violent conflict in Assam, Northeast India', *Political Geography*, 77, p. 102117. https://doi.org/10.1016/j.polgeo.2019.102117.

Dwyer, M. B., Ingalls, M. L., and Baird, I. G. (2016) 'The security exception: development and militarization in Laos's protected areas', *Geoforum*, 69, pp. 207–217. https://doi.org/10.1016/j.geoforum.2015.11.002.

Ethirajan, A. (2013) 'Nepal steps up battle against drug traffickers', *British Broadcasting Corporation*, 4 May, 2013. http://www.bbc.co.uk/news/business21963600 (Accessed: 4 May 2013).

Express News Service. (2015) *Three Army jawans hunt down, skin two peacocks for 'feast'*, probe on. https://indianexpress.com/article/cities/kolkata/three-army-jawans-hunt-down-skin-two-peacocks-for-feast-probe-on/.

Fiott, D. (2014) 'Reducing the environmental bootprint? Competition and regulation in the greening of Europe's defense sector', *Organization & Environment*, 27(3), pp. 263–278.

Fa, J. E., et al. (1995) 'Impact of market hunting on mammal species in equatorial Guinea', *Conservation Biology*, 9(5), pp. 1107–1115. https://doi.org/10.1046/j.1523-1739.1995.951107.x.

Fjeldså, J., et al. (2005) 'Illicit crops and armed conflict as constraints on biodiversity conservation in the Andes Region', *AMBIO: A Journal of the Human Environment*, 34(3), pp. 205–211. https://doi.org/10.1579/0044-7447-34.3.205.

'Forest land identified for development after reorganisation of Jammu and Kashmir' (2020) *Mongabay-India*. Available at: https://india.mongabay.com/2020/01/is-reorganisation-hastening-land-use-change-in-jammu-and-kashmir/ (Accessed: 9 July 2021).

Foundation, T. R. (no date) *India's virus lockdown fuels timber-smuggling in Kashmir forests*, *news.trust.org*. Available at: https://news.trust.org/item/20200612051624-20q88/ (Accessed: 19 July 2021).

Gaynor, K. M., et al. (2016) 'War and wildlife: linking armed conflict to conservation', *Frontiers in Ecology and the Environment*, 14(10), pp. 533–542. https://doi.org/10.1002/fee.1433.

Gazenbeek, A. (2005) 'LIFE, Natura 2000 and the military', *Life Focus—Journal of the European Commission, Environment Directorate General Life III Program (2000–2006)*.

Goldsmith, G. S. (2010) *Environmental impacts of military range use*. ARMY WAR COLL CARLISLE BARRACKS PA.

'Grasslands of grey: The Kaziranga model isn't perfect—but not in the ways you think' (no date) *The Wire*. Available at: https://thewire.in/environment/kaziranga-bbc-shoot-poaching (Accessed: 17 September 2021).

Grima, N., and Singh, S. J. (2019) 'How the end of armed conflicts influence forest cover and subsequently ecosystem services provision? An analysis of four case studies in biodiversity hotspots', *Land Use Policy*, 81, pp. 267–275. https://doi.org/10.1016/j.landusepol.2018.10.056.

Hanson, T., et al. (2009) 'Warfare in biodiversity hotspots', *Conservation Biology*, 23(3), pp. 578–587. https://doi.org/10.1111/j.1523-1739.2009.01166.x.

Havlick, D. G. (2011) Disarming nature: converting military lands to wildlife refuges. *Geographical Review*, 101(2), pp. 183–200.

Henk, D. (2006) 'Biodiversity and the military in Botswana', *Armed Forces & Society*, 32(2), pp. 273–291. https://doi.org/10.1177/0095327X05277907.

Hilaluddin, Khan, D. N., and Abbassi, F. (2011) 'Impact of Hunting on Large Mammal Densities in the West Indian Himalaya', *Indian Forester*, 137(10), pp. 1141–1146.

Hübschle, A. M. (2016) *A game of horns-transnational flows of rhino horn*.

Hübschle, A. M. (2017) 'The social economy of rhino poaching: of economic freedom fighters, professional hunters and marginalized local people', *Current Sociology*, 65(3), pp. 427–447. https://doi.org/10.1177/0011392116673210.

IPCC. (2018) *Summary for Policymakers*. In: Global warming of 1.5°C.

Jentsch, A., Friedrich, S., Steinlein, T., Beyschlag, W., and Nezadal, W. (2009) 'Assessing conservation action for substitution of missing dynamics on former military training areas in Central Europe', *Restoration Ecology*, 17(1), pp. 107–116.

Johnsingh, A. J. T., and Williams, A. C. (1999) 'Elephant corridors in India: lessons for other elephant range countries', *Oryx*, 33(3), pp. 210–214. https://doi.org/10.1046/j.1365-3008.1999.00063.x.

Jorgenson, T. E. A. K. et al. (2010) 'Research articles militarization and the environment: a panel study of carbon dioxide emissions and the ecological footprints'.

Karanth, K. K. et al. (2010) 'The shrinking ark: patterns of large mammal extinctions in India', *Proceedings of the Royal Society B: Biological Sciences*, 277(1690), pp. 1971–1979. https://doi.org/10.1098/rspb.2010.0171.

'Kashmir's forests face the axe' (2019) *The Third Pole*, 12 December. Available at: https://www.thethirdpole.net/en/nature/kashmirs-forests-face-the-axe/ (Accessed: 4 June 2021).

Keith, A., Ahner, D., and Curtis, N. (2019) 'Evaluation theory and its application to military assessments', *The Journal of Defense Modeling and Simulation*, 16(4), pp. 305–322.

Kelly, A. B., and Ybarra, M. (2016) 'Introduction to themed issue: "Green security in protected areas"', *Geoforum*, 69, pp. 171–175. https://doi.org/10.1016/j.geoforum.2015.09.013.

Larsson, K. (2012) *Dynamic route planning in the Baltic Sea: How can the negative impacts of maritime traffic on sensitive or protected areas be reduced?* Världsnaturfonden, WWF.

Lawrence, M. J. et al. (2015) 'The effects of modern war and military activities on biodiversity and the environment', *Environmental Reviews*, 23(4), pp. 443–460. https://doi.org/10.1139/er-2015-0039.

Loucks, C., et al. (2009) 'Wildlife decline in Cambodia, 1953–2005: exploring the legacy of armed conflict', *Conservation Letters*, 2(2), pp. 82–92. https://doi.org/10.1111/j.1755-263X.2008.00044.x.

Lunstrum, E. (2014) 'Green militarization: anti-poaching efforts and the spatial contours of Kruger National Park', *Annals of the Association of American Geographers*, 104(4), pp. 816–832. https://doi.org/10.1080/00045608.2014.912545.

Lunstrum, E. (2016) 'Green grabs, land grabs and the spatiality of displacement: Eviction from Mozambique's Limpopo National Park', *Area*, 48(2), pp. 142–152.

Lunstrum, E., and Givá, N. (2020) 'What drives commercial poaching? From poverty to economic inequality', *Biological Conservation*, 245, p. 108505. https://doi.org/10.1016/j.biocon.2020.108505.

Mahatara, D., Rayamajhi, S., & Khanal, G. (2018) 'Impact of anti-poaching approaches for the success of Rhino conservation in Chitwan National Park, Nepal', *Banko Janakari*, 28(2), pp. 23–31.

Massé, F., Lunstrum, E., and Holterman, D. (2018) Linking green militarization and critical military studies. *Critical Military Studies*, 4(2), pp. 201–221.

Massé, F. (2018) 'Anti-poaching's politics of (in)visibility: representing nature and conservation amidst a poaching crisis', *Geoforum* [Preprint]. Available at: https://doi.org/10.1016/j.geoforum.2018.09.011 (Accessed: 17 September 2021).

Mehta, A. (2021) 'Climate change is now a national security priority for the Pentagon', *Defense News*. Available at: https://www.defensenews.com/pentagon/2021/01/27/climate-change-is-now-a-national-security-priority-for-the-pentagon/ (Accessed: 17 September 2021).

Mendel, L. C. (1999) *Scenery to wilderness: National park development in Tasmania*, 1916–1992 (Doctoral dissertation, University of Tasmania).

Milner-Gulland, E. J., and Bennett, E. L. (2003) 'Wild meat: The bigger picture', *Trends in Ecology & Evolution*, 18(7), pp. 351–357.

Mishra, C., Madhusudan, M. D., and Datta, A. (2006) 'Mammals of the high altitudes of western Arunachal Pradesh, eastern Himalaya: An assessment of threats and conservation needs', *Oryx*, 40(1), pp. 29–35. https://doi.org/10.1017/S0030605306000032.

Mishra, S. (2011) *Army jawans found cooking chinkara meat in Barmer army camp*. https://www.indiatoday.in/india/west/story/indian-armychinkaras-salman-khan-146887-2011-11-26.

Münster, D., and Münster, U. (2012) 'Consuming the forest in an environment of crisis: nature tourism, forest conservation and neoliberal agriculture in south India', *Development and Change*, 43(1), pp. 205–227. https://doi.org/10.1111/j.1467-7660.2012.01754.x.

Muralidharan, R., and Rai, N. D. (2020) 'Violent maritime spaces: Conservation and security in gulf of Mannar Marine National Park, India', *Political Geography*, 80, pp.102–160.

Murillo-Sandoval, P. J., et al. (2020) 'The end of gunpoint conservation: forest disturbance after the Colombian peace agreement', *Environmental Research Letters*, 15(3), p. 034033. https://doi.org/10.1088/1748-9326/ab6ae3.

O'Brien, K., and Barnett, J. (2013) 'Global environmental change and human security', *Annual Review of Environment and Resources*, 38, pp. 373–391.

Ojeda, D. (2012) 'Green pretexts: ecotourism, neoliberal conservation and land grabbing in Tayrona National Natural Park, Colombia', *The Journal of Peasant Studies*, 39(2), pp. 357–375. https://doi.org/10.1080/03066150.2012.658777.

Ordway, E. M. (2015) 'Political shifts and changing forests: effects of armed conflict on forest conservation in Rwanda', *Global Ecology and Conservation*, 3, pp. 448–460. https://doi.org/10.1016/j.gecco.2015.01.013.

Owen, M. (1990) 'The damage-conservation interface illustrated by geese', *Ibis*, 132(2), pp. 238–252.

Parvaiz, A. (2020) *India's virus lockdown fuels timber-smuggling in Kashmir forest*. https://news.trust.org/item/20200612051624-20q88/.

Peluso, N. L. (1993) 'Coercing conservation?: The politics of state resource control', *Global Environmental Change*, 3(2), pp. 199–217. https://doi.org/10.1016/0959-3780(93)90006-7.

Peluso, N. L., and Vandergeest, P. (2011) 'Political ecologies of war and forests: counterinsurgencies and the making of national natures', *Annals of the Association of American Geographers*, 101(3), pp. 587–608. https://doi.org/10.1080/00045608.2011.560064.

Pennaz, A. K., Ahmadou, M., Moritz, M., and Scholte, P. (2018) 'Not seeing the cattle for the elephants: The implications of discursive linkages between Boko Haram and wildlife poaching in Waza National Park', Cameroon. *Conservation and Society*, 16(2), pp.125–135.

Piombo, J. (2013) 'Civil-military relations in an emerging democracy: South Africa', In *The Routledge handbook of civil-military relations* (pp. 255–272). Routledge.

Rangarajan, M. (2001) 'From princely symbol to conservation icon: A political history of the lion in India', *The Unfinished Agenda, Nation Building in South Asia*, pp. 399–442.

Robinson, J. G., and Redford, K. H. (1991) 'The use and conservation of wildlife', *Neotropical Wildlife Use and Conservation*, pp. 3–5.

Sarin, M., and Springate-Baginski, O. (2010) 'India's Forest Rights Act—The anatomy of a necessary but not sufficient institutional reform', IPPG Discussion Paper Series No. 45.

Scarpelli, C. M. (2018) 'Out of the Ashes: examining forest biodiversity conservation efforts in post-conflict Vietnam and Sierra Leone'. Available at: https://scholarsbank.uoregon.edu/xmlui/handle/1794/24095 (Accessed: 17 September 2021).

Shafer, C. L. (1999) 'National park and reserve planning to protect biological diversity: Some basic elements', *Landscape and Urban Planning*, 44(2–3), pp. 123–153.

Sharma, S., Dutta, T., Maldonado, J. E., Wood, T. C., Panwar, H. S., and Seidensticker, J. (2013) 'Forest corridors maintain historical gene flow in a tiger metapopulation in the highlands of central India', *Proceedings of the Royal Society B: Biological Sciences*, 280(1767), p. 20131506.

Sharma, V. (2021) *Man, out to hunt, shot dead by security forces in Latehar*. https://news.trust.org/item/20200612051624-20q88/.

Simlai, T. (2015) 'Conservation 'wars': Global rise of green militarisation', *Economic and Political Weekly*, pp. 39–44.

Simlai, T. and Kazmi, R. (2017) *Are conservation organizations complicit in ethnic discrimination?* https://thewire.in/environment/conservation-organisations-complicit-ethnicdiscrimination.

Smith, C. L., Hooks, G., and Lengefeld, M. (2014) 'The war on drugs in Colombia: the environment, the treadmill of destruction and risk-transfer militarism', *Journal of World-Systems Research*, pp. 185–206. https://doi.org/10.5195/jwsr.2014.554.

Seager J. (1992) 'Operation desert disaster', In C. Peters (Ed.), *Collateral damage: The new world order at home and abroad*. Boston: South End Press.

Smith, G. (2017) *The war and environment reader*. Illustrated edition. Charlottesville, VA: Just World Books.

Stevens, K., et al. (2011) 'Examining complexities of forest cover change during armed conflict on Nicaragua's Atlantic Coast', *Biodiversity and Conservation*, 20(12), pp. 2597–2613. https://doi.org/10.1007/s10531-011-0093-1.

Sundar, N. (2016) *The burning forest: India's war in Bastar*. Juggernaut Books.

'The military's contribution to climate change' (2021) *CEOBS*, 16 June. Available at: https://ceobs.org/the-militarys-contribution-to-climate-change/ (Accessed: 17 September 2021).

The Routledge handbook of civil–military relations (2012) Routledge Handbooks Online. https://doi.org/10.4324/9780203105276.

Thomas, W., and Seager, J. (1994) *Scorched Earth: military's assault on the environment*. Philadelphia, PA: New Society Publishers.

UK Ministry of Defense. (2017–2018) *Annual Report*. https://assets.publishing.service.gov.uk/government/uploads/system/uploads/attachment_data/file/918349/SustainableMOD2018__final_.pdf.

Velho, N., Karanth, K. K., and Laurance, W. F. (2012) 'Hunting: a serious and understudied threat in India, a globally significant conservation region', *Biological Conservation*, 148(1), pp. 210–215. https://doi.org/10.1016/j.biocon.2012.01.022.

Verweijen, J., and Marijnen, E. (2018) 'The counterinsurgency/conservation nexus: Guerrilla livelihoods and the dynamics of conflict and violence in the Virunga National Park, Democratic Republic of the Congo', *The Journal of Peasant Studies*, 45(2), pp. 300–320.

Weladji, R. B., and Tchamba, M. N. (2003) 'Conflict between people and protected areas within the Bénoué Wildlife Conservation Area, North Cameroon', *Oryx*, 37(1), pp. 72–79. https://doi.org/10.1017/S0030605303000140.

Wikramanayake, E., et al. (2004) 'Designing a conservation landscape for tigers in human-dominated environments', *Conservation Biology*, 18(3), pp. 839–844. https://doi.org/10.1111/j.1523-1739.2004.00145.x.

Wild Hope (no date). Available at: https://press.uchicago.edu/ucp/books/book/chicago/W/bo13823467.html (Accessed: 17 September 2021).

Wright, B. (2010) 'Will the tiger survive in India?', In *Tigers of the World* (pp. 87–100). William Andrew Publishing.

Ybarra, M. (2016) '"Blind passes" and the production of green security through violence on the Guatemalan border', *Geoforum*, 69, pp. 194–206. https://doi.org/10.1016/j.geoforum.2015.06.004.

Zentelis, R., and Lindenmayer, D. (2015) 'Bombing for biodiversity—enhancing conservation values of military training areas', *Conservation Letters*, 8(4), pp. 299–305.

Zentelis, R., Lindenmayer, D., Roberts, J. D., and Dovers, S. (2017). 'Principles for integrated environmental management of military training areas', *Land Use Policy*, 63, pp. 186–195.

CHAPTER 9

Policing the High Speed 2 (HS2) Train Line: Repression and Collusion Along Europe's Biggest Infrastructure Project

Andrea Brock and Jan Goodey

Introduction

> If you want a picture of the future, imagine a boot stamping on a human face—for ever
>
> —George Orwell, 1984.

The UK government's High Speed 2 (HS2) rail project is not only Europe's largest infrastructure project (Construction News, 2020) but, at an estimated cost of £200bn and rising (Wolverhampton Express and Star,

A. Brock (✉)
University of Sussex, Brighton, UK
e-mail: a.brock@sussex.ac.uk

J. Goodey
Kingston University London, London, UK
e-mail: J.Goodey@kingston.ac.uk

A. Dunlap and A. Brock (eds.), *Enforcing Ecocide*,
https://doi.org/10.1007/978-3-030-99646-8_9

2017), it is also the largest and most expensive post-war development in the UK. Framed by the UK government as environmentally beneficial and contributing to economic growth, it is strongly resisted by campaigners and residents along its proposed route. They criticise the project for negatively affecting 108 ancient woodlands,[1] destroying irreplaceable nature reserves, ecosystems, chalk aquifers, and waterways, as well as for the lack of economic benefits, and the huge costs for taxpayers (Stop HS2, 2021). Many further criticise the extractive character of the project—while it is discursively positioned as beneficial for Northern communities by the government (McLoughlin, 2015). The true benefits, however, will accrue to London, and the 'City' in particular, as well as wealthy commuters (New Economics Foundation, 2019a; Metz, 2020; Ramchurn, 2013; interviews).

Resistance along the route takes various shapes and forms—from legal challenges, lobbying, demonstrations, to direct action, including the continuous occupation and setup of protest camps along the proposed HS2 pathway to disrupt and slow down its development. This chapter examines the policing, corporate-state collusion, and private-state security collaboration to manage resistance along the HS2 high-speed rail project. We analyse the range of techniques and technologies of control and the violence exercised in the oppression of dissent, by police officers, private security forces, and bailiffs. We seek to uncover the mechanics of clandestine working relationships between police, private security companies, and corporations—in other words, the political (re)actions 'from above' (Geenen and Verweijen, 2017; Verweijen and Dunlap, 2021) to manage resistance. Particular attention will be paid to the formal arrangement between police and HS2 Ltd (the project operator): the Enhanced Police Service Agreement (EPSA).

Our analysis shows that policing is integral to the ecological destruction that HS2 entails. We illustrate how policing upholds the extractive character of the project, contributing to the 'policing for green capitalism' (Brock and Stephens Griffin, 2021) and green extractivism (Dunlap and Brock, 2021; Brock, 2020a) that have become part and parcel of the contemporary political economy. Policing, we show, in line with the other chapters in this edited volume, creates and defends ecological degradation (Brock and Stephens Griffin, 2021). Its technologies can be analysed

[1] With the recent cancellation of one part of the project, the number is now likely to be lower (see the Woodland Trust, 2021).

as counterinsurgency techniques—developed, tried, and tested through colonial policing and refined in the domestic context (see introduction to this volume). While a full overview of policing or counterinsurgency techniques employed in the HS2 context would go much beyond the scope of this chapter, we focus on three areas of policing: (1) Silencing dissent and controlling the narrative; (2) Policing through criminalisation; and (3) Policing through physical coercion—(state) violence on the ground.

Engaging with a set of relevant literatures that examine the policing of protest in the global North, political ecology scholarship of extractivism, and counterinsurgency, we draw on approximately 15 interviews (conducted in 2019 and 2020) and quantitative data (studies, reports, newspaper articles, and blogs) for our analysis. We further look to freedom of information requests from campaigners and journalists which resulted in information on HS2/police collusion as well as numerous non-disclosure agreements from organisations with links to HS2 Ltd. Lastly, this chapter draws on 'observant participation' (Sullivan and Brockington, 2004) and is grounded in personal experiences of policing along the railway and elsewhere.

This introduction is followed by a literature review and theoretical framework that lays out our approach, which is informed by political ecology debates, extractivism, and critical research on the history of policing and counterinsurgency. We introduce the (political economy of the) HS2 project before turning to the policing strategies employed by private security and police forces as well as their collaboration. The analysis is structured into the three areas introduced above: the ways in which dissent is silenced and the narrative controlled by HS2; the role of criminalisation *in* policing; and policing through physical coercion—(state) violence on the ground. Here, we further consider state-sponsored surveillance and its role in inhibiting delays to the building programme, including the use of open-source intelligence. We turn to the work of the private security firm G4S Risk Consulting Limited, which gathers intelligence on protesters through various surveillance techniques in a bid to limit potential risks to the progress of an infrastructure project. Lastly, in our discussion, we argue that policing is integral to enforcing ecological destruction under the name of green capitalism and to serve green extractivism.

Extractivism, Railways, and the Political Ecology of Policing

Much critical (political ecology) scholarship examines the injustices associated with ecological struggles in the global South, especially those caused by (often state-backed) extractive and infrastructure 'development' projects (e.g. Temper et al., 2018; Menton and Le Billon, 2021). The responses 'from above' to local opposition and resistance have garnered increased attention over the last few years, with scholars pointing to the importance of social engineering of consent and repression of resistance by corporations and state forces (Verweijen and Dunlap, 2021; Dunlap, 2019; Wiegink, 2020; Huff and Orengo, 2020; Leifsen, 2020; Kaur, 2021; Jespersgaard Jakobsen, 2020). Indigenous communities and people of colour are particularly at risk from violence by state and corporate forces, and even more harshly policed (see Gobby and Everett, this volume).

At the same time, recent scholarship on the policing and criminalisation of ecological dissent in the global North has laid bare the remarkably similar logics and technologies at play (Brock, 2020a, b; Brock and Dunlap, 2018; Pickard, 2019; Jackson et al., 2019; Monaghan and Walby, 2017; Dunlap, 2020). While the risk of death and serious injury is much diminished, of course, coercion, psychological warfare, and physical violence are part and parcel of the management of dissent in European and Western states. Policing in Britain, where our case study is located, is closely connected to the stigmatisation of protesters (Brock et al., 2022), characterised by increased surveillance, militarisation, and excessive force (Pickard, 2019). Policing in Britain further involves '*a priori* criminalisation' through the use of corporate injunctions (Brock, 2020b), pre emptive house raids and arrests, mass arrests, extended detention, and restrictive bail conditions (Pickard, 2019). It takes place against the well-documented history of undercover surveillance (Lubbers, 2012), with police officers spying on more than 1,000 political groups between 1968 and 2011 (Evans and Lewis, 2013), deceiving female activists into romantic/sexual relationships for years (Stephens Griffin, 2020; Schlembach, 2018). Policing goes hand in hand with harsher criminalisation (Brock et al., 2018) and the application of anti-terror legislation. The idea of 'consensual', nonviolent policing, Joanna Gilmore and colleagues have shown, has always been a myth (Gilmore et al., 2017). Instead, 'the police represent the most direct means by which the state imposes its will on the citizenry. When persuasion, indoctrination, moral pressure, and incentive measures all fail—there are the police ' (Williams, 2007: 29)—and private security services that pick up some of the dirtiest policing work, as our chapter shows. When resistance presents an actual threat to the viability of a project, we argue, this comes to the fore.

Repression is integral to business, but businesses are integral to this repression too; private companies are key to the large-scale surveillance of activists, initiating lawsuits, PR campaigns, greenwashing, and intense lobbying (Lubbers, 2002; 2012). 'Public-private security partnerships' are integral to the policing of extractive projects (Brock, 2020b).[2] The repression and policing of dissent have been analysed as corporate counterinsurgency strategies employed by resource extraction companies and infrastructure developers (often in collaboration with state forces) to deal with so-called 'insurgencies' against their projects, questioning their legitimacy and actions (Brock and Dunlap, 2018; Brock, 2020b; 2018, 2019). Such a framework illustrates how they resemble, and have learnt from, colonial and domestic practices to co-opt and repress resistance. Counterinsurgency is defined by the British army manual as 'military, law enforcement, political, economic, psychological and civic actions taken to defeat insurgency, while addressing the root causes' (British Army, 2009: 1–6). The aim is to maintain political stability as well as governmental and corporate legitimacy. It's a style of warfare that makes use of intelligence networks, psychological operations, media manipulation, and security provision, including social development that seeks to maintain governmental legitimacy (FM3-24 2014; see also Dunlap, 2018). Open-source intelligence gathering, the harvesting of information that is available publicly, is crucial for counterinsurgency.

Counterinsurgency efforts involve militaristic tactics (Brown, 2021) that are composed of 'soft' and 'hard' counterinsurgency tactics (Williams, 2007). The former includes pacification and 'engineering consent' through sponsorship, generational engagement, lobbying efforts, public relations campaigns, and neoliberal social development; whereas the latter involves criminalisation, physical violence, surveillance, and intimidation (Brock and Dunlap, 2018). In the US, fracking companies and their PR specialists work with psy-op experts who have been told to read the US army counterinsurgency manual in order to deal with local opposition

[2] Other examples include the policing of Standing Rock protests, where different police forces collaborated with the private security firm TigerSwan (Brown et al., 2017a, b).

(Javers, 2011); counterinsurgency operations have been documented in the context of the Standing Rock resistance (Brown et al., 2017a, b) and the Rio Tinto operations in Bougainville (Lasslett, 2014). Private security companies with experience in counterinsurgency operations in Iraq are employed to protect fracking operations in the UK (Hope and Collett-White, 2018; Brock, 2020b). Private security firms make huge profits from such conflicts (Brown, 2018), and revolving door relationships[3] (Brown, 2021) ensure smooth collaboration between private and public security actors. Pacification through conservation schemes and other green initiatives is key to these practices (Huff and Orengo, 2020; Brock, 2020b) and integral to divide-and-conquer strategies to co-opt 'moderate' opposition. 'Green counterinsurgency', Alexander Dunlap (2018: 648–9) has argued, involves the use of the green economy 'as a legitimizing device to push through wider projects of control', positioning projects as green, sustainable, or clean. '[T]he "green" in the notion of green grabs and green economy can be read as a larger pacification device to continue land acquisition and industrial development for continued market expansion' (Dunlap, 2018: 648–9).

Drawing on this body of work that explores the policing and counterinsurgency approach to environmental conflicts, we examine the HS2 project and its policing as an extractive operation. Extractivism not only describes mining projects but also a mode of accumulation that involves the extraction and removal of resources, such as forests or industrial monocultures, and has served as 'a mechanism of colonial and neocolonial plunder and appropriation' (Acosta, 2013: 63). Extractivism further provides a lens through which to understand the statist colonial logic and ideology of extraction and exploitation to create value, bound to ideas of progress and growth (Gudynas, 2009). Extractivism involves the creation of profits for national or international business elites as it alters and destroys existing social ecological relationships (ibid., see Introduction to this book). It is characterised by their prioritisation of growth and production over human and ecological health (Preston, 2017), privileging

[3] 'Revolving doors' describes the movement of (influential) individuals between industry and legislators or regulators.

state and corporate actors over communities and their ecosystems.[4] Transportation infrastructure in general, but specifically railways, have been instrumental to facilitating extractive operations and the movement of capital.

Railways and Extractivism

Railway projects have historically been crucial for the facilitation of extraction activities. The first railways in India were built by private British companies in the middle of the nineteenth century to extract and transport resources like coal, iron ore, and cotton (Bagchi, 1982; Habib, 1975). Building railways and irrigation systems

> convert[ed] India into a major supplier of raw materials and foodgrains for Europe and many of its colonies overseas. The route alignments and rate structures of railways were such as to make it cheaper to transport goods from the ports to the interior and back rather than between points in the interior (Bagchi, 1982: 86).

Unlike in Europe, railway construction actually led to *deindustrialisation* in India—resources were processed in British factories—and 'increased the intensity of dominion of advanced capitalist countries' (Bagchi, 1982: 34). Through the increasing importance of export crops and the commodification of land, capitalist social relations were deepened, enhancing the power of landowners, traders, and money lenders over those working the land. The railways thus extended the processes of colonisation and helped Britain to retain India as the principal market of the British industry (Habib, 1975). Drawing on government records, parliamentary reports, and newspapers, Pallavi Das illustrated that 'much of the deforestation seen in the Himalayas today can be traced back to the second half of the nineteenth century when railway construction began in colonial India' (2011: 38). Railway construction led to the depletion of India's natural resources, including its forests (ibid.), and wealth extraction.

[4] For further comments on how the postcolonial concept of extractivism might be useful to understand extractivism in the European context, see (Brock, 2020b).

While today, rail transport is often (rightly) praised as an ecologically conscious alternative to automobility and promoted as 'green transport', high-speed railways often have a huge ecological footprint and trigger major resistance. 'Large, imposed, and useless' infrastructural projects, including a number of high-speed railways, are often opposed by local communities through campaigns, occupations, and blockades (Mauvaise Troupe Collective, 2018; also Best and Nocella, 2006). The ZAD in Notre-Dame-des-Landes (France) and the Sanrizukata movement in Narite (Japan) which both emerged to resist new airports, the No Tav movement in Susa Valley against the Turin–Lyon high-speed railway, and the resistance to the North Dakota Pipeline on (unceded) US land have shown that these struggles are not just *against* a particular development project, but *for* land and autonomy (Ross, 2018). The TAV (Treno ad Alta Velocità/High Velocity Train) in Susa Valley is part of an EU project which plans to connect Lyon to Budapest and Ukraine, and its opposition has come to define the lives of 70,000 people in the Italian Susa Valley for over twenty years (Mauvaise Troupe Collective, 2018). Despite brutal repression and designation of the project as of 'strategic interest' by the state, inhabitants critique the project as useless and unnecessary, profiting private companies while causing ecological and social devastation to the valley, destroying villages and small businesses. Similar to the ZAD in France, which has been fighting '*Against the airport and its world*', the No Tav movement is thus not just a struggle against a railway project but its 'world' too (Mauvaise Troupe Collective, 2018). The 'airworld' that high-speed railways are part of, Ross (2018) describes, is all about global luxury trade; fast transport; smooth connectivity between cities, leaving towns and countryside unserved; '*frictionlessness* – the ability to move people and goods in and out as quickly and effortless as possible'. Here, '[p]eople and things, torn from their living entanglements, are freed to become mobile investments in a world where the fungibility of space is taken as a given' (ibid.). Opposition against this high-speed railway is not just against a trainline but against this world that is grounded in the ideology of growth, markets, and capital; merging life with revolt; and balancing confrontation with community, non-hierarchical resistance (Mauvaise Troupe Collective, 2018). Here, people not only defend their ecosystems and seek to build new social relations with the soil and the land, but with each other too (ibid.), living differently and inspiring people to do so across the world.

Large railway projects are ‘mega-infrastructures – grandiose, landscape-transforming projects, often spanning across several countries or regions, involving multiple public and private actors – [that] reorder capitalist relations across the globe, producing infrastructural spaces wherein rationalities of capital accumulation materialise’ (Lesutis, 2021: 1, drawing on Bouzarovski et al. 2015). They are important state practices of ‘infrastructural territorialisation’ (Lesutis, 2021: 1) in the interest of accumulation and social control. This reordering can occur through ‘bureaucratic land grabs’ where land is designated as development land for a project, with little democratic control, and often against local resistance. This concept puts emphasis on the ‘the procedural legitimization of land theft and examines how unpopular and environmentally destructive development projects are permitted’ (Dunlap, 2020: 112). High-speed railway project facilitates accumulation not only through privatisation and profits for (international) investors, but also because they are subordinated to the overarching goals of growth at huge social and ecological costs. Such infrastructures ‘replicate colonial/statist [extractivist] sociocultural values in the local’ (Leonardi, 2013: 35).

High-speed railway projects’ green extractivist nature thus lies not only in their function within the broader political economy, but also in their role in the extraction of value from regions, places, and spaces, by connecting them (as in the case of HS2) with bigger markets, financial centres, economic hubs, and airports, in the name of ‘sustainability’. This is about mobility in the service of capital. The case of HS2, which is framed as ‘benefitting Northern communities’, illustrates this point particularly well. A thorough analysis by the New Economics Foundation (NEF), based on HS2’s own data, however, has shown that ‘40% of the passenger benefits that underpin HS2’s economic case will accrue to London’; thus deepening existing regional inequalities, rather than ameliorating them (2019a: 4), by investing in and facilitating access to London. In addition to business and other employers in London (including the financial centre), the line will primarily benefit wealthy commuters and shareholders: ‘the HS2 demand model forecasts that its average commuting passenger will be in the top 10% of the income distribution’ (NEF, 2019b). This extends to future investments, Andrew Pendleton argues, ‘intensify[ing] the north-south investment divide’, making HS2 a ‘trickle-down transport policy’ (in NEF, 2019b). The project thus redistributes taxpayers’ money from the bottom up, benefitting the construction industry, wealthy commuters, and travellers, as well as the businesses they work for—in the name of sustainability and economic benefits to the North.

'Extractive infrastructures bind the state-extraction-ecocide nexus together' (Brock, 2020b: 3). 'Green extractivism' (Brock, 2020a, b; Dunlap and Brock, 2021), or 'total extractivism' (Dunlap and Jakobsen, 2019) are thus framed as sustainable and part of a green transition, while continuing the same extractive logics and processes that are part of capitalist modernity and the march for 'progress'. Building on these recent literatures on the management and policing of resistance against extractive projects and contributing to the growing political ecology literature of the Global North (McCarthy, 2002; Schroeder et al., 2006), this chapter thus provides a European case study exploring the policing by corporate and state forces of dissent against green extractivism in the form of high-speed infrastructure, to repress resistance and maintain a clean image. We continue with a brief overview of the HS2 project.

HS2—Too Big to Stop and Too Big to Fail?

The HS2 high-speed rail project is the biggest European infrastructure project since World War II (BBC, 2020). Heralded by the UK government and business leaders as an industrial panacea to reinvigorate the UK economy, drive connectivity, and bridge the gap between the capital and more deprived areas in the north of the country (ibid.; McLoughlin, 2015), its construction started with Phase 1 in 2020 and is slated to continue until at least 2040 (Parsons, 2019). While the roots of the project can be traced back over at least half a century, the current project was established in 2009. Its political inception took the form of a hybrid bill, enacted in 2017. Hybrid bills are introduced to support major infrastructure projects that are deemed to be 'works of national importance'. They are promoted and sponsored by the government and granted automatic development consent (planning permission), which authorises the compulsory purchase of land and land rights, and allows for a wide range of 'ancillary and incidental activities' (Mould, 2017: 3). This is similar to the Declaration of Public Utility (Déclaration d'utilité publique) in France or Projects of Common Interest (Proyectos de interés común) in Spain; the 'bureaucratic land grabs' mentioned above (Dunlap, 2021). This designation makes them uniquely powerful political instruments.

The high-speed project, managed by HS2 Ltd, a private company set up by the government under control of the Secretary of State for Transport, is comprised of two phases: Phase 1 (London to West Midlands) and Phase 2 (West Midlands to Crewe, Phase 2a; and to Leeds, Phase 2b[5]). The UK government has thus been the main driver of the project, not just through providing project funding but also serving as its PR apparatus, promoting HS2 as economically beneficial, a 'boost to the North', and 'making Britain greener'; taking freight and passengers off the roads and out of domestic flights (HS2 Ltd., 2021a, b). In his six-month report to parliament, Transport minister Andrew Stephenson (2021), responsible for HS2, claimed that:

> HS2 remains at the forefront of our long-term investment plan to better connect people and places, boost productivity and create jobs to help rebalance opportunity across the UK. Just as importantly, HS2 will play a pivotal role in creating a greener alternative to regional air and road travel. This is essential if we are to meet our commitment to bring greenhouse gas emissions to net-zero by 2050.

While damage to woodlands and habitats is recognised as 'unavoidable', the Government points to externally verified Environmental Impact Assessments and claims to be able to offset (mitigate and compensate) this destruction through the planting of new trees and the translocation of animals and habitats.[6] A number of initiatives, including the 'Green Corridor Prospectus' are meant to provide environmental benefits and mitigate impacts; to 'create a network of bigger, better-connected, climate resilient habitats and new green spaces for people to enjoy', 'add[ing] benefit over and above committed mitigation and statutory compensation' (HS2 Ltd, 2020). For Phase 2b of the project, HS2 Ltd claims to go beyond the delivery of *No Net Loss* and to move towards *Net Gains* of biodiversity (Stephenson, 2021). While unable to replace ancient woodlands, such schemes are important mechanisms for the discursive 'greening' of extractive projects (Brock, 2020a).

[5] As we are writing this chapter, the government has just confirmed the cancellation of the Eastern leg of phase 2b between Birmingham and Leeds (Construction News, 2021a).

[6] For critiques of biodiversity offsetting, see Carver and Sullivan (2017), Hannis and Sullivan (2012), and Brock (2020a).

Critics

Environmentalists, farmers, and homeowners along the route have long criticised HS2 as a 'self-aggrandising project' that benefits those in positions of power and their friends and networks who stand to gain financially and politically. They typically criticise the project on environmental and economic grounds, as the slogan 'No business case. No environmental case. No money to pay for it'[7] demonstrates.

Many years of research and FOI requests have revealed the shaky grounds on which the project's environmental claims are built, debunking many of the alleged benefits. The 'rail not roads' narrative, for instance, is built on claims that HS2 rail traffic will lead to fewer internal and short haul flights. Yet, potential reductions in these flights are expected to be offset by long haul flights with increased numbers of UK holidaymakers and business executives using the new airport rail terminals (as confirmed by British Airports Authority), and thus actually increasing emissions by several times (Stop HS2, 2015, interviews). In addition, despite claims of reducing flight numbers, several Northern airports are already expanding their capacity in response to HS2 development (Business Live, 2013). Birmingham airport, for one, 'has announced that it will market itself as a fourth London airport with HS2 as the link. The airport said it is currently running at 40% of its capacity and could take 9m extra passengers a year. Birmingham Airport's expansion plans will double the number of flights, resulting in increased emissions' (ibid.). Critics further point to misleading claims about the carbon neutrality of the project, which, according to the company's own estimates, will take over 120 years to achieve (The Times, 2020).

In the meantime, diverse habitats, ancient woodlands, and areas of natural beauty, including some greenbelt land to the north west of London (featured below) which has a confluence of rivers and ancient woodlands, are being destroyed to make way for construction sites and associated developments. These include 'enabling works': spur roads, temporary access roads, roads to the new developments, and new stations. HS2 Ltd. has refused to disclose the number of miles of access roads that are being built for construction—campaigners speculate that this is due to this number amounting to more than the planned number of miles of

[7] The UK-wide Stop HS2 campaign was set up in 2010 with the aim to stop HS2 by persuading the Government to scrap it. It facilitates local and national campaigning.

train tracks. According to the Wildlife Trusts (2020), 33 Sites of Special Scientific Interest (SSSIs) and 21 Local Nature Reserves are at risk of destruction and close to 700 Local Wildlife Sites (LWS) are set to be part-damaged or destroyed by the HS2 route.

Land defenders and ecologists have long documented and flagged numerous instances of local ecological destruction of water sources, woodlands, and other habitats, as well as drinking water pollution[8]—through both legal and illegal activities. Protestors at Denham Ford, for instance, gathered video evidence of rare bats whose habitat is being destroyed, and in Warwickshire, campaigners have documented evidence of HS2 staff destroying birds' nests in contravention of the Wildlife and Countryside Act 1981 (The Independent, 2020). This points to structural flaws in HS2's environmental impact assessments (EIA). In its 2016 response to the High Speed Two Phase 2A EIA and Scope and Methodology Report (SMR), the National Trust, a UK charity and membership group for nature and heritage conservation (historically close to the British government), pointed to serious flaws in the methodology:

> The proposed habitat surveys have some significant omissions, such as the consideration of veteran trees, Brown Hare and deer. Habitat connectivity mapping and modelling, using the method recommended by the HS2 Ecology Technical Group, should be incorporated into the EIA and we would like to see a programme of proactive monitoring of the significant residual impacts of the scheme (National Trust, 2016).[9]

The 'limited approach to reporting, besides apparently failing to make best use of the data acquired' (National Trust, 2016) was likely to underplay the impacts of the development, particularly in respect to connectivity of habitats at a landscape scale and consequent impacts upon movements of species. Others criticise that some destruction would be avoidable by choosing more expensive alternatives. The River Colne at Denham Ford, whose pristine waters are home to critically endangered European eels, bream, and lustrous river weeds, will be devastated by the building

[8] Between August and November 2020, 1,600 tonnes of bentonite, a pollutant, was released into the chalk aquifer at the Chalfont St Peter vent shaft work site. For more details see HS2's own report (Align, 2021).

[9] More details can be found in the actual response document (National Trust blog, 2016) 'National Trust response to: HS2 Phase Two: West Midlands to Crewe EIA Scope and Methodology Report—Draft for Consultation (March 2016)'.

of an access road with a bridge to move a nearby pylon. Campaigners have argued that the pylon could be moved via the nearby Grand Union canal—but HS2 has opted for the cheaper option. The project's offsetting scheme is critiqued for the lack of environmental 'follow up', which has already led to the death of entire new tree plantations. Even the review by HS2 Ltd.'s former chief executive, Douglas Oakervee (2020), which gave the HS2 project its seal of approval, pointed out that ancient woods cannot be replaced by definition, questioning the offsetting rationale.

HS2's economic benefits and viability are equally contested. In July 2020, the government's own Infrastructure and Projects Authority (IPA) gave it the highest risk rating, 'red'. Red ratings are given to schemes when 'successful delivery' appears to be elusive. Independent estimates put the 'value for money' ratio at £0.66 return for each public pound spent (Berkeley, 2020) and £1.30 for each public pound spent (Institute for Government, 2020). This means it ranks as a 'poor value project'. Profits go to private developers, many with close links to government officials and the Conservative Party, and HS2 staff. HS2 chief executive officer Mark Thurston earned over £660,000 in 2019 (Construction News, 2020), making him the government's best paid official (Martin, 2020). The average salary for HS2 workers is over £60,000 a year (Marshall, 2021), with one in four on a six-figure salary (Sky news, 2018).

Meanwhile, campaigners and residents along the route excoriate the company for ongoing local dispossession and expropriation of farmers' and other residents' land and properties. The Parliamentary and Health Service Ombudsman has repeatedly found HS2 Ltd to be 'dishonest, misleading and inconsistent' in dealing with residents; and guilty of maladministration in negotiations of family farm owners (Parliamentary and Health Service Ombudsman, 2021). Many accuse HS2 of lying, blackmailing,[10] pressuring, and providing late payments (Parliamentary and Health Service Ombudsman, 2015). Those with close connections to government officials, however, might benefit from compulsory purchase orders. To illustrate this with two juxtaposed examples: Stanley Johnson, Prime Minister Boris Johnson's father, received £1m for land near his home in Euston (Tominey, 2020). Meanwhile, a local farming family near Birmingham whose land the HS2 line will cut across, with nearly 1000

[10] http://stophs2.org/news/14708-hs2-guilty-maladministration-ombudsman.

trucks passing by every day for years to come, have been told that they are only 'marginally impacted' and thus have no right to compensation.[11]

The ecological and social destruction and dispossession have long triggered resistance. For years, locals fought against the line and the loss of their land through legal and judicial means. The first protest camp was set up along the route in 2018. The camps' aims were to raise awareness of the railway line's damage to the natural ecology, and to stop or delay its construction through direct action. The inequalities and injustices associated with compensation and land ownership have facilitated unusual alliances between landowners and more combative parts of the resistance movements. An example is the local Golf Club at Denham Ford, a redoubt of Conservative Party membership, which allowed protesters onto its land to access its water. Many camps receive(d) field support from the local community sympathetic to the cause; residents whose houses might be in line for compulsory purchase, or those living on nearby council estates where working class values naturally lead to support for the underdog. This resistance is heavily policed, by police forces and private services often working in collaboration. By 2021, an estimated £75m has been spent on policing direct action against ongoing HS2 developments—a small but significant percentage of the official £100–£109bn budget (Topham, 2021). We now turn to policing in the next section.

Policing HS2, Policing Green Capitalism

The policing of protest sites involves a range of public and private actors, with shared and overlapping responsibilities and, often, partnerships. Police forces are at the frontline of policing during the eviction of protest camps and at HS2 sites bordering protest camps, and they enforce corporate injunctions. Their job is to keep protesters in check through threat of arrest, actual arrest, and enforcement of 'bailing off site'. The latter entails the enforcement of bail conditions by re-arrest and the removal of people who are breaking their bail condition to stay away from the camp. The police work closely with the High Court Enforcement Group (HCE Group), the holding company of the National Eviction Team (NET), which operates at a number of protest sites along the route. On its website, NET claims to 'deal with more evictions of trespassers, unwanted

[11] Personal conversation, August 2021. There are many similar stories and examples.

environmental protesters and squatters than anyone else in the country' with 'a very large workforce, enabling [it] to manage even the largest of environmental protests' (NET, 2021). It claims to 'meet and exceed… deadlines to provide [its] clients [including HS2] with the safe and effective removal of trespassers and protesters and to ensure the security of the site afterwards to prevent recurrence'.

> If it is likely that police support is required, we will arrange that on your behalf, *agreeing with the police the number of officers who can be deployed, the timescales, the arrest policy and the primacy policy* (NET, 2021, italics added).

The latter already illustrates the close relationship to police forces, which campaigners frequently witness on the ground.[12] The company works on infrastructure projects that involve Compulsory Purchase Orders (CPOs), legally enforced orders forcing property owners to sell their land, usually when said property is on the route of a planned road, runway, or rail line. By deliberately choosing the High Court moniker, the HCE group plays into the officialdom of nomenclature lending authority and reach, despite it being a private company, subcontracted by HS2 Ltd. People on the receiving end—homeowners, farmers, small businesses facing eviction through compulsory purchase orders (CPO)—automatically assume that these operatives are High Court sheriffs acting in the name of the UK judiciary, when they are not.

The HCE Group carried out a number of high-profile evictions over decades gone by. They include the infamous 2011 eviction of Europe's largest traveller site, Dale Farm (HCE Group, 2021), countless road bypass protest sites in the 1990s, as well as the more recent fracking protest sites at Balcombe in Sussex and Lancashire. It has further tried and tested means of eviction during evictions of travellers and squatters. Its tactics involve using plant machinery and trained climbers, as well as a tunnel team, to gain access to sites and take people off high structures or trees, or evict them from underground. The HCE Group provides training in restraint techniques and employs operatives who Koshka Duff (2021) and others would refer to as 'violence workers'

[12] The NET also supplies to the Ministry of Defence (High Court Enforcement Group HCE, 2021), pointing to further entanglements of their respective political economic interests.

consisting of former police officers, door personnel at night clubs, or city centre security staff. The company currently faces legal proceedings following the injuries of a number of protestors in a spate of separate incidents (Griffin, 2020). The company also faced criticism after a young protester lost consciousness and was hospitalised following an alleged chokehold up a tree in Denham in July 2020 (Swan, 2020). The NET is renowned in environmental circles for assaults on activists. When reported to the police, these assaults are often not investigated for a lack of evidence, or even lead to investigations of the victim (interviews), unless incontrovertible evidence (video footage) is presented in court or pre-trial.

To keep protests contained, HS2 has a number of contracts with other high-profile security firms including G4S Risk Consulting Limited and Servest. G4S is the world's largest security company and works for mining, oil, and gas industries around the world; explicitly targeting the natural resources sector in 'high risk and complex environments' for profit-making (Raphael, 2016). Servest operates in Birmingham, under contracts that entail land and property warrant enforcement (Compulsory Purchase Orders), forcing farmers, businesses, and homeowners to move if they live on the path of the route. It also covers the evictions of protest camps and the patrolling of the various project building sites. Control Risks was awarded a contract for over 42 million GBP (tenders electronic daily, 2017),[13] 'to provide safe and effective incident management and response, proactive area patrolling, close personal protection and management of locked on protesters' (tenders electronic daily, 2017). It was instructed 'to carry out these operations with minimal impact on HS2 Ltd.'s programme, whilst operating at the highest safety and operational levels to minimise any reputational impact on HS2 Ltd and the HS2 project' (HS2 Ltd. in government-online, 2017). Intelligence ('insight') gathering forms 'a significant part of the contract, the remainder being reactive to incident' (tenders electronic daily, 2017).

The private security industry has come to play an important role in the policing of extractive projects in Britain and worldwide. The industry took off with the occupation in Iraq when large contracts were awarded by the British military in order to train the Iraqi police (Raphael, 2016). In the UK, as in many countries, private security personnel vastly outnumber the

[13] https://assets.hs2.org.uk/wp-content/uploads/2018/01/21160810/hs2-current-contract-opportunities.xlsx.

police, with 386,657 licence holders compared to 146,000 police in 2019 (Townsend, 2019). HS2 Ltd. relies on private security firms for most of their policing work, often in collaboration with police.

Policing Along the Route: Public-Private Security Partnerships

Since 2018, HS2 Ltd. has a formal arrangement with the police that is known as an Enhanced Police Service Agreement (EPSA) (Construction News, 2019). It sets out that the British Transport Police 'second' an analyst, a researcher, and a National Police Liaison Officer to this joint operation—these three state employees are thus being paid by a limited company and indirectly managed by corporate executives, although they remain ultimately answerable to the British Transport Police. Similar agreements were used during the London Olympics in 2012 and the major London rail link project Crossrail since 2009 (Construction News, 2019). This partnership enables HS2 and police operatives to coordinate tactics across all 16 police forces along the route and to draft in officers from different forces, with climbing teams and plant machinery. Its aim is to obviate delays and keep costs down. Should major protests arise and people from different camps join forces to resist evictions, police forces and security services can quickly communicate and share security details and intelligence on protesters and camps. The analyst and researcher work alongside open-source intelligence staff who are also on HS2's payroll (as explained below). This all-encompassing approach means that HS2 is cognisant of any surprise tactics and can be in full control of any 'threats' (or 'insurgencies') to their building programme. Yet, this control is not complete, as the company's lack of awareness of the London Euston tunnels proved, when a small group of activists dug and lived in a network of tunnels beneath the HS2 Ltd. groundworks at Euston and kept between 12–25 NET team operatives busy for a month in early 2021 (Melia, 2021). The costs of the operation for policing and operational delays spiralled to over £140,000 (Layton, 2021).

FoI requests reveal that the costs of the Enhanced Police Service Agreement were around £336,000 by 2020; with annual increases from £31,261 in 2016/17 to £125,782 in 2019/20 (Maxey, 2020). From 2016 to 2019, around £230,000 was also paid to various security companies. These security and policing payments form part of the budget for the building work, which has grown from £56bn to £78bn in 2015, £88bn in 2019 and over £109bn in 2020. Between 2018 and 2020,

HS2 further paid out £224,000 on a Home Office scheme to provide a contracted immigration officer—who works for the Home Office but gets paid by HS2—to monitor illegal employment practices and halt any possible exploitation of migrant workers. In the meantime, however, HS2 Ltd. was flying in migrant workers who are paid below living wage to work as security guards, forced to break social distancing measures[14] during the Covid-19 lockdown periods of 2020–2021. At the same time, HS2's practices of sub-contracting controversial security specialists and bailiffs are not made public; there is no mention of the NET partnership on the HS2 website's list of contractors, for instance. The sub-contracting to private security companies and bailiffs—and police forces, through EPSA—means that HS2 effectively has its own private enforcement arm to evict protest camps as well as homes and businesses in the path of the planned route.

HS2 Ltd.'s collaboration with police forces illustrates that the view of the police as protector of public safety was only ever a liberal myth. Rather than protecting the right to protest, police collaborate with corporate security services to further limit this right and to facilitate HS2 development, while ignoring assaults on protesters and wildlife crimes. We now turn to the specific techniques and technologies of policing that are used, focusing on (a) Policing through silencing dissent and controlling the narrative, (b) Policing through criminalisation, and (c) Policing through physical coercion and (state) violence.

How Is This Being Policed?

a. **Silencing dissent and controlling the narrative**

Silencing dissent occurs through non-disclosure agreements (NDAs), the use of 'commercial sensitivity' arguments, pressure on landowners not to speak about their experiences with HS2 Ltd., and redacted planning documents, among others. Between 2012 and 2021, HS2 Ltd. has initiated over 300 NDAs with public and private bodies, including councils, universities, airports, fossil fuel and water companies, wildlife groups, the British Geological Society, individuals, and even the UK government's

[14] Incidents of the flouting of physical distancing regulations have been filmed by protesters in Denham.

Health and Safety Executive (New Civil Engineer, 2020). It took journalists from the New Civil Engineer trade magazine 18 months to win a Freedom of Information (FoI) battle to gain access to a list of the publicly listed companies and public bodies which signed NDAs after the intervention of the Information Commissioner's Office (ICO) which regulates the FoI Act and acts as arbiter in appeals. NDAs help circumvent the transparency that is required by the regular UK planning process which—at least in theory—is meant to protect the public interest (Raynsford, 2019). Their primary purpose is to preserve commercially sensitive details, to stop controversial information on ecological and archaeological impacts from being published, and thus to avoid negative media coverage and project development delays. Enforcing nesting seasons or the relocation of rare species such as great crested newts, campaigners explain, could delay developmental work for weeks and even months. Landowners who had to sell their land to HS2 Ltd. were also threatened with CPOs and pressured to sign NDAs (interviews). They reported blackmailing and threats that supporting the HS2 opposition or camps could lead to reduced or no compensation for their land on the route. Others describe being bullied or 'punished' with late compensation payments, making it impossible for people to buy new land or housing. In effect, NDAs, combined with pressure on individual landowners, silenced opposition and enabled a positive narrative in the mainstream media—of HS2 being 'green', economically beneficial, and a job creator. .

Since its inception, HS2 Ltd., the Department of Transport, and the government more widely, have pushed sustainability narratives and alleged benefits to Northern communities to justify the project (Elledge, 2020, Fig. 9.1). These narratives rest on claims that HS2 benefits the UK's 'low

Fig. 9.1 Benefits of HS2 (HS2 Ltd., nd)

carbon future' (HS2 Ltd, 2021a, 2021b) by getting freight off the roads, displacing diesel emissions, planting seven million trees, and creating new jobs, while ignoring large-scale habitat destruction, the die-back of HS2's tree plantations, drinking water pollution, stress to water levels,[15] and loss of jobs and livelihoods along the line. To create that narrative—to 'build back better' and 'build back green' in its newest (post-Covid) reiteration—HS2 Ltd. employed 17 PR companies and 12 consultancies, as was revealed in 2017 (The Times, 2017). This narrative is further reiterated through parliamentary statements formulated by the Department of Transport and delivered periodically by the Junior Minister in charge of HS2, as well as the heavily redacted development agreements between HS2 and the Secretary of State for Transport (i newspaper, 2017). Their framing of the project as 'bringing prosperity' to the country, but also the lack of public awareness of policing collaborations, violence, and costs, help marginalise criticisms of the project, allowing the mainstream media and HS2 supporters to portray protesters as a minority of militant left-wing agitators (Bigland, 2017). Most British media have adopted these narratives, positioning HS2 as a brand new, bright, shiny infrastructure project to reinvigorate the British economy—a feat of British engineering (Irwin, 2020).[16] HS2's PR strategy thus seems to have been based on two pillars: silencing critique and creating a 'green' image of the project in public/political discourse. Together, they target the hearts and minds of the population and feed the big media outlets.

b. **Policing through criminalisation**

I've never seen anything like it. We were out walking the dogs when seven police vehicles turned up. Two officers said they were just looking around. Then 30 officers descended on the area with the express intention of arresting people. This is the first time in my life that I've felt I'm in a police state. I was scared, intimidated. At one point it was suggested I couldn't leave

[15] The drinking water implications are one of the main planks of opposition: HS2 construction relies upon huge volumes of water to drill its boreholes (an estimated 6.5 m litres a day for 36 months for its boring machines) with aquifer contamination an ever-present danger. (Denham Against HS2: https://www.facebook.com/DenhamAgainstHS2/).

[16] The framing resembles the discourse around the first British motorway in 1959 (the M1) or the M25 in 1986.

the area, or I may be arrested. I was struggling to believe this was the UK (Grand Union Canal boat resident, interview).

Policing further occurs through (*a priori*) criminalisation and deterrence, unwarranted arrests, and the threat thereof. Here, we examine the use of corporate injunctions, policing through bail conditions, and the role of open-source intelligence gathering in mapping, identifying, and criminalising protesters (and their relationships). Once charged with an offence relating to a protest site, police or courts can impose bail conditions that require the defendant to stay away from that named site, or even all HS2 sites.[17]

Corporate injunctions—originally an instrument to stop stalking under the Protection from Harassment Act 1999 (the 'Stalkers Law'/PfHA)—are court orders to stop particular acts, such as resistance against the cutting of trees or the building of access roads. Breaches of injunctions may result in prosecution for contempt of court, and ultimately fines, asset seizing, and prison sentences. They thus act as deterrence mechanisms and operate on the basis of fear. HS2 Ltd. has successfully applied for injunctions against specific individuals as well as against 'persons unknown'—i.e. everyone—to pre-empt protest, including 'lawful' activity. The High Court started to grant interim injunctions as the company formally began the building project and while they can vary in timescale and numbers, they have become increasingly wide-ranging (Taylor and Barkham, 2020). On the Colne Valley building site, for instance, an injunction covered a much larger area than the previous injunction that year, naming 33 individual protesters. Protesters claim that injunctions not only serve to deter protest but also to bar them from monitoring illegal activities. In 2020, for instance, a nine-month injunction was meant to stop protesters from witnessing alleged illegal removal of nesting birds and badgers in a piece of ancient woodland in the West Midlands (interviews). Police have been instrumental in enforcing injunctions on behalf of the company (Taylor and Barkham, 2020), thus facilitating policing through *a priori* criminalisation.

Police further enforce the 'bailing away' of activists and work closely with security personnel at the various building site compounds to ensure there are reinforcements and back-up from local constabularies—including City of London Police or Thames Valley Police in the case

[17] Court conditions remain in place for as long as the court case takes to be heard unless they are successfully contested in preliminary hearings.

of Denham Ford and Hillingdon—when needed. Together, police and security deal with direct actions and blockades through arrests and physical violence. To date there have been approximately 300 arrests and nine prosecutions in relation to the HS2 protests (Guardian 2021). Charges were brought forward under section 12–14 of the Public Order Act (1986) as well as Section 241 of the Trade Union and Labour Relations Act which stops protesters from obstructing workers in their line of duty—normally used in relation to the intimidation of striking workers—among others. Most court cases were later dropped due to lack of evidence. However, the imposition of bail conditions, the emotional and logistical impacts on protesters, and the financial implications still constitute policing mechanisms to deter action. The arrests themselves are significant tools to gather individuals' data and map their networks. Such data complements the open-source intelligence (OSINT) gathered by private firms and is often shared between different policing bodies.

iii. ***Open-source intelligence—policing through data***

Project developers such as construction firms or mining companies often employ firms to collect and process OSINT, as occasionally revealed in court proceedings where such data is presented (e.g. Brock, 2020b). Through physical and online surveillance techniques, OSINT practitioners use a variety of software and programming techniques to build data maps of organisations, groups, and people within those groups, including activists or journalists. Data collection can entail infiltration of private Facebook groups, smart phone technologies, the dark web, and much more. HS2 subcontractors—private security firms who have operatives schooled in OSINT gathering techniques—harvest and analyse OSINT that 'enable[s] HS2 to mitigate risks to the security of HS2 personnel, premises and stakeholders' (Government online, 2017). Their work includes the surveillance of protesters and building legal cases based on the evidence gathered, including 'conspiracy to cause criminal damage', for instance. Embedded analysts and global risk intelligence analysts interpret the data and develop specific 'risk-mitigation strategies' and report on 'situational awareness'. One such contract was advertised to be worth over GBP 1,250,000 and involved not just 'collat[ing] security intelligence, upload intelligence data into an HS2 security system, analys[ing] intelligence data and produc[ing] reports' but also 'general (HS2 wide) and functional (cyber, physical, information, personal, personnel, etc.) intelligence and threat assessments', integrating data with reports from police, local authority, and HS2 internal data sources, and recommending mitigation actions (ibid.). This particular contract (Gov.uk, 2017) was awarded to G4S Risk Consulting Limited, as an FOI request revealed.[18]

Such data facilitates policing through criminalisation, often through legally questionable tactics. Campaigners at Denham Ford have questioned whether the arrest of one protester for alleged trespass as well as breach of bail conditions in August 2020 by undercover police officers was in fact legal, as it was based on surveillance carried out in dubious circumstances. In 2016 it was uncovered in the Sunday Express (Wheeler, 2016) that an internal HS2 Ltd. document stated that the company was looking to gather information on the sexual orientation, sex lives, mental health, criminal records, and political views of protesters, complainants, and litigants including those seeking compensation or objecting to the project as part of the company's Privacy Notice. The information could be volunteered freely but it could also be aggregated from 'doctors, the taxman, lawyers, the courts, security companies and credit agencies'; third parties including 'healthcare, social and welfare advisers or practitioners, HM Revenue and Customs, law enforcement and security agencies and bodies, and relatives, guardians or other persons associated with the individual' (Wheeler, 2016).[19] This illustrates the involvement of different parts of the (welfare) state in policing dissent. British anti-fracking policing also involved this type of operation, where Lancashire police admitted 'passing on details of disabled fracking protesters to the Department for Work and Pensions – who then questioned them about disability benefits claims' (Brock, 2020b: 9).

Police officers are integral to data collection too. Police liaison officers, often seen in and around protests, wear blue hi-viz tabards, and seek to politely chat to and befriend protesters while gathering intelligence on future actions and strategies, as FoI requests by the Network for Police Monitoring have revealed (Netpol, 2014): '[Police Liaison Teams] are

[18] https://www.whatdotheyknow.com/request/intelligence_and_security_organi#incoming-967321.

[19] Hours after the revelation was made plain to HS2 Ltd, the document was withdrawn with 'immediate effect'.

likely to generate high-quality intelligence from the discussions they are having with [protest] group members ... all PLT officers must ensure all intelligence is recorded on Crimint' [a criminal intelligence database] and all intelligence obtained during an event 'is passed to Bronze Intelligence for analysis and dissemination to Silver and the rest of the Command Team (in the same way as any other intelligence)' (Standard Operating Procedures, 2013). In addition, evidence gathering teams record and film activists, gathering data that is key to criminalisation.

At the same time, police rarely take action following reports of illegal behaviour by HS2 Ltd. or their subcontractors. Campaigners and journalists have documented countless instances of refusal to adhere to legal procedures and environmental laws, including tree felling during nesting season, for instance (see StopHS2 website, for instance), and police bodies usually refuse to prosecute such activities, interviewees report. The eviction of Jones Hill Woods in Buckinghamshire serves as an example here. Police did not intervene, neither to stop illegal tree felling, nor to halt the eviction of protesters without having served proper CPO papers to the landowner, nor to prevent violent eviction tactics on the part of bailiffs (interviews). Rather than protecting the safety of protesters and ecologies—including nesting birds, bats, badgers, and foxes—policing forces plainly defend HS2 interests against environmental defenders.

iv. **Policing through physical coercion**

Strategies of violence by police and private security services involve a range of techniques including harassment, physical and verbal abuse, and intimidation. These occur during everyday camp life, camp actions, and evictions, usually outside of public scrutiny, and encompass direct coercion, the use of restraint techniques including chokeholds, pressure point moves, and handcuffing, as well as cutting of (safety) ropes when protesters are on trees or walkways. Occasionally, incidents of violence—especially when recorded by campaigners—make it into the media, with accusations against security forces (Shadwell, 2020) and bailiffs (Courtney-Guy, 2020a) for severely injuring activists, including causing head injuries (Shadwell, 2020) and choking campaigners (Stop HS2, 2020). Yet, they rarely lead to prosecutions or convictions, and usually remain without consequences. Verbal and physical violence is often gendered and sexualised. Female interviewees speak of inappropriate touching during arrests and evictions, unnecessary strip-searches of female activists, sexualised comments shouted at (underage) girls, and sexist comments by private security. Often, such micro-aggressions and abusive behaviours appear to be employed strategically, to trigger aggression by male partners (a practice that is common at protest camps across the UK, see also Brock, 2020a).

Campaigners at Denham Ford, for instance, point to the lack of proper safety measures and concern for protesters' safety during evictions and protest policing, especially when removing protesters from treehouses or walkways.[20] In two separate incidents, police officers cut the ropes of activists who then fell from rope walkways across a small river. Lachlan 'Lazer' Sandford and Samantha 'Swan' Smithson fell 20ft with Sandford actually hitting the shallow riverbed below, his injuries requiring hospital treatment (Metro, 2020). They had spent over 12 hours trying to thwart teams belonging to Thames Valley police heights team, the London Metropolitan Police force, the City of London force, as well as the NET, from cutting down the tree to which their ropeway was attached. In the space of two days, a number of protesters and a police officer at this incident were injured, and seven people were arrested for alleged trespass and assault (Courtney-Guy, 2020b). While an HS2 spokesperson stated that '[t]hese protests are a threat to the security and safety of the public and our workers, and are costly to the taxpayer' (Metro, 2020), activists critiqued them as knowingly endangering the lives of protesters as well as committing ecological harm. Swan recounts:

> I have bruises and carpet burns from the rope swinging around my arm. Nothing major, more trauma, mental trauma to be honest with you. I'm very angry with the police. HS2, I'd expect it from, and that day they continued to chainsaw the tree down even though we were connected to it, so they had no care for our health and well-being, our safety… But then this is usual… And aside from my personal safety there was also the safety of the other people on the lines but also under the trees – we had peaceful protesters stood below and we had bailiffs standing in the tree and they were standing on their [protestors'] fingers and trying to stop them peacefully protesting under the tree. Pushing them around in the water… There were three different police forces plus the ambulance which

[20] Rope walkways are ropes slung between trees and used as walkways.

> we called because they didn't. The Thames Valley climb team and the Met Police and the City of London police and the Gold, Silver and Bronze commanders from Thames Valley..., so they treated it like an operation and their main thing was preventing any of the protesters obstructing HS2. When we questioned that they said their main concern was our safety. If that was the case, why did they cut my line. They insisted...I should get off the line...which I refused to, because I was peacefully protesting my rights.

She tells us how police officers willingly put her in danger:

> My line [was cut] without speaking to me without advising me what to do. [It was] ... loosely tied on to another line and they waited for me to fall and [I] ended up dangling a metre above the water. And we [had] warned them that this 10mm line had been used before, so its strength was questionable and essentially, they trusted that line with my life. They had no control of it when that line went slack – they couldn't lower me. They also didn't know that the rope would be short enough to stop me hitting the ground or the water. It was purely a guesstimate and a very lucky guess at that... And I was disappointed because that was the police, I'd expect that from HS2 or NET because on all the other scenarios all the time I've been on sites they've been acting despicably. Obviously, it's wrong on every level... and we're trying to expose that but I did not expect that from the police and it was very clear to me from that moment in time that the police were not only there facilitating HS2 in the development but were totally complicit in the process, assisting and basically taking HS2's side and did not care for the public health and safety. If they did then they would have stopped the chainsawing [of the tree Swan was defending] by HS2 but in reality, after I came down and got arrested the chainsaw continued on the tree.

Other incidents of violence resulted in broken fingers through police officers stamping on them, broken noses through security services, activists losing consciousness after being choke-held high up in tree canopies (interviews), or inability to move after a bailiff stamped on an activist's head (Morning Star, 2021). In October 2020, NET bailiffs were under police investigation for allegedly assaulting protesters while off duty (Courtney-Guy, 2020a). Four NET operatives were suspended after assaulting three activists in a hotel/pub car park in Kenilworth, Warwickshire, where the activists were trying to investigate how many security operatives would be attending the camp's eviction planned for the next day, by counting NET vehicles in the carpark. One activist was hospitalised with a broken jaw (Courtney-Guy, 2020b). They report getting assaulted when trying to leave in their car:

> Eventually someone opened the car, punched me in the face at least three times and broke my jaw… I'm still in shock. I've seen the NET hurt my friends before, but I never expected to get attacked outside of a protest. I went to the hotel out of concern for the people I'd met at Rugby Road Protection Camp. I wanted to give them a heads up about how many [bailiffs] would come in the morning – exactly because of this kind of violence (Courtney-Guy, 2020a).

In a subsequent incident, activists supporting the disruption of a crane being transported for HS2 work near Aylesbury, Buckinghamshire, were arrested and 'pinned down' by police as their vehicle stalled (Courtney-Guy, 2020b). Video footage further shows the violent arrest of an activist, with three officers pinning him down, lying face down on the ground. A police officer kneeling on the protester's leg then hits him three times. 'They dismissed the punching as "distraction strikes" although I could see they were not looking at it' (Courtney-Guy, 2020b). Charges against the protester were dropped, but the case against the police officers was not prosecuted. This incident was preceded by the filming of another HS2 bailiff allegedly kneeling on the neck of an activist for up to four minutes, after the same activist had thrown carrots onto a construction site. In October, William Harewood, a black person in a group in Wendover, near Aylesbury, had his neck and back allegedly knelt on by a police officer after being forced to the ground (Griffin, 2020). Thames Valley Police refuted that the officer's knee was placed on his neck despite video evidence. These instances of coercion are part and parcel to policing along the HS2 line to deter and intimidate resistance and enforce the project.

Other strategies of violence against protesters are subtler. Camp residents criticise the physical and mental torture resulting from lights, generators, and radios kept on all night to inhibit sleeping, interfering not only with protesters' sleep patterns, but with the lives of nocturnal creatures and the natural habitat in general. Intimidation of activists and landowners takes place every day (interviews). Residents who provide water and other amenities to the camps, interviewees explain, have been pressured into no longer providing such access. At times, access to water from nearby standpipes by landowners sympathetic to the cause is simply blocked. Any means that can be employed to make life difficult for the camp will be put in place with the aim to close the camp down quickly and effectively, according to interviewees.

Discussion

The policing of anti-HS2 protest is integral to the 'success' of the project, deterring and repressing resistance, avoiding negative media coverage through NDAs, and securing a positive 'green' narrative, however weak and contested. It lays bare, once again, that policing is not about protecting public safety, but about protecting industrial interests, selectively enforcing some, and ignoring other pieces of legislation, particularly those which seek to defend the local ecology and wildlife within it, such as the Wildlife and Countryside Act 1981. Wildlife laws are badly implemented and often ignored, not only by HS2 contractors but more generally in the UK; legislation is 'inadequate to the task of wildlife protection, subject to an equally inconsistent enforcement regime... that fails to address the specific nature of wildlife offending' (Nurse, 2013, 4).

Large extractive/infrastructure projects are intimately tied to state power and thus particularly violently enforced (Meehan, 2014). Their financial and political profitability to the political elites and their networks (e.g. banks and corporations) mean that they must take precedence. Infrastructures are not only integral to state building and state legitimacy, but also key for pacification and domestication of subjects—whether we look at early water infrastructures and their importance to the very first state building exercises and emergence of social hierarchy (Gelderloos, 2017), or the importance of 'lower-carbon' energy infrastructures to the legitimation of 'green' capitalism and pacification of ecological resistance.

The analysis shows not only how 'green' extractive projects require similar levels of policing to enforce them, but also how ecologically and socially destructive these projects can be—serving to create a green image rather than meaningful ecological and political change. This is facilitated by their framing as projects of 'national significance' that override all other interests, as well as their positioning as innovative examples of industrial excellence and the apogee of competitivity. This further justifies enormous costs to the regular population, especially working classes, who—like in the case of HS2 or similar projects—will not actually be able to make use of them for financial reasons. On top of the harsh policing and criminalisation, stopping such projects is made nearly impossible regardless of how many laws and regulations they breach. Subsuming of protesters' rights and the ecological wellbeing to the rights of developers is thus not about the *abuse* of police/security powers. Protecting extractive and infrastructure projects—upholding private property rights and industrial interests—is the *aim* of policing . Repression and ecological harm are inherent to policing as a system of social control and an illustration of the collusion between the police and private company concerns.

The UK's road building programme of the 1990s and plans to expand London's Heathrow Airport illustrate this point conclusively. These major infrastructure projects once again prioritised extractive growth over ecological or social justice, at the expense of environmental degradation. They were/are often decided in secrecy (Goodey, 2012). The facilitating of legal protest by comparison is paid lip service and, as is presented in the case study above, protesters' rights are trampled upon, literally and figuratively. Internally, resistance is framed as a threat and security risk that needs to be carefully managed while minimising reputational impacts (see 4.2). Yet, policing language—uncritically adopted by mainstream media and tabloids—celebrates ideas of 'balancing' rights and interests, as this quote by a Thames Valley Police spokesperson illustrates: 'At all times, suitably trained officers worked in partnership to ensure appropriate risk assessments were carried out, while striving to balance the rights of those protesting, against those of HS2 to conduct their lawful business' (Courtney-Guy, 2020c). The police seek to remove themselves as perpetrators of violence, drawing on 'rotten apple' narratives[21] to explain particularly violent incidents, and instead frame themselves and HS2 security staff as victims of violence from protesters (Burnett, 2021).

The narrative perpetuated by the government is that any form of resistance is counter to the economic fortunes of the nation state, drawing on patriotic tropes: including of HS2 be(com)ing the 'backbone of our rail network'; 'help[ing] level-up the country'; and 'putting Britain on track to a net zero carbon future' (HS2 Ltd., nd). Policing thus becomes important to protect the country's national economic interest (especially important in times of Brexit) and to defend 'British values' such as the right to work in the teeth of 'eco-mob violence' which harks back to the UK miners' strikes of the 1980s.

[21] In the 'bad apple narrative' corruption and problematic behaviour by police officers—including misogynist or racist violence—is framed as the exception, rather than a systemic issue.

In effect, policing HS2 involves winning hearts and minds of the population, while employing a range of (counterinsurgency) strategies to repress dissent. On-the-ground policing illustrates how policing protects and militantly enforces regional ecological harm that accumulates with all the other industrial acts of ecological degradation. As the project has gone forward, the confidence of the those who are employed to provide 'security' and surveillance for HS2 has grown, as with the full backing of the police and state apparatus they see themselves as invincible, courtesy of this additional authority. A febrile, aggressive atmosphere often pervades the policing of sites, with no accountability for violence committed by police and security officers (interviews). At the same time, framing the project as environmentally beneficial hides its destructive and carbon-intensive (Webster, 2020) nature.

Conclusion

As people across the world were making great sacrifices to stop the spread of the Covid-19 virus and protect themselves and others in early 2020, HS2 Ltd. used the distraction generated by the pandemic to enforce evictions and violently remove protesters and journalists[22] to make way for its high-speed rail project—with little to no media scrutiny. With the full weight of the government behind it—despite critical voices questioning the economic costs and benefits—it has moved forward, creating facts on the ground. Policing has been integral to these efforts.

Unlike it is often assumed, the police (and private security services) are not upholding the right to protest and defending public safety. They enforce the development-at-all-cost political philosophy of capitalism that is driving biodiversity collapse and environmental catastrophe. Police forces work alongside HS2's private security operatives to avoid delays to the project and subsequent financial penalties.

The policing and shaping of the narrative around HS2 as environmentally and economically beneficial are just as important, however. These narratives require the silencing of dissent through NDAs and bullying and intimidation of landowners, but also investment into PR work and advertising that paint the project as 'green' and job creating. Drawing on a rich and successful history of direct action in the UK—such as major

[22] See for instance The Canary (2021).

road protests of Twyford Down, Newbury, the M11 and Solsbury Hill, anti-GMO actions, and anti-fracking resistance more recently—will this campaign have the same impact? Those earlier protests effectively stopped two-thirds of a roads programme in the 1990s; the establishment of GMO agriculture (despite recent signs that the fight against it is not over); and the development of a fracking industry.

Could anti-HS2 campaigning, in addition to the public backlash to inefficiency and lack of cost effectiveness, be as effective? HS2 supporters with pro-HS2 transport and travel blogs believe that activists with 'limited legal knowledge and back-up' will be crushed as a minor irritant, having underestimated the sheer size of the HS2 project. They state that this is not merely some local bypass campaigners are trying to stop after all, pouring scorn on even their arboreal knowledge (Bigland, 2021). According to some industry sources, however, the answer to this question could be affirmative (Construction News, 2021b). The later phase of the project, the eastern leg to Leeds, is now officially shelved (Daily Mail, 2021). Public opinion is shifting, and even former HS2 directors are now critiquing the scheme and pushing for alternatives. Anti-HS2 protest constitutes the number one 'core risk' to the HS2 schedule and delivery, according to HS2 Ltd. (Horgan, 2021). This is why repressive policing to create facts on the ground is crucial to the continuation of a project that is of poor value financially, and an ecological catastrophe, environmentally and socially. Yet the question remains as to whether it will ever be fully realised, while it still yields profits to developers, construction firms, and hundreds of wealthy individuals and subcontractors all the way down the line.

Bibliography

Acosta, A. (2013). Extractivism and neoextractivism. In *Beyond development*, ed. M. Lang and D. Mokrani. Rosa Luxemburg Foundation. https://www.rosalux.de/publikation/id/8051/beyond-development/.

Align, (2021). Analysis on Bentonite loss during Chalfont Saint Peter D-wall. Excavation Document no: 1MC05-ALJ-GT-NOT-CS02_CL04–000001. https://www.hs2facts.org/wp-content/uploads/2021/09/201202_Analysis_on_Bentonite_Loss_during_Chalfont_Saint_Peter_D_wall_Excavation_C03-02_12_2020_redacted.pdf.

Bagchi, A. K. (1982). *The political economy of underdevelopment*. Cambridge: Cambridge University Press.

BBC, (2020). HS2: Six reasons why the rail route is so expensive. *BBC News*. https://www.bbc.com/news/business-51415590.

Berkeley, T. (2020). A review of high speed 2: Dissenting report by Lord Tony Berkeley. https://theengineer.markallengroup.com/production/2020/01/Lord-Berkeley-HS2-Review-FINAL.pdf.

Best, S. and A.J. Nocella, (2006). *Igniting a revolution*. Oakland: AK Press.

Bigland, P. (2017). A busy week for HS2. Travel blog. https://paulbigland.blog/2017/07/20/a-busy-week-for-hs2/.

Bigland, P. (2021). Crazy anti-HS2 campaigner of the week. No 29. https://paulbigland.blog/2021/02/28/crazy-anti-hs2-campaigner-of-the-week-no-29/.

Bingley, L. (2020). Public support for HS2 is falling; will protests rise? *Construction News*. https://www.constructionnews.co.uk/hs2/public-support-for-hs2-is-falling-will-protests-rise-08-09 2020/.

Bouzarovski, S., Bradshaw, M., & Wochnik, A. (2015). Making territory through infrastructure: The governance of natural gas transit in Europe. *Geoforum, 64*, 217–228.

British Army, (2009). Field manual volume 1 Part 10 countering insurgency (Vol. 1).

Brock, A. (2020a). Securing accumulation by restoration—Exploring spectacular corporate conservation, coal mining and biodiversity compensation in the German Rhineland. *Environment and Planning E: Nature and Space*. https://doi.org/10.1177/2514848620924597.

Brock, A. (2020b). 'Frack off': Towards an anarchist political ecology critique of corporate and state responses to anti-fracking resistance in the UK. *Political Geography* 82: 102246.

Brock, A. and A. Dunlap, (2018). Normalising corporate counterinsurgency: Engineering consent, managing resistance and greening destruction around the Hambach coal mine and beyond. *Political Geography* 62: 33–47.

Brock, A. and N. Stephens Griffin, (2021). Policing environmental injustice. IDS Bulletin. https://doi.org/10.19088/1968-2021.1.

Brock, A, A. Huff, J. Verweijen, J. Selby, D. Ockwell, and P Newell, (2018). Fracking democracy, criminalising dissent. The Ecologist. https://theecologist.org/2018/oct/18/fracking-democracy-criminalising-dissent.

Brock, A, C. Stephenson, N. Stephens-Griffin, and T Wyatt, (2022). Go home, get a job, and pay some taxes to teplace a bit of what you've wasted: Stigma power and solidarity in response to anti-open-cast mining activism in the coalfields of rural county Durham, UK. *Sociological Research Online*. https://doi.org/10.1177/13607804211055486.

Brown, A. (2018). From North Dakota to Puerto Rico, controversial security firm profits from oil protests and climate disasters. The Intercept. https://theintercept.com/2018/03/12/tigerswan-dapl-private-security-climate-disaster-response/.

Brown, A. (2021). Oil company official overseeing crackdown on pipeline resistance cut teeth at Amazon and Exxon. The Intercept. https://theintercept.com/2021/09/17/enbridge-line-3-pipeline-amazon-security-exxon/.

Brown, A., W. Parrish and A. Speri, (2017a). The battle of treaty camp. The Intercept. https://theintercept.com/2017a/10/27/law-enforcement-descended-on-standing-rock-a-year-ago-and-changed-the-dapl-fight-forever/.

Brown, A., W. Parrish and A. Speri, (2017b). Leaked documents reveal counterterrorism tactics used at standing rock to “defeat pipeline insurgencies”. The Intercept. https://theintercept.com/2017b/05/27/leaked-documents-reveal-security-firms-counterterrorism-tactics-at-standing-rock-to-defeat-pipeline-insurgencies/.

Brown, G. (2013). HS2 makes Birmingham Airport ‘viable for expansion’. Business Live. https://www.business-live.co.uk/economic-development/hs2-makes-birmingham-airport-viable-6415045.

Burnett, T. (2021). HS2 staff attacked near Wendover construction site with nine left injured—Thames Valley Police say. Buckinghamshire Live. https://www.buckinghamshirelive.com/news/buckinghamshire-news/hs2-staff-attacked-near-wendover-5240350.

Business Live, (2013). HS2 makes Birmingham Airport ‘viable for expansion’. https://www.businesslive.co.uk/economic-development/hs2-makes-birmingham-airport-viable-6415045.

Carver, L. and S. Sullivan, (2017). How economic contexts shape calculations of yield in biodiversity offsetting. *Conservation Biology* 31(5): 1053–1065.

Challenor Gardiner, (2021). Bat conservation—High court ruling averts costly delay to HS2 rail project. https://www.challenor-gardiner.co.uk/bat-conservation-high-court-ruling-averts-costly-delay-to-hs2-rail-project/.

Chandra, B. (1991). Colonial India: British versus Indian views of development. Review (Fernand Braudel Center).

Clark, T. (2020). Revealed: The 253 companies and public bodies to sign HS2 ‘gagging orders’. New Civil Engineer. https://www.newcivilengineer.com/latest/revealed-the-253-companies-and-public-bodies-to-sign-hs2-gagging-orders-16-11-2020/

Clifton, K. (2021). Secret network of tunnels built near Euston station by anti-HS2 protesters. https://metro.co.uk/2021/01/27/hs2-protesters-dig-100ft-tunnel-network-under-euston-station-park-13970805/.

Construction News, (2019). HS2: National police group established to secure project. https://www.constructionnews.co.uk/civils/hs2-national-police-group-established-secure-project-23-04-2019/.

Construction News, (2020). HS2 boss’s bonus boosts salary to £660,000. https://www.constructionnews.co.uk/hs2/hs2-bosss-bonus-boosts-salary-to-660000-22-07-2020/.

Construction News, (2021a). HS2 eastern leg cancelled. https://www.constructionnews.co.uk/hs2/hs2-eastern-leg-cancelled-18-11-2021a/.

Construction News, (2021b). Major projects body brands HS2 phase 2b 'unachievable'. https://www.constructionnews.co.uk/hs2/major-projects-body-brands-hs2-phase-2b-unachievable-19-07-2021b/.

Courtney-Guy, S. (2020a). HS2 bailiffs investigated for 'breaking protester's jaw' while off duty. Metro. https://metro.co.uk/2020a/10/09/hs2-bailiffs-investigated-for-breaking-protesters-jaw-while-off-duty-13395738/.

Courtney-Guy, S. (2020b). Police officer accused of repeatedly punching restrained HS2 protester. Metro. https://metro.co.uk/2020b/11/24/police-officer-accused-of-repeatedly-punching-restrained-hs2-protester-13646460/amp/?ito=cbshare&__twitter_impression=true&s=09.

Courtney-Guy, S. (2020c). Eco-activists injured in 20-foot drop from trees when contractors cut their ropes. Metro. https://metro.co.uk/2020c/07/27/eco-activists-injured-20-foot-drop-trees-when-contractors-cut-ropes-13037213/.

Cruz, M. and D. Luke, (2020). Methodology and academic extractivism: the neo-colonialism of the British university. *Third World Thematics: A TWQ Journal* 5(1–2): 154–170.

Daily Mail, (2021). HS2's eastern link hits the buffers as the line to Leeds is shelved amid soaring costs—Despite Boris Johnson's 'levelling up' pledge. https://www.dailymail.co.uk/news/article-9845605/HS2s-eastern-link-hits-buffers-line-Leeds-shelved-amid-soaring-costs.html.

Dalton, J. (2020). HS2 workers accused of destroying birds' nests and failing to maintain social distancing during coronavirus pandemic. https://www.independent.co.uk/climate-change/news/hs2-bird-nest-trees-woodland-coronavirus-social distance-a9484141.html.

Das, P. (2011). Colonialism and the environment in India: Railways and deforestation in 19th century Punjab. *Journal of Asian and African Studies* 46(1): 38–53.

Dept Transport, (2018). Development agreement between secretary of state for transport and HS2 Ltd. https://assets.publishing.service.gov.uk/.../916509/hs2-development-agreement-document.pdf.

Duff, K. (ed) (2021). *Abolishing the police*. London: Dog Section Press.

Dunlap, A. (2018). Counterinsurgency for wind energy: The Bíi Hioxo wind park in Juchitán, Mexico. *The Journal of Peasant Studies* 45(3): 630–652.

Dunlap, A. (2019). 'Agro sí, mina NO!' the Tía Maria copper mine, state terrorism and social war by every means in the Tambo Valley, Peru. *Political Geography* 88: 10–25.

Dunlap, A. (2020). Bureaucratic land grabbing for infrastructural colonization: Renewable energy, L'Amassada and resistance in Southern France. *Human Geography* 13(2): 109–126.

Dunlap, A. (2021). Spreading 'green' infrastructural harm: Mapping conflicts and socioecological disruptions within the European Union's transnational energy grid. *Globalizations*: 1–25. https://doi.org/10.1080/14747731.14742021.11996518.

Dunlap, A. and A. Brock, (2021). When the wolf guards the sheep: Green extractivism in Germany and Mexico. In: Springer, S., Locret, M., Mateer, J., et al. (eds) *Anarchist political ecology* Volume 3. London: Rowman & Littlefield, 91–123.

Dunlap, A. and J. Jakobsen, (2019). *The violent technologies of extraction.* London: Palgrave.

Elledge, J. (2020). Why everyone—And Especially Greens—Should still support HS2. New Statesman. https://www.newstatesman.com/politics/environment/2020/10/why-everyone-and-especially-greens-should-still-support-hs2.

Evans, R. and P. Lewis, (2013). *Undercover.* London: Faber and Faber.

FM3-24, (2014). *Insurgencies and countering insurgencies.* http://fas.org/irp/doddir/army/fm3-24.pdf.

Geenen, S. and J. Verweijen, (2017). Explaining fragmented and fluid mobilization in gold mining concessions in eastern Democratic Republic of the Congo. *The Extractive Industries and Society*, 4(4): 758–765.

Gelderloos, P. (2017). Worshipping power: An anarchist view of early state formation.

Gilmore, J., W. Jackson and H. Monk, (2017). 'That is not facilitating peaceful protest. That is dismantling the protest': Anti-fracking protesters' experiences of dialogue policing and mass arrest. *Policing and Society* 29(1): 36–51.

Goodey, J. (2012). Road building programme gets green light by stealth. The Ecologist. https://theecologist.org/2012/oct/05/road-building-programme-gets-green-light-stealth.

Government online, (2017). Provide security intelligence services to HS2. https://www.government-online.net/provide-security-intelligence-services-to-hs2/.

Gov.uk, (2017). HS2 Ltd Information Fusion (previously advertised as Intelligence Fusion) service—Birmingham. https://www.contractsfinder.service.gov.uk/Notice/32f00b16-1933-41fe-8bf6-997ccbc5a94d?p=@NT08=UFQxUlRRPT0=NjJ.

Griffin, J. (2020). Police accused of unprovoked assault on HS2 protester. The Guardian. https://www.theguardian.com/uk-news/2020/oct/10/chilling-police-tactics-used-on-hs2-protesters-claims-report.

Gudynas, E. (2009). Diez tesis urgentes sobre el nuevo extractivismo. In J. Schuldt, et al. (Eds.), *Extractivismo, politica y sociedad* (pp. 187–225). CLAES, Quito: CAAP.

Habib, I. (1975). Colonialization of the Indian economy, 1757–1900. *Social Scientis* 3(8): 23–53.

Hannis, M. and S. Sullivan, (2012). Offsetting nature? Habitat banking and biodiversity offsets in the English land use planning system. Technical Report. Green House, Weymouth, UK.

HCE group, (2021). Services: Traveller eviction. https://hcegroup.co.uk/.

Hope, M. and R. Collett-White, (2018). Investigation: Facebook surveillance of fracking protesters. *Desmog*. https://www.desmog.co.uk/2018/06/15/investigation-fracking-company-facebook-surveillance-environmentalprotesters-key-information-and-documents#Injunctions.

Horgan, R. (2021). Five 'core risks' to HS2 schedule and budget. New Civil Engineer. https://www.newcivilengineer.com/latest/five-core-risks-to-hs2-schedule-and-budget-21 07 2021/.

HS2 Ltd., (2020). HS2 green corridor. https://www.hs2.org.uk/building-hs2/hs2-green-corridor/.

HS2 Ltd., (2021a). Carbon—Putting Britain on track to a net zero carbon future. https://www.hs2.org.uk/why/carbon/.

HS2 Ltd., (2021b). 'We're taking an extra 24,000 lorries off Buckinghamshire roads…' https://twitter.com/HS2ltd/status/1425056856599371787?s=19.

HS2 Ltd., (nd). Why HS2? https://www.hs2.org.uk/why/.

Huff, A. and Y. Orengo, (2020). Resource warfare, pacification and the spectacle of 'green' development: Logics of violence in engineering extraction in southern Madagascar. *Political Geography* 81: 102195.

i Newspaper (Weekend) (2017). HS2 Cover Up, 29–30/07/2017.

Institute for Government, (2020). *High Speed 2 costs*. At https://www.instituteforgovernment.org.uk/explainers/high-speed-2-costs.

Irwin, D. (2020). The major HS2 projects taking shape across the region in the coming months. Birmingham Mail. https://www.birminghammail.co.uk/news/midlands-news/majorInstitute for Government, 2020. High Speed 2 costs. At https://www.instituteforgovernment.org.uk/explainers/high-speed-2-costs-hs2-projects-taking-shape-18750091.

Jackson, W., J. Gilmore and H. Monk, (2019). Policing unacceptable protest in England and Wales: A case study of the policing of anti-fracking protests. *Critical Social Policy* 39(1): 23–43.

Javers, E. (2011). *Oil executive: Military-style "psy ops" experience applied*. https://www.cnbc.com/id/45208498

Jespersgaard Jakobsen, L. (2020). Corporate security technologies: Managing life and death along a Colombian coal railway. *Political Geography* 88: 102273.

Kaur, R. (2021). Nuclear necropower: The engineering of death conditions around a nuclear power plant in south India. *Political Geography* 88: 102315.

Lasslett, K. (2014). *State crime on the margins of empire*. London: Pluto Press.

Layton, J. (2021). Police spent more than £140,000 guarding HS2 protesters' tunnels in Euston. Metro. https://metro.co.uk/2021/03/11/police-spent-more-than-140000-on-hs2-protest-in-london-euston-14227116/?ico=related-posts.

Leifsen, E. (2020). The socionature that neo-extractivism can see: Practicing redistribution and compensation around large-scale mining in the Southern Ecuadorian Amazon. *Political Geography* 88: 102249.

Leonardi, E. (2013). Foucault in the Susa Valley: Tthe NO TAV movement and struggles for subjectification. *Capitalism Nature Socialism* 24(2): 27–40.

Lesutis, G. (2021). Infrastructural territorialisations: Mega-infrastructures and the (re) making of Kenya. *Political Geography* 90: 102459.

Lubbers, E. (2002). *Battling big business: Countering greenwash, infiltration, and other forms of corporate bullying*. Green Books.

Lubbers, E. (2012). *Secret manoeuvres in the dark: Corporate and policy spying on activists*. London: Pluto.

Marshall, J. (2021). Minister discloses HS2 staff are picking up an average of £60k a year. *Building*. https://www.building.co.uk/news/minister-discloses-hs2-staff-are-picking-up-an-average-of-60k-a-year/5112648.article.

Martin, D. (2020). Boss of delayed and over-budget HS2 project rakes in colossal salary of £625,000... four times higher than Boris Johnson's. *Daily Mail*. https://www.dailymail.co.uk/news/article-7922959/Boss-delayed-budget-HS2-project-rakes-colossal-salary-625-000.html.

Mauvaise Troupe Collective, (2018). *The Zad and No Tav*. Translated/edited/preface by K. Ross. London/New York: Verso Books.

Maxey, L. (2020). Policies on policing HS2, FOIA request. https://www.whatdotheyknow.com/request/policies_on_policing_hs2#incoming-1606878.

McCarthy, J. (2002). First world political ecology: Lessons from the wise use movement. *Environment and Planning* A 34: 1281–1302.

McLoughlin, P. (2015). HS2 and the northern powerhouse, department for transport. https://www.gov.uk/government/speeches/hs2-and-the-northern-powerhouse.

Meehan, K. M. (2014). Tool-power: Water infrastructure as wellsprings of state power. *Geoforum* 57: 15–224.

Melia, S. (2021). Railing against HS2. *The Ecologist*. https://theecologist.org/2021/feb/02/railing-against-hs2.

Menton, M. and P. Le Billon, eds. (2021). *Environmental defenders*. Taylor & Francis Ltd.

Metro, (2020). Video: Eco-activists fall 20 feet and suffer injuries after rope cut. Metro. https://metro.co.uk/video/hs2-protester-falls-20-feet-breaks-collarbone-rope-cut-2217206/.

Metz, D. (2020). HS2: true value to cities in the north of England has not been made clear. *The Conversation*. https://theconversation.com/hs2-true-value-to-cities-in-the-north-of-england-has-not-been-made-clear-130918.

Monaghan, J. and K. Walby, (2017). Surveillance of environmental movements in Canada: Critical infrastructure protection and the petro-security apparatus. *Issues in Criminal, Social, and Restorative Justice* 20(1): 51–70.

Morning Star, (2021). Anti-HS2 protesters 'violently assaulted' during eviction. *Morning Star*. https://morningstaronline.co.uk/article/b/anti-hs2-protesters-violently-assaulted-during-eviction.

Mould, T., QC (2017). Introduction to hybrid bills. https://www.landmarkchambers.co.uk/resources/bank/introduction-to-hybrid-bills/.

National Trust, (2016). National trust response to: HS2 phase two: West Midlands to Crewe EIA scope and methodology report—Draft for consultation (March 2016). https://ntplanning.wordpress.com/2016b/05/13/national-trust-response-to-hs2-phase-2a-consultation-still-more-to-do/.

NEF, (2019a). A rail network for everyone: Probing HS2 and its alternatives. New Economics Foundation. https://neweconomics.org/uploads/files/A_Rail_Network_for_Everyone_WEB.pdf.

NEF, (2019b). HS2 will serve wealthier passengers and deliver more benefits to London than the North. New Economics Foundation. https://neweconomics.org/2019b/03/hs2-will-serve-wealthier-passengers-and-deliver-more-benefits-to-london-than-the-north.

NET, (2021). National evictiont team. https://www.nationalevictionteam.co.uk/.

Netpol, (2014). Police liaison officers https://netpol.org/police-liaison-officers/.

New Civil Engineer, (2020). Revealed: The 253 companies and public bodies to sign HS2 'gagging orders'. https://www.newcivilengineer.com/latest/revealed-the-253-companies-and-public-bodies-to-sign-hs2-gaggingorders-16-11-2020/.

Nurse, A. (2013). Species justice: The future protection of wildlife and the reform of wildlife laws. *The Green Criminology Monthly* 6: 1–11.

Oakervee, D. (2020) Oakervee review. https://www.gov.uk/government/publications/oakervee-review-of-hs2.

Parliamentary and Health Service Ombudsman, (2015). Report on an investigation into complaints about High Speed Two Limited. HC 620. Presented to Parliament pursuant to Section 10(4) of the Parliamentary Commissioner Act 1967.

Parliamentary and Health Service Ombudsman, (2021). An investigation into HS2's failure to communicate with a family about acquiring their home. HC 211. Presented to Parliament pursuant to Section 10(4) of the Parliamentary Commissioner Act 1967.

Parsons, R. (2019). HS2 may not arrive in Yorkshire until 2040 and will be £25bn over budget, government admits. *Yorkshire Post*. https://www.yorkshirepost.co.uk/news/transport/hs2-may-not-arrive-yorkshire-until-2040-and-willbe-ps25bn-over-budget-government-admits-493730.

Pickard, S. (2019). Excessive force, coercive policing and criminalisation of dissent. Repressing young people's protest in twenty-first century Britain. *Revista Internacional de Sociologia* 77(4): e139.

Preston, J. (2017). Racial extractivism and white settler colonialism: An examination of the Canadian Tar Sands mega-projects. *Cultural Studies*, *31*(2–3), 353–375.

Ramchurn, R. (2013). Scrap HS2, and solve the North-South divide while you're at it. *The Independent*. https://www.independent.co.uk/voices/comment/scrap-hs2-and-solve-north-south-divide-while-you-re-it-8851459.html.

Raphael, S. (2016). *Mercenaries unleashed. The brave new world of private military and security companies*. London: War on Want.

Raynsford, N. (2019). Raynsford review into planning. https://www.gov.uk/government/organisations/ministry-of-housing-communities-and-local-government.

Ross, K. (2018). Preface: Making a territory. In: *The Zad and NoTAV: Territorial Struggles and the Making of a New Political Intelligence*, Mauvaise Troupe Collective and K. Ross. Verso Books.

Schlembach, R. (2018). Undercover policing and the spectre of 'domestic extremism': The covert surveillance of environmental activism in Britain. *Social Movement Studies* 17 (5): 491–506.

Schroeder, R.A., K. St Martin and K.E. Albert (2006). Political ecology in North America: Discovering the third world within? *Geoforum* 37: 163–168.

Shadwell, T. (2020). HS2 protester filmed with blood dripping from head 'after being assaulted by worker'. Mirror. https://www.mirror.co.uk/news/uk-news/hs2-protester-filmed-blood-dripping-21971339.

Sky News, (2018). One in four HS2 workers on six-figure salaries. https://news.sky.com/story/quarter-of-hs2-workers-on-pay-deals-over-100k-11465148.

Standard Operating Procedures, (2013). https://www.whatdotheyknow.com/request/170321/response/422052/attach/19/BLOWE%20SOP.doc.pdf.

Stephens Griffin, N. (2020). 'Everyone was questioning everything': Understanding the derailing impact of undercover policing on the lives of UK environmentalists. *Social Movement Studies* 20(4): 459–477.

Stephenson, A. (2021). HS2 6-monthly report to parliament: March 2021. https://www.gov.uk/government/speeches/hs2-6-monthly-report-to-parliament-march-2021.

Stop HS2, (2015). HS2–reduce flights or increase flights? http://stophs2.org/news/14390-hs2-reduce-flights-increase-flights.

Stop HS2, (2020). Video on Twitter. https://twitter.com/stophs2/status/1254361506579447808?s=20.

Stop HS2, (2021). Case against HS2. http://stophs2.org/.

Sullivan, S. and D. Brockington, (2004). *Qualitative methods in globalisation studies*. Centre for the Study of Globalisation and Regionalisation.

Taylor, D. and P. Barkham, (2020). Rise in injunctions against HS2 protesters. *The Guardian*. https://www.theguardian.com/uk-news/2020/jun/13/rise-in-injunctions-against-hs2-protesters.

Temper, L., F. Demaria, A. Scheidel, D. Del Bene and J. Martinez-Alier, (2018). The Global Environmental Justice Atlas (EJAtlas): Ecological distribution conflicts as forces for sustainability. *Sustainability Science* 13: 573–584.

Tenders electronic daily, (2017). United Kingdom-Birmingham: Security services 2017/S 014–022145. https://ted.europa.eu/udl?uri=TED:NOTICE:22145-2017:TEXT:EN:HTML.

The Canary, (2021). UK police just arrested two journalists in one week. https://www.thecanary.co/exclusive/2021/02/01/uk-police-just-arrested-two-journalists-in-one-week/?fbclid=IwAR31OwoM4R0eNi3_jLdKSM9OQeC3JdVdNwpxPyZ4vd90z-NupmmJ_qSXbTA.

The Times, (2017). HS2 hires 17 firms to make its case. https://www.thetimes.co.uk/article/hs2-hires-17-companies-to-make-its-case-3gh6bv2rd.

The Times, (2020). HS2 will never be carbon neutral in 120-year lifespan. https://www.thetimes.co.uk/article/hs2-willnever-be-carbon-neutral-in-120-year-lifespan-wk3s29pkl.

The Independent, (2020). HS2 workers accused of destroying birds' nests and failing to maintain social distancing during coronavirus pandemic. https://www.independent.co.uk/climate-change/news/hs2-bird-nest-trees-woodlandcoronavirus-social-distance-a9484141.html.

Tominey, C. (2020). HS2 spends a FORTUNE buying up £1m homes—Then Boris's father makes £2.5m profit from it.*Daily Express*. https://www.express.co.uk/news/uk/843344/HS2-Stanley-Johnson-Boris-Johnson-father-government-scheme-homes/amp.

Topham, G. (2021). HS2 asks government to help it deal with rising number of protests. *The Guardian*. https://www.theguardian.com/uk-news/2021/jun/24/hs2-asks-government-help-rising-number-protests.

Townsend, M. (2019). Death on the door: How security industry risk has grown as police numbers fall. *The Guardian*. https://www.theguardian.com/uk-news/2019/jan/05/private-security-industry-risk-grows-as-police-numbers-fall.

Verweijen, J. and A. Dunlap, (2021). The evolving techniques of the social engineering of extraction: Introducing political (re)actions 'from above' in large-scale mining and energy projects. *Political Geography* 88: 102342.

Webster, B. (2020). HS2 will never be carbon neutral in 120-year lifespan. *The Times*. https://www.thetimes.co.uk/article/hs2-will-never-be-carbon-neutral-in-120-year-lifespan-wk3s29pkl.
Wheeler, C. (2016). Rail bosses wanted to spy on sex lives of people who opposed controversial route. *Express*. https://www.express.co.uk/news/uk/734205/HS2-rail-chiefs-controversial-route-spy-sex-lives-mental-health.
Wiegink, N. (2020). Learning lessons and curbing criticism: Legitimizing involuntary resettlement and extractive projects in Mozambique. *Political Geography* 88: 102192.
Wildlife Trusts, (2020). What's the damage? Why HS2 will cost nature too much. https://www.wildlifetrusts.org/news/hs2-exorbitant-cost-nature.
Williams, K. (2007 [2004]). *Our enemies in blue: Police and power in America*. Cambridge: South End Press.
Wolverhampton Express and Star, (2017). HS2: Claims high-speed rail link could cost £200bn. https://www.expressandstar.com/news/politics/2017/02/20/hs2-claims-high-speed-rail-link-could-cost-200m/.
Woodland Trust, (2021). HS2 rail link. https://www.woodlandtrust.org.uk/protecting-trees-and-woods/campaign-with-us/hs2-rail-link/.

CHAPTER 10

Ecological Terror and Pacification: Counterinsurgency for the Climate Crisis

Peter Gelderloos

Introduction

In April of 2009, representatives of the 28 member states of the North Atlantic Treaty Organization (NATO), gathered in Strasbourg and Kehl, neighboring cities divided by the French–German border. Climate change was a central point on the agenda of the largest military alliance on the planet. This was a time when politicians in most NATO member states were at best inactive and at worst actively denying the reality of the problem. In the lead-up to the conference, top generals and policy experts were circulating proposals for greater border security as well as introducing biometric IDs and more surveillance for their domestic

This chapter is a modified version of Chapter 2 of Peter Gelderloos (2022), *The Solutions Are Already Here: Strategies of Ecological Revolution from Below*, London: Pluto Press.

P. Gelderloos (✉)
Oslo, Norway

A. Dunlap and A. Brock (eds.), *Enforcing Ecocide*,
https://doi.org/10.1007/978-3-030-99646-8_10

populations, reemphasizing their goal of normalizing the domestic use of military forces in urban environments by 2020, and increasing cooperation with non-governmental organizations (RTO/NATO 2003; Monroy 2008). These proposals by top NATO planners systematically linked climate change with "refugee problems" (Naumann et al. 2007, p. 34). An earlier strategy paper said plainly that "Security challenges are predominantly socio-economic, not military-technical, in character" (Sens 2006). The paper goes on to make clear that "socio-economic" threats include those posed by poor people, "illegal migration and shortcomings in the social integration of immigrant communities [that] can create a racially and religiously defined underclass," and "climate change and environmental degradation [that] will precipitate an increase in environmentally-induced conflict." As such, the paper argues, NATO "must incorporate political and economic programmes into military planning and operations" and "establish consultative groups combining international staff, government officials, academia, industry, intelligence, NGOs, and expatriates." This translates into greater border militarization, privacy intrusions, policing, and confinement (Monroy 2008; Dunlap and Fairhead 2014). Said differently, authoritarianism, coercion, and control is the proposed mitigation pathway to climate change.

Anyone who is shocked by this response, or sees it as some kind of *non sequitur*, has failed to appreciate the true nature or scope of the ecological crisis. World governments, particularly those in the Global North, understand the crisis as a security issue and they have since NATO began studying the problem in the 1960s. They know the problem is real, and they know growing deserts and rising sea levels will force hundreds of millions of people from their homes in search of their very survival. Their solution is to further militarize the borders—those borders of Fortress Europe and the American southwest that are most definitely "designed to kill"—so that people fleeing for their lives will be dissuaded by the very real possibility of dying in their journey (CrimethInc 2017).

To put it simply, the major militaries of the world, already among the greatest producers of toxic waste and greenhouse gases, propose killing even more people to shield themselves from the consequences of the crisis they are in large part responsible for (Smith 2017; Belcher et al. 2020). As we shall see, they are also oriented toward increasing repression against their own citizens, most likely aware that the lower classes everywhere will bear the brunt of the crisis, and rebellions are to be expected in the North as well as in the South.

While the rich and powerful rubbed shoulders in Strasbourg, thousands of anarchists and other anticapitalists took to the streets, battling with police and burning down banks as they tried to disrupt the summit.[1] Though the media condescendingly portrayed them as a mindless horde bent on evil deeds, they were perfectly aware of what NATO was proposing and the stakes for all the rest of us. Accordingly, their response might be characterized as the most reasonable and intelligent, at the very least if we compare it to that of the scientists and NGOs who continue to dialogue with the same governments sponsoring the summit, producing report after policy paper with no appreciable change of course.

This chapter offers an overview of repressive campaigns and different repressive techniques used against ecological resistance movements, in both the so-called Global North and South. It thus serves three primary purposes. First, I demonstrate the extent of resistance to defend the territory and social fabrics from ecological degradation, and how systematically repression is carried out to maintain and enforce ecological degradation. Whereas democratic logics portray political violence in the Global South as endemic failings of democracy, thus reinforcing an ultimately white supremacist and colonial North–South divide, I show that repression to maintain the structures responsible for the ecological crisis is globally integrated, and that a counterinsurgency lens shows that the same repressive strategies are used in the North and the South, with differentiation of techniques based on differing circumstances. Finally, I point to the role that academics and NGOs considered to be a part of the climate movements can and sometimes do play in these counterinsurgency campaigns.

[1] A perfect example of where the university system sits in regards to the faultline of counterinsurgency, academic citation guidelines do not take the reality of state repression into account. At best, anonymous interviews require a clear separation between a legal, legitimate, and consequence-free researcher, and an anonymous interviewee. Such a format does not account for repression using conspiracy or illegal organization laws, that can be used against participants in illegal social movements. For all these reasons, I can attest that the information in this paragraph is accurate, but for reasons of self-incrimination I cannot state whether I am the source or whether the source is someone I have communicated with.

Repression as a Guarantor of Ecological Degradation

Around the world, people are defending the land against infrastructural and extractive projects that result in ecological degradation. These movements of resistance, which can be more or less formal, evoke a consciousness of the interdependence between human communities and a specific territory, as well as a consciousness of the planet as a globally interconnected living system that we are a part of and need to take care of. Also around the world, governments—as well as corporations carrying out private acts of governance with the permission and complicity of the sovereign state power in a given territory—systematically respond to such movements with a full gamut of repressive techniques. These include misrepresentation and demonization; techniques of bribe, divide, and rule; the weaponization of poverty and forced economic dependence on degrading productive activities; extensive legal harassment and lengthy imprisonment; and the widespread application of bodily harm including the use of lethal force, which is to say, murder (see Dunlap 2020). Different techniques are chosen in accordance with the democratic/colonial divide between Global North and Global South. State violence in the Global North is portrayed as simply the application of law, of the free market, or at worst as accidental excesses necessitating inconclusive internal reviews, and the objectively greater violence in the Global South is portrayed as an inherent defect of unstable governments that are putatively struggling with imported concepts of democracy and civilization, an insinuation that is constantly made in media portrayals of violence in such countries, in line with the crypto-racist narrative of development (Mignolo 2005; Moraña et al. 2014; Patel 2020). In actual fact, many of the actors behind repressive violence in the Global South are governments and corporations of the Global North, and state and private actors in zones of opacity in the Global North also make easy recourse to supposedly anti-democratic repressive techniques like murder.

Ecological Resistance & The Politics of Killing in the Global South

There can be no doubt that repressive violence used to maintain and facilitate ecologically destructive practices in the Global South results in an elevated body count, given the frequent recourse to paramilitary and military force. Yet it is also true that movements in the Global South often accomplish much more in terms of winning land autonomy. The oft-cited figure about biodiversity being disproportionately located on Indigenous lands should provide a major clue about the importance of autonomy for ecological resistance. An autonomous territory, by definition, is one in which the forces of state repression and corporate extractivism do not have a decisive foothold; they must organize a process of reinvading such territory. Given the scaled response that is at the center of counterinsurgency methodology (Galula 1964; Kilcullen 2010), a counterinsurgency lens explains the relation of repressive force and the ecological crisis much better than implicitly racist arguments about the health of democracy in the Global South. And as we shall see in the next section, indubitably democratic governments in the Global North seem most likely to use lethal force against their citizens precisely in the context of those struggles that are in the process of winning land autonomy.

Repression on the island of Borneo provides a clear example of the connections between autonomy, counterinsurgency, and globally integrated economic practices responsible for the ecological crisis, in this case the global palm oil, logging, and paper industries that cause massive deforestation and pollution. One activist from Borneo who works on agrarian issues and Indigenous land struggles, particularly with Dayak communities that have halted several palm plantations and maintain a high degree of land autonomy, describes to me how the state neutralizes resistance. "In the worst cases, the state murders its opponents." The state, they continued, also "criminalizes and arrests Dayak activists or cultivators [...] to cover up the actions of the palm oil companies[...]. Sometimes, arrested activists die in police custody." The activist described one case of a dissident journalist who was tortured and mutilated after writing critically of the business oligarchy that controls politics on the island. The journalist subsequently died in questionable circumstances. "I could name other cases of repression, but it's a bit psychologically exhausting" (Gelderloos 2022, p. 68).

China has been the site of a growing number of "mass incidents" or riots, many of them by rural people against development projects, pollution, unsafe working conditions, and corruption between Party bosses and the new industrialists. A staggering 65% of the 180,000 annual "mass incidents" are rural conflicts triggered by land grabs, with government officials and private developers forcibly requisitioning village lands, sometimes without even offering compensation (Economy 2012; Forsythe 2011).

Communication between these different sites of resistance has become increasingly difficult over the last ten years as the Chinese government consolidates its control over the internet. Signs of the underlying tension still boil over despite the censorship, as when a dispossessed villager killed four with a homemade bomb at a government office near Guangzhou this past March, amidst a conflict with farmers being dispossessed with little or no compensation for a major development project (Guardian 2021). Meanwhile, the Chinese government has been documented using ecological degradation to bring nomadic territories further under control, foreclosing the possibilities of autonomy by contaminating the land as a conscious byproduct of rare earth minerals mining, and covering it with wind turbines (Klinger 2017).

The changing balance of power also means the overturning of earlier environmental victories, such as the defeat of a planned hydroelectric dam on the Yangtze River at Tiger Leaping Gorge. The dam, which would have displaced 100,000 people, was halted by locals in 2006, but now the government is resurrecting plans for the dam, reflecting their new control over the situation, as well as the opportunities that "green energy" and "carbon neutral" policies afford governments for neutralizing environmental movements (Liu 2013; Standaert 2020). In another example of the same dynamic, three conservationists in the Ningxia Hui Autonomous Region were arrested and accused of extortion, making threats and "provoking trouble" in September 2020 after filing whistleblower complaints regarding pollution from a paper mill and destruction of protected gazelle habitat for the construction of wind farms (Li 2020).

Repressive violence makes no distinction for renewable energy projects or "green" industries. During the very week that I was researching campaigns of repression in Latin America, two Indigenous land defenders were killed in Honduras. José Adan Medina and Félix Vazquez, of the Tolupan and Lenca peoples, respectively, were assassinated for opposing hydroelectric dam construction and other land abuses (Aljazeera 2020).

According to the NGO Global Witness (Global Witness 2020), between 2002 and 2018, over 1,700 land defenders—people trying to stop the destruction of the environment or human rights abuses related to resource extraction—were murdered in 50 countries. Some 40% of the dead were Indigenous and the killers, the vast majority of whom committed their deeds with impunity, were police , military, and paramilitaries working at the direction of the state or of multinational companies, many of them based in the North (Butt and Menton 2019). A record number of land defenders were murdered in 2019, with 212 deaths recorded and many more unrecorded (Guy 2020). Colombia, the Philippines, and Brazil topped the list for murders (with the caveat that extractivist violence in many African countries is underreported, in part for reasons we will look at shortly).

Violence in Brazil and Colombia underscores how globally integrated counterinsurgency strategies are, and not the result of endemic democracy problems. In those countries, paramilitary death squads created with the support of the United States attack communities that stand in the way of the large economic interests that destroy the rainforest to establish mines, logging operations, and plantations (Cuellar 2005; Grajales 2013). In Colombia, corporations from the US like Dole Food Company and Chiquita Brands International paid paramilitaries several million dollars to protect their interests. The companies would pass the names of labor organizers, community activists, and suspected leftists to the paramilitaries, who would then execute them (Smith 2009). Between 1997 and 2004, paramilitaries committed 4,335 homicides and 1,306 forced disappearances in communities near Chiquita plantations (Alsema 2018). And while it was on the Chiquita payroll, the AUC (*Autodefensas Unidas de Colombia*) paramilitary group forcibly evicted 60,000 people from their homes in the banana growing region of just one state. Even after the group was declared a terrorist organization by the US State Department (a move that came only after decades of public pressure), Chiquita was sending them money and smuggling in thousands of assault rifles for the AUC using the company's international distribution facilities (Cohen 2015). They were punished with a small fine after a 2007 legal action, in which they were protected from serious consequences by attorney Eric Holder, whom President Obama would soon appoint as Attorney General.

Across the Atlantic in 1995, Ogoni writer and activist Ken Saro-Wiwa was executed by the Nigerian state, together with eight other activists who had been protesting the destruction of Ogoni lands in the Niger Delta by the oil industry, particularly Royal Dutch Shell. As a nonviolent activist, Saro-Wiwa received substantial attention from human rights and environmental NGOs in the Global North, and his prison memoirs were printed by a major publisher. In December 1998, a youth conference of the neighboring Ijaw people committed to a direct action campaign against the oil industry, also expressing a commitment to nonviolence. The Nigerian military invaded the region with over ten thousand troops and opened fire on a protest march with machine guns, killing at least three. They again opened fire on a march demanding the release of those arrested, killing at least three more, and then invaded communities, terrorizing the people and raping women and girls. In January 1999, a hundred soldiers directly attached to a Chevron facility attacked two Ijaw villages, killing and disappearing dozens (Online Nigeria 2005). Despite this extreme level of violence, the Ijaw did not give up, and one might say they *could* not give up, as the oil industry makes their lives impossible. Frequent leaks and explosions poison fields, destroy forests, rivers, and coastal waters, and kill workers and neighboring villagers. Since Dutch, British, and US companies began oil extraction in 1956, they have leaked one and a half million tons of oil into the land, or a current rate of a quarter of a million barrels every year (Kadafa 2012). Increasingly, local Ijaw and Ogoni people cannot feed themselves from the fields, forests, and fisheries on which they had traditionally depended. This is a perfectly acceptable state of affairs within the sort of development framework favored by the World Bank and similar institutions: people *should not* feed themselves from the land. Subsistence agriculture is dismissed as a primitive activity. Rather, they should get jobs and then pay for food from industrialized and globalized producers (relying, of course, on oil to power their machinery, manufacture their chemicals, and transport their product). However, this is also not an option for the peoples of the Niger Delta, as the profits of oil extraction go to foreign companies and the central government.

International media attention quickly waned. Eventually, the Ijaw took up arms to defend their lives and formed groups like the Movement for the Emancipation of the Niger Delta (MEND). MEND sabotaged oil facilities and kidnapped highly paid North American and European oil workers. Their actions and those of similar militant groups led to major reductions in Nigerian oil production, for example shutting down 10% of production with a single attack in 2008 (BBC 2008) and leading to widespread reductions over the better part of a decade. The Nigerian military responded with bloody repression, including aerial bombardments of Ijaw communities. The US military also stepped in to help the government better defend its oil platforms in the name of fighting piracy (Maupin 2016), helping oil production in the Niger Delta to grow again. This was complemented by corporate-led, "soft" counterinsurgency operations integrated within the international framework of emissions reductions and putatively sustainable extractivism (Dunlap and Fairhead 2014, 952).

The state-sanctioned murder of activists using what the West deems to be legitimate (nonviolent) tactics to defend their lands and their livelihoods frequently drives communities to take up arms just to survive. Subsequently, they are often become more able to attack and stop the companies responsible for poisoning them. However, when they are then massacred by their governments with weapons from Europe and North America, human rights NGOs no longer count their deaths—or those of the community members supporting them and giving them shelter—in their lists of land defenders and environmental activists killed each year. So, the above-cited figure of 1,700 murdered land defenders is woefully inadequate. The figure does indicate quite clearly, however, what type of resistance movements will receive resources from the Global North, conditioning those movements in what are legitimate and what are illegitimate forms of resistance.

For all the hundreds of millions of dollars in charitable donations that go to saving the Amazon or stopping hunger in Africa, little or no money goes to the organizations that take up tactics that are actually effective in halting the forms of structural harm they face. This means that such organizations have to finance themselves within the economies they have at hand, leading to less than emancipatory results, whether in Colombia or with the case of the MEND.

Social War in the Global North

Though the intensity of resistance tends to be much more tame, brutality against land defenders is also a problem in the Global North. As the European Union promotes itself as a model for the green capitalist future, it too profits off gory acts of repression. In the last few years, seven Romanian park rangers have been murdered and over 650 have been assaulted or threatened for going up against logging interests connected to an Austrian company, with the line between legal and illegal logging "increasingly blurred" (Larsson 2020). And environmental activists have been threatened or have had to go into hiding due to attempts on their lives after they investigated logging, hydroelectric dams, and other industries. Incidentally, these hydroelectric dams and these forests-turned-logging-plantations are the ostensible green energy sources, carbon sinks, and nature preserves that lie at the heart of the European Union's model for profitable environmental action.

At anticapitalist and ecological protests across North America and Europe, police routinely attack protesters with less lethal weaponry, regardless of whether they are being peaceful or combative, occasionally causing deaths and more frequently leaving people permanently injured. Hundreds of people have had eyeballs shot out or suffered brain damage from high impact munitions. At the protests against the Free Trade Area of the Americas in Miami in 2003—a major neoliberal initiative that would have accelerated ecocidal practices across both continents and which was defeated thanks in part to fierce opposition from the streets—police raped detainees. Sexual assault by police during arrest, especially against women, nonbinary, and trans people, is common (see note 1). In 2016, water protectors resisting new oil and gas pipelines crossing Dakota territory were met with an extremely militarized police force and private security personnel with counterinsurgency experience in Iraq. They brutalized hundreds and came close to killing several protestors, causing critical injuries, while imprisoning many of the survivors (Estes 2019).

The government classifies people opposing pollution, deforestation, and global warming as "extremists" and treats them accordingly. In 2002, shortly after the September 11th attacks and in the midst of the new War on Terror, the US government declared radical environmentalists to be the number one domestic terrorism priority, at a time when lethal white supremacist attacks were already on the rise and groups like Earth First!, the Earth Liberation Front (ELF) and the Animal Liberation Front (ALF) were not even accused of having committed bodily harm (Potter 2011).

In what has become known as the Green Scare, the US government brought its full repressive weight against these movements (Best and Nocella 2006). To name just a few examples out of dozens arrested and entire movements subjected to police surveillance and the threat of imprisonment, Jeff "Free" Luers was given a 22 year prison sentence for setting fire to three SUVs. Eric McDavid was sentenced to 20 years in prison for conspiracy to bomb a dam as part of a plot that was funded, organized, and pushed ahead by a paid FBI informant who had become romantically involved with McDavid as part of her infiltration activities. In 2006, six members of Stop Huntingdon Animal Cruelty, a group that coordinates aggressive boycott actions against Huntingdon Life Sciences, one of the largest animal testing companies in the world, were sentenced to three to six years in prison for sending large numbers of faxes to HLS or "internet stalking," convicted under the Animal Enterprise Terrorism Act, a law that allows the government to grant special punitive protections to companies that abuse and exploit animals (Green Is the New Red 2021). Meanwhile, the UK government stepped in to help the company secure financing when they were on the verge of bankruptcy due to the boycott campaign (Harrison 2001). William "Avalon" Rogers committed suicide in jail while being threatened with literally hundreds of years of imprisonment. In 2009, Marius Mason was sentenced to 20 years in prison for an arson attack against a university lab involved in genetic engineering research funded by Monsanto, arson of logging trucks, and attacks against the construction of luxury homes. Like other Green Scare prisoners before him, Mason is being held in a Communications Management Unit, an opaque corner of the federal prison system that uses extreme isolation and a prohibition on communication with the outside world as a form of psychological torture.

The FBI and Department of Homeland Security currently list "racial and environmentally themed ideologies" as primary motivators for domestic terrorist attacks. As for those "racial" ideologies, the FBI has focused a great deal of attention on so called "Black Identity Extremists," its way of discrediting those who are outraged by racist police murders. The property damage of antiracists and environmentalists merited such attention while white supremacists were carrying out hundreds of attacks, many of them lethal, including full-scale massacres. Activists who carried out civil disobedience against oil pipelines—simply turning off valves—were placed on the same "extremism" lists as those responsible for mass shootings in Black churches (Federman 2020). At the time of writing, two land defenders in the US were recently in court on terrorism charges for simply shutting off the valves on oil pipelines and then turning themselves in; one was given 8 years in prison. They are not the only people to face terrorism charges for acts of resistance against new pipelines (Johnson 2020).

As Andrea Brock (2020a) has documented, similar politics apply in Europe. Those who set up encampments to defend forests are treated as enemies of society and met with militarized repression and marginalization by interlocking consortia of public and private security, elected officials, and corporate board members. Farmers and anticapitalists occupying the land and opposing the construction of an airport at Notre-Dame-des-Landes, France, for instance, were repeatedly attacked by a militarized police force, with one young person having their hand blown off by a flash bang grenade (Mauvaise Troupe Collective 2016).[2] German police defending the expansion of an expanding coal mine that is destroying the Hambach Forest regularly brutalize protesters (Brock and Dunlap 2018). Numbering in the thousands and backed by military grade weaponry, they have cut protesters down from six meter high tree sits, shot them with high pressure water cannons in below freezing temperatures, struck people with vehicles, attacked them with dogs, and kicked their teeth out (Brock 2018).

Italian police shot anarchist Carlo Giuliani in the head, killing him, during heavy protests against the G8 in Genoa in 2001, where they also raided a protest convergence point, brutally assaulting and torturing dozens of people trapped there (CNN 2001; BBC 2015). At the G8 protest in Evian, France, in 2003, police cut down an activist suspended from a bridge, causing him to fall twenty meters and almost killing him (Millar and Langley 2003). In 2014, French police killed an environmental activist during heavy repression against a campaign to stop deforestation associated with a new dam (Penketh 2014). The campaign was eventually successful, and the dam was canceled.

Campaigns of ecological resistance regularly have to deal with legal cases, police raids, and imprisonment, with occupations against coal mining, highways, or energy infrastructures being evicted with violent police action from the UK and Germany to France, Italy, and Spain.

To preserve the illusion of democracy and the beliefs of privileged citizenry that the institutions of power are on their side, police forces in the Global North use a higher proportion of "soft" counterinsurgency techniques (Dunlap 2020; Nomad 2016), but these come with their own forms of violence. Many participants in the George Floyd Rebellion in 2020 associated counterinsurgency with discourses of reformism and nonviolence that facilitated the isolation of participants in the uprising, exposing them to the violence of the police, National Guard, and prison system, all of which could be naturalized if they were seen as being used against a criminal minority and not against the movement itself (Rodríguez 2020; Marshall 2020). While the more radical participants of

[2] For documentation of a similar experience of effective direct action and land occupation against extractivist infrastructure and repressive police violence, see A. Dunlap (2020), Bureaucratic Land Grabbing for Infrastructural Colonization: Renewable Energy, L'Amassada and Resistance in Southern France. *Human Geography* 13: 109–126.

the movement were being neutralized by hundreds of court cases, lethal violence by police and non-uniformed white supremacists, and smear campaigns or conspiracy theories promoted by the media, nominally antiracist NGOs were hard at work shifting the widespread popular rejection of policing in its entirety to an electoral demand for "defunding" the police. In many cities, this institutionalization of the uprising was further co-opted and distorted into a refunding that got police departments more money for sensitivity trainings or somehow improved crowd control weaponry. This interplay illustrates how soft and hard counterinsurgency methods go hand in glove (Brock and Dunlap 2018).

Effective soft counterinsurgency requires massive intelligence gathering. Spying on politically active people in the northern democracies, even to an extent that violates the applicable laws, is systematic (see also Goodey and Brock, this edited volume). This includes the long term infiltration of anticapitalist and ecological direct action movements (Brock 2020b). In the UK, around 150 police agents went on deep infiltration assignments against anticapitalist, environmental, and other leftwing movements since 1968, living under assumed identities and participating in those movements for several or even a dozen years (Schlembach 2018). Agents routinely began romantic and sexual relationships with female activists they were informing on, using the forms of manipulation they had been trained in. Several of these undercover cops impregnated activists, committed to forming a family with them, and when their stint was up, disappeared (Evans and Lewis 2013; Woodman 2018; Lubbers 2012).

Though the details that have become public are horrible, we should assume the truth is far worse: police were caught destroying a large quantity of documents they had been ordered to preserve concerning the infiltration program since the 90s, when it expanded and shifted largely to environmental and anarchist groups (Lewis and Evans 2020). Since the scandal broke in the news, the UK has introduced the Covert Human Intelligence Source Act (2021) to officially allow domestic police to carry out illegal acts while undercover, with support from both major parties.

Corporate involvement in such repression is another salient feature. It was recently revealed that agrochemical giant Monsanto runs an "intelligence fusion center" to compile information on and conduct disinformation and harassment campaigns against dissident journalists, academics, and activists who threaten the company's financial interests through their research or organizing. "Fusion center" is the same term the FBI uses for its counterterrorism centers. In just one example, Monsanto targeted a Reuters journalist investigating the carcinogenic effects of the company's star product, glyphosate, or Roundup. Their campaign included coordinating "third parties" to post negative reviews of the book, hiring scientists to cast doubt on the book's conclusions, pressuring the journalist's editors at Reuters "very strongly every chance we get" in the hope "she gets reassigned," covering up their financial relationship with scientists claiming their product was safe, accusing the journalist of being a "pro-organic capitalist" activist, as though there were big bucks to be made in opposing some of the world's largest chemical companies, and contracting search engine optimization (SEO) experts to make sure that their alternative facts, their negative reviews, and their various slanders of said journalist would appear in search engines above results showing how Roundup causes cancer (Levin 2019). In another case, in 2020, an academic publisher abruptly canceled the publication of a book that showed how Canadian mining companies benefited from the genocide in Guatemala, moving in to stake their claims sometimes even before the death squads had left. The publishers expressed fears of lawsuits for defamation, though they refused to point out what part of the book, which received favorable peer reviews, might be considered defamation (Nolin and Russell 2020).

The new media environment is also particularly hostile for people trying to defend the land. Facebook has provided an important platform for the growth of the extreme Right, while they regularly ban anarchist organizations not connected to harm against any living person. When Facebook finally agreed to tamp down on misinformation related to climate change denial, they actually gave the boot to Indigenous and environmentalist pages that were fighting climate change (Milman 2020; Indybay 2020). For its part, Hollywood churns out a steady stream of movies and TV series reinforcing the trope of "eco-terrorists." Even progressive mainstream media like the *Guardian* lend their support to policing operations against environmental movements by portraying anti-capitalists who have always been at the heart of the ecological resistance as *outside agitators* (Willsher 2019; Gelderloos 2013).

Direct police interventions also make the internet an unequal terrain for movements fighting for liberation. Articles I used in my research on this very topic became unavailable after Dutch police raided and seized computer servers for several anticapitalist news sites (Enough 14 2021). For us, this is a common occurrence, almost completely erased from the mainstream, where right-wing pundits and transphobic authors, many of them millionaires with enormous platforms, control the debate around free speech and complain of being "canceled."

Making radical movements disappear through selective media coverage or wearing them down through contrition is vital to establishing the preeminence of movements that are reconcilable with the dominant power structures and existing forms of exploitation. These include the corporate climate movement that is centered around media events like the COP conferences, and of citizen movements more broadly, whether the issue is support for model immigrants within the overall border and deportation apparatus, or quixotic attempts to reform the police. Such movements respond positively to conditioning by media, police, and charitable foundations, they mimic the tactics and objectives of corporate-financed NGOs, they center governmental action and campaigns catering to the media and to the notion of middle class sensibilities the media produce, and they avoid intersectional critiques and practices. This constitutes a necessary part of the process that invisibilizes climate refugees, Indigenous liberation movements, anticapitalists, community organizers struggling for the commons, and others who have a realistic view of the ecological crisis beyond the narrow confines of climate- and emissions-centered discourse.

Borders Internal and External: Repression as Ongoing Colonialism

One of the most important sites of human resistance and State violence related to the unfolding ecological crisis is the border regime. Environmental factors are a growing reason for people to migrate in search of a better life. The border regime is a form of continuous repression designed to prop up the (neo)colonial appropriation of global resources by a handful of wealthy countries (Walia 2021). The extractive, accumulative processes behind this ongoing appropriation are a chief cause of the ecological crisis, and one's chances for surviving that crisis are closely connected to one's citizenship status.

The State deals with migrants with the same apparatus it brings to bear against social movements. Intelligence gathering, surveillance, demonization and scapegoating in the media, imprisonment, torture, targeted killings, and not least of which, depoliticization. Just as those fighting for clean water, a healthy relationship with the land, or a livable planet for future generations are deemed extremists, people who go through the trauma of leaving their homes behind are denied all reason and presented as criminals. NATO planners and member states routinely present migrants as a security threat (see for example, Naumann et al. 2007; RTO/NATO 2003; Sens 2006; Heise 2021).

The economies of the US, Canada, and the European Union depend on immigrant labor. As Harsha Walia (2021) documents, the purpose of the border regime is not to stop immigration but to control and terrorize it. When Germany decided in 2015 to take in over a million refugees from the Syrian civil war, it was only because the largest association of German employers had just declared that the national economy faced a shortfall of millions of skilled laborers.[3] But at no point did the German government allow direct flights from Turkey or Lebanon, where so many of the refugees were warehoused. Instead, they obligated refugees to make the expensive and perilous journey over the Aegean Sea, through the Balkans, under and over razor wire fences, through police truncheons and tear gas, past violent, xenophobic crowds, so that finally they would arrive with almost nothing, willing to accept any labor conditions and bureaucratic controls. It was a journey that cost on average several thousand euros, on top of the steep psychological price, effectively ensuring that primarily only members of the university-educated middle class would be able to make it.

Migrants from Africa are made to cross the open sea on flimsy rafts, left in the water or even deliberately rammed by the coast guards of the various Mediterranean states: between 2014 and early 2020, over 20,000 have died in the crossing (Migrant Project 2020). Once across, hundreds of thousands of migrants are either held in detention centers—sometimes for years—or left to live in crowded shanty towns, often without access to running water, heating, toilets, or medical facilities and at the mercy of arson attacks by fascists often connected to the police and operating with impunity, from Spain to Greece (Karakoulaki 2019; Council of Europe 2020).

In the US, the border regime has been designed to force migrants to cross in the most dangerous regions of the Sonoran Desert. Between 1998 and 2019, Border Patrol recovered the bodies of 7,800 migrants who died while crossing (Giartellia 2019). Many more have died on the

[3] The *New York Times* has a paywall, so I cannot access the article at the time of preparing this chapter. https://www.nytimes.com/2018/05/09/opinion/germans-secret-labor-experiment.html.

Mexican side of the border, or crossing through Guatemala and southern Mexico, where the Mexican government has increased enforcement in line with US demands. Border control, one of the branches of police with the strongest far Right sympathies, often brutalize and occasionally kill those who are crossing, as do right-wing militias. They also destroy caches of water left in the desert by volunteers trying to decrease the number of deaths (CrimethInc 2017). In detention, people are held in cells or cages with temperatures near freezing. People are left to die of medical neglect, or denied attention in the case of pregnancies, leading to frequent miscarriages; hundreds and possibly thousands have been involuntarily sterilized, leading some prisoners to compare the centers to "an experimental concentration camp" (Shen 2020; Manian 2020). Torture is systematic and "inherent" in the US immigrant detention system, according to a new report, and many are brutalized and forced to sign their own deportation papers, even though it means going back to a country where their lives are at risk (Kladzyk 2021; Borger 2020).

Australia is another terrible case of institutionalized racism that reveal borders to be an active measure of ongoing colonialism. The country legalized the buying and selling of enslaved Aboriginal laborers into the 1970s, and had an immigration policy explicitly designed to only allow white immigrants, also lasting until the 1970s. In 2018, the government was forced to pass a new law outlawing slavery again after finally acknowledging the tens of thousands of Pacific Islanders working in the agricultural sector, getting paid only $10 a week after deductions by their employers.[4] Determined to keep out "unskilled" immigrants from non-white countries, the Australian government obliges its poorer neighbors like Papua New Guinea to imprison asylum seekers indefinitely in facilities that have been compared to concentration camps (Charles 2016).

[4] Thalia Anthony and Stephen Gray ("Was there slavery in Australia? Yes, it shouldn't even be up for debate," *The Conversation*. 12 June 2020. https://theconversation.com/was-there-slavery-in-australia-yes-it-shouldnt-even-be-up-for-debate-140544) describe the process but incorrectly identify the enslavement of Aborigines as ending in the 1950s. For continuity through the '70s see Sarah *Collard*, "Class action launched against West Australian Government over Indigenous stolen wages," *ABC News*. 18 October 2020. https://www.abc.net.au/news/2020-10-19/wa-government-faces-class-action-over-stolen-wages/12737046; Norman Hermant, "Seasonal farm workers receiving less than $10 a week after deductions, investigation reveals," *ABC News*. 26 February 2016. https://www.abc.net.au/news/2016-02-25/seasonal-farm-workers-receiving-as-little-as-$9-a-week/7196844.

Parallel to state violence, humanitarian aid constitutes one of the mechanisms by which migrants are surveilled, controlled, and maintained in abject precarity and poverty [**source?]. In contrast, direct action and solidarity has helped migrants win housing and gain resources on their own terms, pass clandestinely through strict border regimes, and survive the crossing. In Europe, people are increasingly being legally required to report on undocumented people and prohibited from giving them food or water; squats and autonomous spaces created for and by migrants are targeted for eviction; autonomous transportation infrastructure that provides mobility to undocumented people is criminalized and shut down; and those aiding migrants have been charged with trafficking, which carries a sentence of up to ten years imprisonment. Ironically, European governments and their private partners, who cynically paint migrants and border activists as traffickers, have funneled millions of euros to the Libyan militias that actually control both the imprisonment and the trafficking of refugees at a key point on the southern Mediterranean coast (Dadusc and Mudu 2020, pp. 25–26, 30). In the US, people can be imprisoned for years for giving migrants a map or giving them a ride to the hospital, and Border Patrol has increasingly been raiding a first aid camp set up by volunteer group No More Deaths. Migrants who speak out about abuses frequently get targeted with speedy deportations.

In July 2019, as the Trump administration was gearing up to carry out major raids targeting undocumented people across the country, anarchist Willem Van Spronsen set fire to an ICE vehicle at the Northwest Detention Center in Tacoma, Washington. His action and other widespread acts of resistance succeeded in getting ICE to dramatically scale back the raids, but it came with a heavy price. Police discovered him in the parking lot lighting the fire, and they took advantage of the situation and executed him on the spot (CrimethInc 2019).

Though the border regime is designed to brutalize racialized people and uphold the privileges of whiteness, these killings show that whiteness is less a question of skin color and more a question of alliances. Those who fight against the alliances represented by white supremacy and colonialism may face the very apparatus of annihilation designed to control racialized people.

Ongoing wars against Indigenous communities are fully tied up in internal border regimes, ongoing processes of state formation and whiteness, and neocolonial double standards around citizenship (Granovsky-Larsen 2021; Brown 2021). When Water Protectors tried to prevent the construction of the Dakota Access Pipeline transporting shale oil through Lakota territory, occupied by the United States, local, state, and federal police mobilized a huge militarized force complemented by the private security firm TigerSwan, complete with helicopters, drones, tanks, infiltration, and surveillance, constituting "the largest mobilization of cops and military in the state's history since 1890, when […] the military was deployed to crush the horseless and starving Ghost Dancers at Standing Rock" (Estes 2019, p. 54). TigerSwan agents, coming out of counterinsurgency operations in Iraq and Afghanistan, called Indigenous Water Protectors "terrorists" engaging in a "jihadist insurgency" (Estes 2019, p. 251). Private and public police forces carried out brutal raids, beat, shot, and gassed Water Protectors, and soaked them with powerful water cannons in subzero temperatures miles from any facility where they might receive adequate care for hypothermia. They used attack dogs, rubber bullets, concussion grenades, tear gas, and sound cannons. No one was killed in any of the multiple acts of police repression only thanks to good luck and well organized medical support. However, a Native woman lost her sight after being shot in the face with a tear gas grenade at close range and a solidarity protester had her arm partially blown off by a concussion grenade (Tolan 2016; Fontaine 2016). Many others received traumatic head wounds from police projectiles. In one single attack on 20 November 2016, 200 people were injured (Estes 2019, p. 55). Multiple people were imprisoned with the aid of police infiltrators. Throughout the resistance, over 830 people were arrested and given state charges, while five people—all of them Indigenous—were given federal charges. One Indigenous and Chicano protester, Steve Martínez, is locked up at the time of this writing for refusing to testify before a grand jury (Fatica 2021). He could be kept in jail for up to 18 months in an attempt to force him to give information against other people in the movement.

It has been pointed out time and again that when far Right groups carry out armed occupations, even when they shoot and murder political opponents in the course of white riots, the police stand back and let it happen. Often, they participate directly. Subsequently, prosecutors are always more lax in bringing charges. Fascists in the street, extractivist profiteers in corporate board rooms, and the criminal justice system are all facets of a white supremacist, colonial system that is destroying the planet.

Because repression transcends formal exercises of state power, we should also consider the connection between pipeline construction, shale oil extraction, and Man Camps, the itinerant labor camps of often highly paid, mostly white male workers associated with increasing rates of drug and alcohol usage and violence against Indigenous women and Two Spirit people. Going back to mining incursions in the Black Hills gold rush at the end of the nineteenth century or the US Army Corps of Engineers damming rivers in Lakota territory in the middle of the twentieth, settler profiteering has always been connected to violence against Indigenous communities, whether or not the perpetrators wear a uniform.

In Canada, progressive prime minister Justin Trudeau came into office on pledges of taking climate action and improving relations with the dozens of Indigenous nations whose stolen lands the Canadian state is built on. In fact, Trudeau took advantage of his conservative predecessor having left the Kyoto accords to go all in for the exploitation of the Athabasca tar sands, a huge deposit of extremely dirty oil that takes an enormous amount of energy to extract and process, leaving an immense expanse of the Canadian interior denuded of all forests and topsoil in the process. To capitalize on the oil boom, Trudeau's government sank billions of dollars into new pipeline construction despite fierce Indigenous resistance across the continent. When the Wet'sewet'en nation effectively blocked the construction of the Coastal GasLink pipeline on their own lands, the RCMP, Canada's federal police created initially as a paramilitary force against First Nations, carried out several militarized raids on Wet'sewet'en encampments and were given permission to shoot to kill. In their raids, the RCMP used boats, helicopters, drones, and intensive physical and social media surveillance of land defenders, in addition to large numbers of agents with armored vehicles and military grade weaponry. They also considered the use of social services to take custody of any children arrested, fully in line with Canada's long and violent history of using so-called residential schools targeting Indigenous children as a tool for genocide (Starblanket 2018). Canada's 2015 Anti-Terrorism Act "sanctions the criminalization of Indigenous environmentalists by enhancing surveillance and legal powers against any potential interference with Canada's 'critical infrastructure' or 'territorial integrity'" while RCMP surveillance documents identify land defenders as "aboriginal extremists" and prepare for the use of lethal force (Dhillon and Parrish 2019; Crosby and Monaghan 2018).

Effective Counterinsurgency: Containing Resistance with Mercenaries and NGOs

In December 2011, global private security firm Stratfor was hacked. Stratfor maintains high level relations with intelligence, political, and military officials from the US to Israel and provides consulting and intelligence services to private companies and governments around the world to aid in public relations, counterinsurgency, disruption, and regime change efforts. They have described themselves as a "shadow CIA" (Gibson and Horn 2013). The resulting data dumps revealed widespread surveillance and illegal actions by government and private entities. We only have access to this information thanks to the hack carried out by Anonymous. The anarchist hacker Jeremy Hammond was sentenced to ten years in prison for the action. Similarly, we only know about the FBI's bloody COINTELPRO program—which resulted in bloody campaigns against Black and Indigenous resistance movements in the '60s and '70s—because a group of revolutionaries broke into an FBI field office and stole the documents. In other words, the only way we ever know what our governments are actually doing is thanks to illegal direct action, leaving any discussion of government accountability rather hollow.

In the Statfor leaks, we learned how governments view social movements, and how they try to neutralize us. What is important to emphasize is that Stratfor, like governments all over the world, use the lens of counterinsurgency when dealing with the dissent of their subjects. With one exception, they do not make a distinction between the middle class high school student going on strike on Fridays to raise awareness about climate change, the journalist researching the cancerous effects of widely used pesticides, the anarchist setting fire to construction equipment to save a forest from being bulldozed, and a Native elder leading a ceremony in an encampment seeking to regain treaty lands from a settler state. They view all these people as enemies. The only distinction they make is how to turn different kinds of dissidents against one another in order to buy out some and marginalize the rest. At the heart of counterinsurgency is the belief that all of a government's subjects are potential enemies.

According to a framework used by Stratfor and other security consultants, there are four types of dissidents and a three-step method to neutralizing them. There are the radicals, who:

> want to change the system; have underlying socio/political motives' and see multinational corporations as 'inherently evil,' These organizations do not trust the … federal, state and local governments to protect them and to safeguard the environment. They believe, rather, that individuals and local groups should have direct power over industry … I would categorize their principal aims … as social justice and political empowerment. (Duchin 2013)

There are the idealists who "want a perfect world" but because of their "intrinsic altruism" can be made to sympathize with the interests of industry and shy away from conflictive positions. There are opportunists who will seize "the opportunity to side with the powerful for career gain" and there are the realists who are "willing to work within the system" and are "pragmatic. The realists should always receive the highest priority in any strategy dealing with a public policy issue."

The three steps are rather simple:

> First, isolate the radicals. Second, 'cultivate' the idealists and 'educate' them into becoming realists. And finally, co-opt the realists into agreeing with industry.
>
> 'If your industry can successfully bring about these relationships, the credibility of the radicals will be lost and opportunists can be counted on to share in the final policy solution'[.] (Duchin 2013)

It is more than a little ironic that those who consider themselves realists are considered the easiest to manipulate.

This is why it is neither sectarianism nor an excess of zeal when we declare that NGOs and humanitarian activists are part of the State's counterinsurgency campaign. We can find this spelled out in US field manuals utilized in Iraq, particularly in the shift in methods that accompanied the increasing effectiveness of the US occupation around 2008 (Kilcullen 2010). Following the lead of renowned counterinsurgency expert David Galula (1964), state and private counterinsurgency campaigns recognize the need to position themselves as the benefactors of the general population. Of course, soldiers and paramilitaries are not the best equipped personnel to go around knocking on doors, asking people what their grievances are and offering solutions. That's where the NGOs come in. Some NGOs are *astroturfed*, directly created by corporations or governments to pursue their interests (Brock and Dunlap 2018). But even putatively independent NGOs often produce the same results. Dependent on the grants of governments and private foundations, they fool themselves into thinking they are making the world a better place, responding to real problems, carrying out harm reduction, when in reality they are pawns in the immensely funded, expertly prepared plans of the very institutional forces responsible for the original problems (Choudry and Kapoor 2013).

Even this might be too charitable a view, given the countless times major conservation NGOs have directly administered paramilitary forces to evict and even murder Indigenous inhabitants from wilderness preserves they have appropriated on multiple continents [Subcommittee 2021; Duffy 2016).

Thus, we have numerous cases of NGOs that make their money positioning themselves against human trafficking and that systematically cooperate with the deportation machinery of the Department of Homeland Security and Immigration and Customs Enforcement in the US, and similar agencies in other countries (Reisenwitz 2021); we have homeless charities aiding deportation patrols (Taylor 2018); NGOs getting government money to imprison migrant children (London No Borders 2012; Kotch 2019); environmental NGOs protecting logging companies from protest in exchange for forest conservation agreements that look good in the media, but in the fine print actually open much more territory to logging than they protect (Paley 2011); Indigenous communities criticizing NGOs for cutting them out of agreements that directly affect their livelihoods, in exchange for backroom deals with the major power brokers (Stainsby and Oja Jay 2009); and many other examples of conservationist complicity with continuing forms of colonial violence (Menton and Gilbert 2021). From a public relations standpoint, smaller NGOs can be seen as laundering legitimacy, receiving money from such dubious institutions as the US military, ICE, the Ford Foundation, or major banks, and then presenting themselves as independent actors. The Ford Foundation, for one, is a major donor to the migrant rights and environmental justice movements, but can be rightly analyzed as cultivating the "soft power" of the US government and capitalists, making sure social movements remain pliable and reform oriented; the Foundation also has a long history of working with the CIA (Chaparral 2013).

The role of Sanctuary Cities deserves special mention, as they represent a point of intervention in which progressive movements, NGOs, and municipal governments—ostensibly representing a more humane scale of state intervention—are frequently lauded as a major success for taking in climate and other refugees, often couple with policies of green urban planning. However, the Sanctuary City model "contributes to a hostile asylum regime by indefinitely deferring" any resolution to the precarity experienced by undocumented people (Bagelman 2013). Another common practice in Sanctuary Cities is for public authorities to hand *undesirables* over to the deportation machinery, further the racist and classist divide between good and bad immigrants (Nair 2015). Sanctuary Cities have also constituted a cynical bait and switch. Multiple European cities declared themselves Sanctuary Cities, praising themselves for taking in a mere 100 asylum seekers each at the same time as the European Union pulled out all the stops from their "push back" policy in which migrants were systematically rounded up and deported with no legal procedures whatsoever, making centuries of immigration protections effectively worthless (Foerster 2019). In a similar move, humanitarian NGOs in Greece congratulated themselves for taking the *problem* of refugee seekers in hand in 2017, making no mention that this was the very moment when the Greek state moved with an iron fist against a multitude of self-organized spaces where migrants provided themselves with dignified housing (International Rescue Committee 2017). Meanwhile, the new asylum centers administered by the NGOs were effectively prisons.

As reported in outlets from the *New Yorker* to *Vice*, "Greece's anarchists are taking better care of refugees than the government" and NGOs, but this is something of a distortion (Crabapple 2017, 2020). Greece's anarchists were sharing resources and tools of collective self-defense to help migrants take care of themselves. The initiatives they supported were illegal direct actions that expropriated entire buildings or other territories from the capitalist market and turned them into self-organized living spaces where people—migrants and citizens living together—could define and secure their own needs. In other words, exactly the kind of initiatives that governments criminalize, that NGOs silence, that academics ignore or attempt to tame and make more palatable, and that police respond to with truncheons and tear gas. Some proponents of the NGO-municipality alliances represented by these "cities of refuge" even gave credit for the dignified housing created by anarchist initiatives to the very public-private partnerships that shut them down (Turam 2021).

The role of NGOs in counterinsurgency ranges from the clownish to the brutal, depending on context. In a European Union increasingly supporting the transition to green capitalism, money floods in faster than local governments and organizations know how to spend it, and NGOs and corporations line up for handouts (Riley 2021). Because the priority is to spend budgeted funds before the earmark expires, the result is often projects that are useless or even counterproductive, and very corporations responsible for ecological devastation are often the prime beneficiaries of climate emergency funds together with the environmental charities that promise some kind of quality control but really are just legitimizing the new wave of extractivism (Binnie 2021; Barteczko 2021). In Oaxaca's Isthmus of Tehuantepec, on the other hand, the rush to build industrial scale wind factories has seen NGOs engaging in "conflict stabilization" and pacification in a way that actually makes it easier for murderous paramilitaries connected to financial interests to carry out their work, leaving social movements on the ground defenseless (Dunlap and Correa 2021). NGOs also continue and accelerate earlier colonial processes of increasing the hierarchization of traditionally stateless Indigenous societies by choosing, promoting, and financing representatives.

Whether dealing with migrants or more conventional environmental movements, NGOs play an important role in deciding which members of society are legitimate arbiters, which demographics are deserving of forms of charity that are frequently victimizing or demobilizing (if not outright carceral), and which ones are undesirable, potentially disruptive, deserving of deportation or other forms of repression (Walther 2021). This habit of dividing target populations into pliable and hostile segments is as clear a sign of counterinsurgency thinking as any.

We can also see this when we look back at resistance to the oil industry in Nigeria. When Ken Saro-Wiwa was led up to a scaffold and executed, his death was counted. He was a poet and a nonviolent protest leader, after all. But what happened to his peers who learned the lesson, who valued their own lives, who took up arms or gave shelter to those who did, and *who actually hurt the profits and cut off the production of the oil industry*? When they were murdered by British or US munitions fired from helicopters dispatched by the Nigerian government or Shell Oil, those very well meaning NGOs did not include them in their list of murdered land defenders. Their lives are neither counted nor valued. The very NGOs who evince a concern for human rights are crucial to this counterinsurgency operation, celebrating the activists who use tactics deemed legitimate by the government. By not celebrating every bit as vociferously the lives and the resistance of those who make the hard choice to resist with illegal and effective methods, these NGOs are signaling who is legitimate and who is subhuman, they are creating the division between good actors and bad actors that the militaries, death squads, and counterinsurgency experts rely on, and they are doing so in a way that most people who think they care about the environment or human rights will find credible. Likewise, they are signaling to the rest of society what kinds of tactics (legal, peaceful, less effective) will be honored, and what kinds of tactics (illegal, transformative) are associated with dangerous, frightening elements who seem to inevitably—in the moral arc of capitalism's universe—end up in unmarked graves (Gelderloos 2013, 2018).

In order to outmaneuver these counterinsurgency strategies, we need to start by becoming aware of them. There can be no effective struggle for healthy lives on a healthy planet unless we are aware of the systematic repression leveled against those who seek this most basic goal, and stand in solidarity with those who are most affected. Because when we support one another and fight back, we can win.

Conclusion

The chapter has demonstrated how state and paramilitary repression are part and parcel of the ecological crisis. They are the systematic violence that protect extractivist industries, environmental racism, and border regimes from people's resistance. Even as leading governments struggle to make the most minimal commitments to international agreements for emissions reductions, they have developed apparatuses for the repression of environmental and Indigenous movements and for the control of climate refugees by leaps and bounds. The ecological crisis is enforced. Not only is it not accidental, we are, on a global scale, being forced to cooperate with the various institutions and economies that produce this crisis, and when we try to obstruct them or directly address the crisis without intermediaries, we are neutralized by the full range of oppressive violence and the most up-to-date counterinsurgency strategies available to the institutions of power, both public and private.

A major part of state and corporate counterinsurgency strategies to pacify the resistance relies on environmentalist NGOs to define respectable and unacceptable resistance. The criterion that divides the two corresponds to the needs of government to rule their subject populations and the needs of the private sector to continue extracting profits, and not to the needs of human communities and ecosystems to guarantee their continued survival.

Given that this is an urgent question of collective survival with obvious culprits, those who are studying facets of the ecological crisis should consider breaking with respectability politics and polite neutrality, or channeling their proposals for solutions through the very institutions that have spent decades undermining their own credibility.

Bibliography

Al Jazeera, 2020. "Second Indigenous activist killed in Honduras this week," *Al Jazeera*. 12 December 2020. https://www.aljazeera.com/news/2020/12/31/honduras-sees-second-indigenous-activist-killed-in-in-past-week.

Allen, G. and A. G. Sens, 2006. "Riga and beyond: The political transformation of NATO," *NATO Review*. https://www.nato.int/docu/review/articles/2006/12/01/riga-and-beyond-the-political-transformation-of-nato/index.html.

Alsema, A. 2018. "Colombia charges 13 former Chiquita executives over hundreds of murders," *Colombia Reports*. 1 September 2018. https://colombiareports.com/terror-for-profit-colombia-charges-14-former-chiquita-executives/.

Bagelman, J. 2013. "Sanctuary: A politics of ease?" *Alternatives: Global, Local, Political* 38(1): 49–62.

Barteczko, A. 2021. "Businesses and NGOs urge Poland to bet on climate in recovery plan," *Reuters*. 8 February 2021. https://news.trust.org/item/20210208145915-c0brm/.

BBC, 2008. "Nigerian attack closes oilfield," *BBC*. 20 June 2008.

BBC, 2015. "Italian police 'tortured' Genoa G8 protester, says ECHR," *BBC*. 7 April 2015. https://www.bbc.com/news/world-europe-32211364.

Belcher, O., P. Bigger, B. Neimark, et al. 2020. "Hidden carbon costs of the 'everywhere war': Logistics, geopolitical ecology, and the carbon boot-print of the US military," *Transactions of the Institute of British Geographers* 45(1): 65–80.

Best, S. and A. J. Nocella, 2006. *Igniting a Revolution*, Oakland: AK Press.

Binnie, I. 2021. "Iberdrola seeks EU recovery funds for Spanish clean power projects," *Reuters*. 8 February 2021. https://news.trust.org/item/20210208122833-5fy1q/.

Borger, J. 2020. "US ice officers 'used torture to make Africans sign own deportation orders,'" *The Guardian*. 22 October 2020. https://www.theguardian.com/us-news/2020/oct/22/us-ice-officers-allegedly-used-torture-to-make-africans-sign-own-deportation-orders.

Brock, A. 2020a. "Enforcing ecological catastrophe at all costs," *New Internationalist*. 19 November 2020. https://newint.org/features/2020a/11/19/enforcing-ecological-catastrophe-all-costs.

Brock, A. 2020b. "'Frack off': Towards an anarchist political ecology critique of corporate and state responses to anti-fracking resistance in the UK." *Political Geography* 82: 102246.

Brock, A. 2018. "The battle of Hambacher Forest," *Red Pepper*. 20 September 2018. https://www.redpepper.org.uk/the-battle-of-hambacher-forest/.

Brock, A. and A. Dunlap, 2018. "Normalising corporate counterinsurgency: Engineering consent, managing resistance and greening destruction around the Hambach coal mine and beyond," *Political Geography* 62(1): 33–47.

Brown, A. 2021. "Corporate counterinsurgency: Indigenous water protectors face off with an oil company and police over a Minnesota pipeline," *The Intercept*. 7 July 2021. Available at: https://theintercept.com/2021/07/07/line-3-pipeline-minnesota-counterinsurgency/.

Butt, N. and M. Menton, 2019. "More than 1,700 activists have been killed this century defending the environment," *The Conversation*. 5 August 2019. https://theconversation.com/more-than-1-700-activists-have-been-killed-this-century-defending-the-environment-120352.

Chaparral 2013. Chaparral respects no borders, "Beware the funders of immigrants' rights," Tucson: 2013. chaparralrespectsnoborders.blogspot.com

Charles, S. 2016. "Our detention centres are concentration camps and must be closed," *Sydney Morning Herald*. 4 May 2016. http://www.smh.com.au/comment/our-detention-centres-are-intentionally-cruel-and-must-be-closed-20160504-golr04.html.

Choudry, A. and D. Kapoor, 2013. *NGOization: Complicity, Contradictions and Prospects*, London: Zed Books.

CNN, 2001. "Hundreds mourn G8 protester," *CNN*. 25 July 2001. https://edition.cnn.com/2001/WORLD/europe/07/25/italy.funeral/index.html.

Cohen, S. 2015. "The supreme court needs to decide: Can victims sue Chiquita for sponsoring terrorism," *New Republic*. 10 April 2015. https://newrepublic.com/article/121506/supreme-court-should-take-chiquitas-terrorism-sponsorship-case.

Council of Europe, 2020. "Council of Europe's anti-torture committee calls on Greece to reform its immigration detention system and stop pushbacks," Council of Europe. 19 November 2020. https://www.coe.int/en/web/cpt/-/council-of-europe-s-anti-torture-committee-calls-on-greece-to-reform-its-immigration-detention-system-and-stop-pushbacks.

Crabapple, M. 2017. "Greece's anarchists are taking better care of refugees than the government," *Vice*. March 2017.

Crabapple, M. 2020. "The attack on Exarchia, an anarchist refuge in Athens," *The New Yorker*. 20 January 2020.

CrimethInc, 2017. "Designed to kill: Border policy and how to change it," *CrimethInc.*, zine updated into book, CrimethInc., *No Wall They Can Build*, 2017.

CrimethInc, 2019. "On Willem Van Spronsen's action against the Northwest detention center in Tacoma," *CrimethInc.* 14 July 2019. https://crimethinc.com/2019/07/14/on-willem-van-spronsens-action-against-the-northwest-detention-center-in-tacoma-including-the-full-text-of-his-final-statement.

Crosby, A. C. and J. Monaghan, 2018. *Policing Indigenous Movements: Dissent and the Security State*, Black Point: Fernwood Publishing.

Cuellar, F. R. 2005. *The Profits of Extermination: How US Corporate Power Is Destroying Colombia*, Monroe: Common Courage Press.

Dadusc, D. and P. Mudu, 2020. "Care without control: The humanitarian industrial complex and the criminalisation of solidarity," *Geopolitics* 17: 25–26, 30.

Dhillon, J. and W. Parrish, 2019. "Exclusive: Canada police prepared to shoot Indigenous activists, documents show," *The Guardian*. 20 December 2019. https://www.theguardian.com/world/2019/dec/20/canada-indigenous-land-defenders-police-documents.

Duchin, R. 2013. quoted in Steve Horn, "Here's how the corporations defeat political movements," *Popularresistance.org*. 29 July 2013. https://popularresistance.org/heres-how-the-corporations-defeat-political-movements/.

Duffy, R. 2016. "War, by conservation," *Geoforum* 69(1): 238–248.

Dunlap, A. 2020. "Wind, coal, and copper: The politics of land grabbing, counterinsurgency, and the social engineering of extraction," *Globalizations* 17(4): 661–682.

Dunlap, A. and J. Fairhead, 2014. "The militarisation and marketisation of nature: An alternative lens to 'climate-conflict," *Geopolitics* 19(4): 937–961.

Dunlap, A. and M. Correa Arce, 2021. "'Murderous energy' in Oaxaca, Mexico: Wind factories, territorial struggle and social warfare," *The Journal of Peasant Studies*. 23 February 2021.

Economy, E. C. 2012. "A land grab epidemic: China's wonderful world of Wukans," Council on Foreign Relations. 7 February 2012.

Enough 14, 2021. "Dutch police seized servers of NoState tech collective – North Shore counter-info, Montreal counter-info, 325 and act for freedom now," Enough 14. 31 March 2021. https://enoughisenough14.org/2021/03/31/dutch-police-seized-servers-of-nostate-tech-collective-north-shore-counter-info-montreal-counter-info-325-and-act-for-freedom-now-down/.

Estes, N. 2019. *Our History Is the Future: Standing Rock Versus the Dakota Access Pipeline, and the Long Tradition of Indigenous Resistance*, New York: Verso.

Evans, R. and P. Lewis, 2013. *Undercover*, London: Faber and Faber.

Fatica, R. 2021. "Steve Martínez jailed again after refusing to testify before a federal grand jury," *Perilous Chronicle*. 3 March 2021. https://perilouschro

nicle.com/2021/03/03/steve-martinez-jailed-again-after-refusing-to-testify-before-a-federal-grand-jury/.

Federman, A. 2020. "Revealed: US listed climate activist group as 'extremist' alongside mass killers," *The Guardian*. 13 January 2020. https://www.theguardian.com/environment/2020/jan/13/us-listed-climate-activist-group-extremists.

Foerster, A. 2019. "Solidarity or sanctuary? A global strategy for migrant rights," *Humanity and Society* 43(1): 19–42.

Fontaine, T. 2016. "Woman hurt in latest Standing Rock confrontation in 'serious' condition," *CBC*. 22 November 2016. https://www.cbc.ca/news/indigenous/standing-rock-serious-injuries-1.3861705.

Forsythe, M. 2011. "China's spending on internal police force in 2010 outstrips defense budget," Bloomberg. 6 March 2011. https://www.bloomberg.com/news/2011-03-06/china-s-spending-on-internal-police-force-in-2010-outstrips-defense-budget.html.

Galula, D. 1964. *Counterinsurgency Warfare: Theory and Practice*, Praeger Security International, Westport, CT.

Gelderloos, P. 2013. *The Failure of Nonviolence*, Seattle: Left Bank Books.

Gelderloos, P. 2018. "Riots and remembrance on the streets of Barcelona: The collective learning of subversive techniques," in Alissa Starodub and Andrew Robinson (eds.), *Riots and Militant Occupations: Smashing a System, Building a World—A Critical Introduction*, London: Rowman & Littlefield International.

Gelderloos, P. 2022. *The Solutions Are Already Here*, London: Pluto Press.

Giartellia, A. 2019. "Deadly crossing: Bodies of 300 illegal immigrants found on US-Mexico Border during fiscal 2019," *Washington Examiner*. 11 December 2019. https://www.washingtonexaminer.com/news/deadly-crossing-bodies-of-300-illegal-immigrants-found-on-us-mexico-border-during-fiscal-2019.

Gibson, C. and S. Horn, 2013. "Exposed: Globally renowned activist collaborated with intelligence firm stratfor," *Occupy.com*. 2 December 2013. https://www.occupy.com/article/exposed-globally-renowned-activist-collaborated-intelligence-firm-stratfor.

Global Witness, 2020. "Annual report 2019: Protecting people and planet," *Global Witness*. May 2020. https://www.globalwitness.org/en/about-us/annual-report-2019-protecting-people-and-planet/.

Grajales J. 2013. "State involvement, land grabbing and counter-insurgency," *Development and Change* 44: 211–232.

Granovsky-Larsen, S. and L. Santos, 2021. "From the war on terror to a war on territory: Corporate counterinsurgency at the Escobal mine and the Dakota access pipeline," *Canadian Journal of Latin American and Caribbean Studies/Revue canadienne des études latino-américaines et caraïbes*: 1–25.

Green Is the New Red 2021. "SHAC 7," *Green Is the New Red*. http://www.greenisthenewred.com/blog/tag/shac-7/.

Guy, J. 2020. "Record number of environmental activists killed in 2019," *CNN*. 29 July 2020. https://edition.cnn.com/2020/07/29/world/global-witness-2019-defenders-report-scli-intl/index.html.

Harrison, D. 2001. "Minister set up deal to save animal lab," *The Telegraph*. 21 January 2001. https://www.telegraph.co.uk/news/uknews/1318759/Minister-set-up-deal-to-save-animal-lab.html.

Heise, R. 2021. "NATO is responding to new challenges posed by climate change," *NATO Review*. 1 April 2021. https://www.nato.int/docu/review/articles/2021/04/01/nato-is-responding-to-new-challenges-posed-by-climate-change/index.html.

Indybay, 2020. "Facebook bans multiple anarchist and antifascist pages," *Indybay*. 24 August 2020. https://www.indybay.org/newsitems/2020/08/24/18836084.php.

International Rescue Committee, 2017. "Transitioning to a government-run refugee and migrant response in Greece: A joint NGO roadmap for more fair and humane policies," International Rescue Committee. 13 December 2017. https://www.rescue-uk.org/report/transitioning-government-run-refugee-and-migrant-response-greece-joint-ngo-roadmap-more-fair.

Johnson, G. 2020. "Two face federal terror charge over train track interference," *AP News*. 1 December 2020. https://apnews.com/article/bellingham-washington-british-columbia-seattle-7afd76ecb1ee9bccac7d0bcb7f9ae1b7.

Kadafa, A. A. 2012. "Environmental impacts of oil exploration and exploitation in the Niger Delta of Nigeria," *Global Journal of Science Frontier Research* 12(3).

Karakoulaki, M. 2019. "The invisible violence of Europe's refugee camps," *Al Jazeera*. 22 October 2019. https://www.aljazeera.com/opinions/2019/10/22/the-invisible-violence-of-europes-refugee-camps/.

Kilcullen, D. 2010. *Counterinsurgency*, New York: Oxford University Press.

Kladzyk, R. 2021. "ICE detainees face systemic torture at El Paso area immigration facility, new report says," *El Paso Matters*. 5 January 2021. https://elpasomatters.org/2021/01/05/ice-detainees-at-el-paso-area-immigration-facility-face-systemic-torture-new-report-says/.

Klinger, J. M. 2017. *Rare Earth Frontiers: From Terrestrial Subsoils to Lunar Landscapes*, Ithaca: Cornell University Press.

Kotch, A. 2019. "These nonprofits and businesses are making millions from detaining immigrant children," *Sludge*. 3 July 2019. https://readsludge.com/2019/07/03/these-nonprofits-and-businesses-are-making-millions-from-detaining-immigrant-children/.

Larsson, N. 2020. "Killed for defending the planet: Murder of environmental activists reaches record high," *Independent*. 5 August 2020. https://www.

independent.co.uk/news/world/europe/environmental-activists-deaths-murder-climate-change-colombia-romania-a9651826.html.

Levin, S. 2019. "Revealed: How Monsanto's 'intelligence center' targeted journalists and activists," *The Guardian*. 8 August 2019. https://www.theguardian.com/business/2019/aug/07/monsanto-fusion-center-journalists-roundup-neil-young.

Lewis, P. and R. Evans, 2020. "Secrets and lies: Untangling the UK 'spy cops' scandal," *The Guardian*. 28 October 2020. https://www.theguardian.com/uk-news/2020/oct/28/secrets-and-lies-untangling-the-uk-spy-cops-scandal.

Liu, J., 2013. "How tiger leaping Gorge was saved," *China Dialogue*. 19 April 2013. https://chinadialogue.net/en/nature/5923-how-tiger-leaping-gorge-was-saved/.

Li You, 2020. "Ningxia conservationists detained for 'Picking Quarrels'," *Sixth Tone*. 11 September 2020. https://www.sixthtone.com/news/1006168/ningxia-conservationists-detained-for-picking-quarrels.

London No Border's, 2012. "Barnado's campaign: Further reading," London No Borders. 8 April 2012. http://london.noborders.org.uk/barnardos/articles.

Lubbers, E. 2012. *Secret Manoeuvres in the Dark: Corporate and Policy Spying on Activists,* London: Pluto.

Manian, M. 2020. "Immigration detention and coerced sterilization: History tragically repeats itself," ACLU. 29 September 2020.

Marshall, Y. G. 2020. "Black liberal, your time is up," *Al Jazeera*. 1 June 2020. https://www.aljazeera.com/opinions/2020/6/1/black-liberal-your-time-is-up/.

Maupin, C. 2016. "The Niger delta avengers declare war on Western oil giants," 8 June 2016. https://www.mintpressnews.com/niger-deltans-plenty-avenge-yet-us-media-ignores-context-terrorism-oil-companies/217030/.

MauvaiseTroupe Collective, 2018 [2016]. *The Zad and NoTAV: Territorial Struggles and the Making of a New Political Intelligence*, New York: Verso Books.

Menton, M. and P. R. Gilbert, 2021. "BINGOs and environmental defenders: NGO complicity in atmospheres of violence and the possibilities for decolonial solidarity with defenders," *Environmental Defenders*, New York: Routledge, 2021.

Mignolo, W. D. 2005. *The Idea of Latin America,* Malden: John Wiley & Sons.

Migrant Project 2020. "Migrant deaths in the Mediterranean exceed 20,000 since 2014," The Migrant Project. 9 March 2020. https://www.themigrantproject.org/mediterranean-deaths-2/.

Millar, S. and A. Langley, 2003. "G8 protester to prosecute Swiss police," *The Guardian*, 9 June 2003. https://www.theguardian.com/uk/2003/jun/09/g8.world.

Milman, O. 2020. "Facebook suspends environmental groups despite vow to fight misinformation," *The Guardian*. 22 September 2020. https://www.theguardian.com/environment/2020/sep/22/facebook-climate-change-environment-groups-suspended.

Monroy, M. 2008 "Social movements against the global security architecture!" Gipfelsoli, 7 September 2008. https://digit.site36.net/2008/10/07/social-movements-against-the-global-security-architecture/.

Moraña, M., E. Dussel, and C. A. Jáuregui, 2014. *Coloniality at Large: Latin America and the Postcolonial Debate*, Durham: Duke University Press Books.

Nair, Y. 2015. "How to make prisons disappear," in Eric A. Stanley and Nat Smith (eds.), *Captive Genders: Trans Embodiment and the Prison Industrial Complex*, Oakland: AK Press.

NATO. 2009. "Strasbourg / Kehl Summit Declaration," *North Atlantic Treaty Organization*, 4 April 2009. https://www.nato.int/cps/en/natolive/news_52837.htm.

Naumann, General K., General J. Shalikashvili, Field Marshal The Lord Inge, et al., "Towards a grand strategy for an uncertain world: Renewing transatlantic partnership," Noaber Foundation, 2007.

Nolin, C. and G. Russell, 2020. "Public letter," 11 November 2020. Available at https://mailchi.mp/rightsaction/public-letter-publication-upcoming-book-blocked?fbclid=IwAR0ehaY2JNu05DBHgtoesuljK8FCFGOOT-tKxnTxnaHg0pI1pXUK9Y9r-gs.

Nomad, T. 2016. "Policing as counterinsurgency: An interview with Tom Nomad," Submedia.TV. Available at: https://submedia.tv/stimulator/2016/08/04/policing-counterinsurgency-interview-tom-nomad/.

Online Nigeria 2005. "Ijaw Tribe," *Online Nigeria*. 19 December 2005. https://onlinenigeria.com/finance/?blurb=669.

Paley, D. 2011. "Fracturing Solidarity," *Briarpatch*. 1 March 2011. https://briarpatchmagazine.com/articles/view/fracturing-solidarity/.

Patel, K. 2020. Race and a decolonial turn in development studies. *Third World Quarterly* 41(9): 1463–1475.

Penketh, A. 2014. "French socialists and greens at odds following death of Sivens dam protester," *The Guardian*. 28 October. https://www.theguardian.com/world/2014/oct/28/french-socialists-greens-sivens-dam-protestor-remi-fraisse-grenade-autopsy.

Potter, W. 2011. *Green Is the New Red: An Insider's Account of a Social Movement Under Siege*, San Francisco: City Light Publishers.

Reisenwitz, C. 2021. "The truth about Polaris," Sex and the State, 15 January 2021. https://cathyreisenwitz.substack.com/p/the-truth-about-polaris.

Riley, B. 2021. "From Brussels with love: The secret to spending the recovery fund," *Euractiv*. 16 April 2021. https://www.euractiv.com/section/energy/opinion/from-brussels-with-love-the-secret-to-spending-the-recovery-fund/.

Rodríguez, D. 2020. "Reformism isn't liberation, it's counterinsurgency," *Black agenda report*. 21 October 2020. https://blackagendareport.com/reformism-isnt-liberation-its-counterinsurgency.
RTO/NATO. 2003 "Urban operations in the year 2020," RTO Technical Report 71. Neuilly-Sur-Seine CEDEX, France: Research and Technology Organisation/North Atlantic Treaty Organisation.
Sens, A. 2006. "Riga and beyond: The political transformation of NATO," *NATO Review*. 1 December 2006. https://www.nato.int/docu/review/articles/2006/12/01/riga-and-beyond-the-political-transformation-ofnato/index.html.
Schlembach, R. 2018. "Undercover policing and the spectre of 'domestic extremism': The covert surveillance of environmental activism in Britain," *Social Movement Studies* 17(5): 491–506.
Shen, A. 2020. "Why did it take a sterilization scandal to retrigger our outrage over ICE?" *Slate*. 18 September 2020. https://slate.com/news-and-politics/2020/09/ice-sterilization-scandal-outrage-abuse-hysterectomies.html.
Smith, G. 2017. *The War and Environment Reader*, Washington, DC: Just World Books.
Smith, J. 2009. "Colombia: Ex-paramilitary implicates two U.S. companies in murder of trade unionists," *NACLA*. 14 December 2009. https://nacla.org/news/colombia-ex-paramilitary-implicates-two-us-companies-murder-trade-unionists.
Stainsby, M. and D. Oja Jay, 2009. "Offsetting resistance: The effects of foundation funding and corporate fronts," July 2009. http://www.offsettingresistance.ca/.
Standaert, M. 2020. "With activists silenced, China moves ahead on big dam project," *The Third Pole*. 5 November 2020. https://www.thethirdpole.net/2020/11/05/with-activists-silenced-china-moves-ahead-on-big-dam-project/.
Starblanket, T. 2018. *Suffer the Little Children: Genocide, Indigenous Nations and the Canadian State*, Atlanta: Clarity Press.
Subcommittee 2021. "Subcommittee on water, oceans, and wildlife, oversight: Protecting human rights in international conservation," Natural Resources Committee US House of Representatives. 26 October 2021. https://naturalresources.house.gov/hearings/protecting-human-rights-in-international-conservation.
Taylor, D. 2018. "Homeless charity aided deportation patrols in search for rough sleepers," *The Guardian*. 5 March 2018. https://www.theguardian.com/uk-news/2018/mar/05/st-mungos-homeless-charity-helped-target-rough-sleepers-to-deport.
The Guardian, 2021. "China bomb attack kills four in suspected protest over development," *The Guardian*. 23 March 2021. https://www.theguardian.

com/world/2021/mar/23/china-bomb-attack-kills-four-in-suspected-pro test-over-development.

Tolan, S. 2016. "Wounded on the front line at standing rock, a protester refuses to give up her fight," *Los Angeles Times*. 22 December 2016. https://www.latimes.com/nation/la-na-standing-rock-wounded-20161222-story.html.

Turam, B. 2021. "Refugees in borderlands: Safe places versus securitization in Athens, Greece," *Journal of Urban Affairs* 43(6): 756–780.

Walia, H. 2021. *Border and Rule: Global Migration, Capitalism, and the Rise of Racist Nationalism*, Chicago: Haymarket Books.

Walther, E.-M. 2021. "Brokers in straitjackets: The moral quandaries of refugee support organisations in Slovakia," *Cultural Dynamics*. https://doi.org/10.1177/09213740211011190.

Willsher, K. 2019. "Anarchists hijack climate march on day of violent protests in Paris," *The Guardian*. 21 September 2019. https://www.theguardian.com/world/2019/sep/21/paris-on-high-alert-as-protesters-try-to-revive-gilets-jaunes-movement.

Woodman, C. 2018. "The Infiltrator and the movement," *Jacobin*. 23 April 2018. https://jacobinmag.com/2018/04/uk-infiltration-secret-police-mi5-special-branch-undercover.

PART IV

Looking Forward

CHAPTER 11

Demilitarize for a Just Transition

Matthew J. Burke and Nina L. Smolyar

Introduction

According to the Costs of War project at Brown University (Watson Institute, 2021), the United States (U.S.) military is the world's single largest institutional consumer of fossil fuels and a major contributor of greenhouse gas emissions. Yet these facts do not imply simply decarbonizing the military. A truly just transition to a post-fossil-fuel economy, a transition that prioritizes people and planet, must leave militarism behind. This chapter contributes to the challenging and necessary project of demilitarization by proposing a just transition of military institutions, centering on the needs and contributions of working people, their communities, and the environment.

M. J. Burke (✉)
University of Vermont, Burlington, VT, USA
e-mail: matthew.burke@uvm.edu

N. L. Smolyar
Gund Institute for the Environment, Burlington, VT, USA
e-mail: nina.smolyar@uvm.edu

A. Dunlap and A. Brock (eds.), *Enforcing Ecocide*,
https://doi.org/10.1007/978-3-030-99646-8_11

We begin by briefly addressing the common yet unconvincing assertion that the U.S. military will lead the way to a low-carbon future (see also de Vries, this volume). Much of the existing narrative around the future of the U.S. military assumes not only its continued existence post-fossil fuels, but also envisions and encourages its central role in advancing the end of the fossil-fuel regime (see for example, Conger et al., 2019; Lieven, 2018; MacNeil, 2020). The general argument seems to rely on three claims, each highly problematic: that the military recognizes climate change as a serious problem; that this recognition drives a meaningful push for alternative energy sources; and that this push will then transform the rest of the economy.

The first claim regarding the military's appreciation of the gravity of a warming world overlooks the fact that the concern of the military is not about climate change per se. From the military's perspective, the overriding question concerns how to maintain global dominance in a world becoming increasingly uncertain and unstable due to climate change. The priority focus remains firmly on adapting the military to heightened risks of disruption to military operations and infrastructure as associated with climate change, especially over the coming decades (Department of Defense, 2019), rather than on mitigating the underlying systemic drivers of climate catastrophe.

Prioritizing control in the face of catastrophe then reveals the problem with the second claim—that the military will meaningfully advance alternative energy systems. Although the U.S. military may explore alternative energy technologies, it is fundamentally committed to fossil fuels and nuclear power. Military operations require nearly unrestricted access to these high-density forms of energy, lest mission readiness and warfighting capabilities be compromised (Deutch, 2020). Despite reported concerns and at-best contradictory commitments, fossil-fuel dependence of military operations continues to grow, even without accounting for classified operations (Distel, 2020). An agenda of global military dominance and energy dominance go hand in hand (Schneider and Peeples, 2018). Incremental technological innovations may enable a diversification of energy sources for some limited set of applications, but such techno-fixes do not constitute the end of reliance upon fossil fuels. Military dominance requires extraction of everything, and more of it.

This point then leads to the problem with the third claim—that a military powered by alternative energy sources will somehow transform energy systems and economies generally. While there may be some transfer of technology to non-military uses, these technological shifts fail to ensure the needed structural economic change. There is no basis for expecting that a military technological transfer would happen within the time needed to meet pressing climate targets, nor that the associated techno-innovations will be of the type that people and the planet actually need—again, geopolitical dominance is the goal, not well-being. Such interventions have little to offer for changing socio-economic organization and patterns of production and consumption, and may very well deepen dependence on expensive, unnecessary, and harmful technologies. To meet its strategic and political goals, the military may be able to achieve marginal improvements in fuel economy, shift to nuclear and biofuels for tanks, aircraft carriers, and warplanes, or even compel research and development of broadly applicable technologies such as digital and communications technologies, large-scale solar installations and batteries, or synthetic materials. Yet clearly such efforts fail to reduce the extractive character of arms manufacturing and military technologies, nor take seriously the core concern for restructuring economies to meet human needs within ecological limits (Steichen and Koshgarian, 2020). Importantly, left unexamined is the central question of who actually gets to make the decisions in this context. In short, fundamental structural changes are far more necessary than militarized technologies.

These issues demonstrate that the U.S. military cannot be expected to lead an energy transition. Yet neither will the military industrial complex simply remain sidelined or inconsequential to these changes. The U.S. military is at once a major source and enabler of fossil-fuel dependence, and the nation's largest employer with nearly 4 million active-duty military, reserve members, and civilian employees working in contracted companies and across roughly 800 U.S.-controlled military bases worldwide (Steichen and Koshgarian, 2020; Vine, 2019). An agenda for transition must directly address the U.S. military due to the scope and significance of this institution for both energy and labor.

Arguments and proposals for demilitarization through economic conversion, i.e., converting from a military to a peace economy, have been around for decades and longer. Encapsulated by the expression "Let us beat swords into plowshares", such economic conversion is a technical, political, and economic process operating at multiple scales, to redirect defense firms, manufacturing, personnel, laboratories, and infrastructure to maximum extent, toward meeting social and community needs (Peattie, 1988; Melman and Dumas, 1990; Bischak, 1991;

Dumas, 1995; Touchton and Ashley, 2019; Pemberton and Hartung, 2020). While considerable attention has been given to conversion of the military and the plight of its workers and their communities, only recently have these questions been explicitly taken up from the perspective of a just transition, that is, taking seriously the combined ecological and social imperatives of conversion (see for example Eisenscher, 2014; Pemberton and Hartung, 2020; Steichen and Koshgarian, 2020).

Here, we focus on the role of military workers and their communities as key agents and beneficiaries (Green and Gambhir, 2020) of a just transition to peaceful and post-fossil-fuel futures. Just transition refers to a set of strategies that shift communities and societies from an extractive economy driven toward the accumulation and concentration of wealth and power, to cooperative and regenerative economies that bring about justice, ecological restoration, social equity, and community resilience (Movement Generation, 2016). Just transition framing is key for simultaneously confronting the climate crisis and the impacts of the response upon workers and their communities, while helping overcome the obduracy and influence of the military.

This chapter examines the relationship of militarism and environmental destruction and the need to overcome militarism through a just transition. Specifically, the chapter presents a case for a more coordinated, proactive, and public-sector approach to planning demilitarization in a manner that remains responsive to growing environmental calamities and the needs of workers and communities. The chapter first considers more closely the role that the U.S. military plays in contributing to ecological and social harms while preventing a just transition, then proposes how the military itself could be transformed. Through this work, we demonstrate that although the U.S. military is a major obstacle to the realization of a just and ecological society, it can—and must—be systematically repurposed to contribute to a just transition.

The U.S. Military Contributes to Ecological Catastrophe and Prevents the Realization of an Ecological Society

Military institutions not only fail to offer meaningful and realistic leadership in response to ecological catastrophe, but remain a key obstacle to a peaceful, socially just, and ecologically sustainable economic transition. The operations of the military are ecologically destructive. Its organizational structures and systems, and its civilian support apparatus, are harmful to working people and communities. The military and its wide-reaching agenda monopolize the federal government's budget and geopolitical priorities, while a militarized response to climate change only serves to deepen the crisis.

Militarism Undermines the Ecological Basis of Well-Being

The U.S. military contributes significantly to ecological catastrophe and prevents the realization of a more ecological society. First, we consider the direct environmental impacts. The U.S. Department of Defense (DoD) is the single largest consumer of energy and the largest institutional consumer of oil in the world, using billions of gallons of fuel annually (Eisenscher, 2014; Steichen and Koshgarian, 2020). Energy use for daily operations of the DoD, approaching 400,000 barrels of oil, exceeds that of any other private or public organization as well as more than 100 nations (Crawford, 2019; Reisch and Kretzmann, 2008; Warner and Singer, 2009). Military vehicles, from tanks, fighter jets and bombers, to battleships and aircraft carriers consume petroleum-based fuels at extraordinarily high rates; for example, four to eight miles per gallon of diesel fuel for HUMVEEs, just over half a mile per gallon for tanks, 134 barrels per hour for non-nuclear aircraft carriers, four gallons of jet fuel per nautical mile for B-2 Bombers and 25 gallons per minutes for F-15 fighter jets (Crawford, 2019; Eisenscher, 2014). Indirect applications would further account for fuel consumed by private contractors and weapons manufacturing. While U.S. average per capita energy use is already among the highest in the world [https://ourworldindata.org/grapher/per-capita-energy-use], the per capita rate for active-duty military and civilian personnel is 35% higher still than this average. The corresponding emissions amount to 59 million metric tons of greenhouse gases annually, the single largest volume of produced emissions in the world, even exceeding that of industrialized countries such as Sweden, Denmark, and Portugal. Since 2001, the U.S. military has emitted 1.2 billion metric tons of greenhouse gases and is one of the highest emitters historically (Crawford, 2019).

Further, it would not be possible for fossil-fuel-dependent global production and consumption levels to keep growing and to remain so high without the U. S. military war machine deploying missions to protect oil supplies. Up to one half of all interstate wars since 1973 have been driven by this purpose, though often not officially (Steichen and Koshgarian, 2020). Intimidation and threat of military action and violence undergird the dynamics of uneven power relations between, on the one hand, Global North state interests to secure their levels of consumption, and on the other hand, communities often in the Global South, whose lands and livelihoods are devastated by the extractive industries and infrastructure protected by the military.

Compounding and obfuscating the situation is the absence of military greenhouse gas emissions in global climate negotiations and agreements, resulting in a highly inaccurate depiction of the scale of the problem that these enormous international efforts seek to solve (Steichen and Koshgarian, 2020). This absence is not an accidental oversight. In 1997, the U.S. team at the United Nations climate talks in Kyoto, Japan, successfully lobbied for exemptions from requirements in emission reductions for the military. Even though the U.S. did not even ratify the Kyoto Protocol, military exemptions remained for all the nations that did sign, and continue to present day (Buxton, 2015).

Many U.S. military practices are also ecologically destructive and have wreaked havoc on the local environments where they operate. The very notion of ecocide, as the mass destruction and extermination of ecosystems, arose in the context of the U.S. military's devastating use of the chemical weapon Agent Orange in Vietnam (Falk, 1973; Higgins, 2015). More recently, the toxicity accumulated from using depleted uranium munitions and burning trash in open pits during military operations in Afghanistan and Iraq has resulted in high incidences of crippling birth defects and cancer among civilians in both places (Webb, 2017; Hussain, 2019).

Even if we were to accept the rather dubious notion that military operations have strengthened the cause of democracy in these parts of the world and solidified U.S. security, their devastated environments result in huge losses to local livelihoods. A cascading effect is that these governments and societies become further entrapped and dependent on international aid while instability is perpetuated. The financial instruments arising from these foreign aid regimes are structured by geopolitically powerful countries to extend their neocolonial power over those already enduring ongoing dispossession, extraction, and marginalization (Hickel, 2018).

Militarism Harms Working People and Communities

Militarism brings harms and devastation to working people and communities, in complex and lasting ways. A culture of violence training and normalization, following of orders in the rigid chain of command, and manufactured loyalty to often misguided and unjust missions, lead to negative impacts on the mental and physical well-being of people and communities. These impacts have only grown in the past decades, correlating with the increasing scale of the U.S. military's protracted involvements in Afghanistan and Iraq. Relative to previous conflicts, service members have been deployed more times, for longer durations, and with shorter times at home in between deployments. Increased deployments of women, of parents with young children, reserve and National Guard troops are also accompanying trends (Institute of Medicine, 2013). In addition to about 7,000 fatalities and 50,000 wounded as direct life and health impacts of the wars, many veterans returning from these operations have a number of difficulties readjusting to civilian life. The gaps in care for these service members and their circles of communities were so significantly large as to necessitate allocation of national funding in 2008 for a full assessment of the deployments' multi-dimensional effects (Ibid.).

The military's own data has indicated a concerning rise in suicides among soldiers (Alvarez, 2009). Overall suicide rates for veterans are 1.5 times higher than in the general population (Department of Veteran Affairs (VA), as cited by Hooper and Hardey, 2020). The picture for female veterans is even worse: the rate is 2.5 times higher than for non-veteran women and represents an increase of 2.9 per 100,000 between 2001 and 2014 (Ibid.). Older veterans are also more likely to commit suicide (Ibid.). These tragic outcomes are often preceded by the co-occurrence of substance and alcohol abuse, depression, anxiety, post-traumatic stress disorder (PTSD), and traumatic brain injury stemming from deployment experiences (Ibid.). Estimates cite that about half of Afghanistan and Iraq veterans have a mental health diagnosis (Ibid.), affecting not only these individuals but their loved ones and wider communities as well.

Accordingly, a higher incidence of Intimate Partner Violence—physical, sexual, emotional abuse—among military families compared to the civilian population is also well documented. Some reports cite a rate three times higher than among non-military families (Gourley, 2016, 5). A 2009 report from the VA Office of Public Health and Environmental

Hazards found a rate 14 times higher than in the civilian population, of post 9/11 PTSD-suffering veterans engaging in a severely violent act, such as domestic violence, murder, and child abuse and neglect (Bannerman, 2017). The unaddressed mental and emotional anguish from extended and traumatic deployments can also manifest in deadly violence toward military colleagues (Associated Press, 2020; Fernandez, 2019). While it is a positive development that as of 2018, more funding is being allocated to serve these needs of the veterans and their circles (VA, as cited by Hooper and Hardey, 2020), that funding is also thus diverted from community care and wellness supports that could be provided instead, if mental illness, violence, and suicide among veterans had not become such dire issues in the first place.

Militarism Diverts Money from Other Needs

Such diversion of funding is a significant matter. As critical as the aforementioned direct environmental and social impacts, are the multiple ways that the U.S. military diverts the necessary funding and resources from meeting other pressing needs. Most obvious here is the relentlessly large military budget. According to the National Priorities Project (2021), for Fiscal Year 2021, military expenses including the Department of Defense budget and related military spending such as nuclear weapons amount to an estimated $740 billion. This figure is significantly higher than the roughly $600 billion budgeted for all other essential expenditures combined, such as education, healthcare, housing, environment, agriculture, transportation, and so on (Swanson, 2021a). The actual annual costs of militarism are higher still, well over $1 trillion, when accounting for spending on veterans' benefits, related interest on the national debt, intelligence agencies, and other "security" expenditures across the Departments of Energy and Homeland Security (D'Agostino and Rynn, 2019). This militarized budget takes up 64% of discretionary spending.

Importantly, this level of spending continues year after year, having increased sharply following 9/11 and remaining elevated since (Crawford, 2019), even as the Pentagon has failed all three of its independent audits, unable to account for millions of dollars in its $700 + billion budget (Chappell, 2021). The U.S. currently spends more than $2000 per person per year on militarism, many times more than any other country in the world, contributing to the poor performance of the U.S. among wealthy nations across various measures including health, education, poverty prevention, environmental sustainability, prosperity, economic mobility, democracy, and freedom (Swanson, 2021b).

Militarized Responses to Climate Change Make Matters more Catastrophic

A militarized response to climate change and related ecological calamities would serve to reinforce rather than transform the very political and economic institutions that have led to these crises, and are already making matters worse. Because military operations are highly dependent upon fossil fuels, merely deploying the U.S. military to respond to climate-induced crises and disasters including climate refugee crises, storms, and wildfires, increases greenhouse gas emissions.

The deeper concern raised here is of a more fundamental and systemic quality, however, in that the specific modes of response are themselves damaging and destructive to workers, their communities, and the environment. The U.S. military has recognized climate change since 2003, yet in only the narrow, self-serving sense as a threat multiplier and thus, a major and increasing risk to national security (Steichen and Koshgarian, 2020; Buxton, 2018). Public documents detailing military planning to respond to climate change bear out a troubling scenario. Key priorities in the military response to climate change include safeguarding its own vast built infrastructure, both domestic and abroad, from extreme weather events; training people to become better fighters; and reducing personnel and equipment loss due to heavy reliance on oil while protecting oil supplies (Steichen and Koshgarian, 2020; Buxton, 2017).

Another priority is preparing for climate-related crises response, which is then used to further justify continued increases in military spending (Johnson, 2020). These costly militarized climate responses include pre-emptive defensive infrastructure in combatant U.S. operations, and militarizing borders to control climate refugee migration from less resilient parts of the world. It is no small matter that these communities are made less resilient by the same socio-economic forces that are now shutting the door to safer and better resourced parts of the planet. An indicative example of the humanitarian tragedy intrinsic to a militarized approach to natural disaster is the 2005 Hurricane Katrina in New Orleans, when

National Guard troops ended up shooting rather than rescuing trapped civilians (Buxton, 2017).

As climate change converges with and accelerates political and economic crises, an overdeveloped military capacity risks defining and dominating the state's policy response. Rather than easing tensions, such a response drives an upsurge of open-ended, global counter-insurgency, or what journalist Christian Parenti and others describe as an "armed lifeboat" approach (Parenti, 2011; Ross-Brown, 2013). Recognizing the efforts of the Pentagon and its allies abroad to plan for and implement a globally militarized climate adaptation, Parenti warns of a so-called green authoritarianism, emerging among wealthy nations of the world, that responds to these converging catastrophes through ever-more and ever-lasting escalation of armaments, exclusion, segregation, repression, policing, and violence to protect these nations from the chaos and disasters experienced across the Global South. This is not to say that such an approach, however horrific, could ever work as planned. Indeed, as Parenti recognizes, "no amount of walls, guns, barbed wire, armed aerial drones, or permanently deployed mercenaries will be able to save one half of the planet from the other" (Parenti, 2011). It is to also emphasize the point that it is imperative to prevent this strategy from being pursued at all (Siddique, 2021).

Militarism Shapes Geopolitics and Undermines Global Cooperation

Despite shortcomings, the Paris Agreement reflects a clear recognition of the need for global cooperation in the collective response to climate change. The fact that the military is already legitimized as a key actor in global climate governance means that the oft-underappreciated role of the military in setting the global climate agenda deserves greater scrutiny (Jayaram and Brisbois, 2021). Militarism and related agendas of nationalism, imperialism, and extractivism undermine the conditions for supporting the necessary global cooperation in various ways (Schwartzman, 2020). The U.S. military influences the domestic and foreign policy agendas of the U.S. This is not to say that U.S. military personnel are incapable of serving in a variety of roles in policymaking (Bechtel, 2017). Rather as related to ecological catastrophes, there are important questions regarding the military taking up civilian roles in climate governance (Jayaram and Brisbois, 2021), especially when required actions may compromise the relative position of the U.S. military as a global power.

The outsized influence of military actors can serve to delegitimize and disempower civilian actors, reinforce existing global power imbalances, and constrain the ability of communities to adapt to climate and environmental change (Jayaram and Brisbois, 2021). Further, so-called strategic interests are then translated into policy and persuasion overseas, such as pressuring members of the North Atlantic Treaty Organization (NATO) to increase military spending in their domestic budgets. Coupled with a commitment to military dominance relative to perceived adversaries, the effect is to drive military spending and its related impacts upward among allies and adversaries alike. Relatedly, the U.S. military must also then ensure continued access and security for sites of extraction and global supply chains for fossil fuels and critical minerals (Belcher et al., 2019; Schwartzman, 2020). These logistical and policing functions heighten tensions in regions with longstanding conflicts, including the Middle East and the South China Sea, and increasingly across less accessible locations, such as the Arctic, the deep sea, and outer space.

More fundamentally, there is a conflict between military logics that frame the world in terms of "national security" and the logic of cooperation and shared-yet-differentiated responsibilities required to respond equitably to ecological catastrophe (Jayaram and Brisbois, 2021). As mentioned previously, military use of fossil fuels has historically been excluded from climate targets and reporting (Belcher et al., 2019), exempting militaries from accountability and undermining the pursuit for greater transparency. As dominant perspectives on the global scene, militarized logics can exert substantial influence on climate action as more people experience the direct impacts of climate change. Military modes of operation escalate the problems rather than strategize their prevention and effective remediation, which require cooperation, diplomacy, and sustained coalition building. The focus on national interests preempts possibilities for building real trust and alliances with other nations and incentivizes each nation to adopt an isolationist policy in their own foreign relations, even if public statements attempt to assert differently.

True collective security arises from social and political peace and well-resourced infrastructure and institutions to meet the real human security needs for healthy food and environment, meaningful, sufficiently compensated livelihood, quality housing, education, healthcare, and self-determination. These social needs are currently deprived of the resources they need, because the funds are diverted into increasing militarization, which generates more conflict while drastically reducing capacity to meet human needs. Decarbonization must mean demilitarization, and demilitarization in turn requires a just transition.

The Military must be Converted and Repurposed

This set of deeply intertwined issues underscores the need for converting the U.S. military to civilian purposes and control. This section turns to the specific options and opportunities for a just transition of the U.S. military. This transition requires nothing less than a complete dismantling and restructuring of this institution of war, firmly rooted under civilian control, toward the goal of creating an equitable and ecological society. There are three broad strategies to pursue as part of this just transition: spending reductions and reallocation; economic conversion to a peace economy; and assistance, retraining, and civilian repurposing for workers and veterans (D'Agostino and Rynn, 2019).

Spending Reductions and Reallocation

Demilitarization and just transition require decreases in military spending and a green fiscal shift toward meeting urgent social and environmental needs. Often it is a budget reduction that catalyzes deeper, more abiding conversions, yet such a curtailment must be combined with targeted reinvestments (Pemberton and Hartung, 2020). Military spending creates many fewer jobs than if those same funds were spent on healthcare, education, and clean energy and infrastructure, and jobs in these other sectors are on average equally well or better compensated (Peltier, 2019; Pollin and Garrett-Peltier, 2011). Shifting military spending to green manufacturing can support a just transition for regions of the U.S. experiencing declines in manufacturing, while scaling up lower-carbon energy infrastructure and public transportation, better preparing communities for impacts of climate change, and improving and expanding jobs in care work (Steichen and Koshgarian, 2020).

A growing consensus is emerging around proposals to cut military spending. For example, a Defense Spending Reduction Caucus has been formed in the U.S. Congress, while an amendment was further introduced in 2020 to cut the Pentagon budget by 10% (Lee, 2020). This reduction is especially urgent in view of the need for and popularity of

Covid-19 economic relief payments, and changes in rules to allow significant cuts to the DoD independently of non-defense spending. Even at just 10%, this reinvestment could achieve significant benefits across any number of priorities: housing half a million people, dramatically expanding coronavirus testing, bridging the funding deficit for majority non-white school districts, providing renewable energy to nearly every U.S. household, transitioning nearly every worker in conventional coal, oil, and gas sectors, hiring close to a million public school elementary educators, and so on (Steichen and Koshgarian, 2020). At a global level, the opportunities for shifting funding become even greater, given $2 trillion annually in global expenditures (half from the U.S.) on war and war preparations. Small percentages of this amount could end starvation or provide drinking water worldwide, reducing unrest, and improving well-being (World Beyond War, 2020).

Economic Conversion

In coordination with these shifts in budgetary priorities, a just transition requires converting from a military to a peace economy, including especially industrial planning and restructuring. Economic conversion refers to "political, economic and technical measures for ensuring the orderly transformation of labor, machinery and other economic resources now being used for military purposes to alternative civilian uses" (Melman and Dumas, 1990). Such efforts have been proposed and tested for decades, emphasizing the use of sector- and community-based planning for workers and communities dependent on defense and defense industries (D'Agostino and Rynn, 2019). Beginning with moves toward nuclear disarmament and the end of the Cold War in the 1980s and 90s, advocates and organizers seriously examined the necessary elements of conversion to a peace economy (Melman and Dumas, 1990; Pemberton and Hartung, 2020). In addition to spending reductions and reinvestment, successful economic conversion depends upon several key conditions. These conditions include availability of technical and financial assistance to firms, workers, and communities; supportive civil society measures including research, education, and legislation; and well-designed industrial policies and planning. Many examples of successful conversion can be traced at the level of firms, communities, bases, laboratories, and sectors since the end of the second world war. However, the size and complexity of the network of militarism in the U.S. now require a large-scale industrial policy to achieve full disarmament and demilitarization (Pemberton and Hartung, 2020).

Much has already been learned about scaling up economic conversions. Especially since the end of the Cold War, the work of Seymour Melman and colleagues, along with efforts of labor and trade unions, has provided a foundation for this comprehensive approach to economic conversion and industrial planning. These lessons reflect the fact that for decades, military priorities have served as the de facto plan for the U.S. economy, and markets are incapable of making the needed shift to peaceful economies (Eisenscher, 2014; Melman and Dumas, 1990). Supportive legislation nearly passed as early as 1963, while legislation such as the Defense Economic Adjustment Act has been introduced in various forms since the early 1990s (Swanson, 2010). More recently, Miriam Pemberton and others have put forward comprehensive strategies for defense transition that operate at all levels and are federally funded and resourced, and state- and locally coordinated and implemented (Pemberton, 2018).

The scale and pace of change needed now demand new models for economic conversion. Demilitarization and conversion to a peace economy will require explicit industrial planning in public and private sectors, a challenge in the U.S. context to say the least. Nevertheless, several tools may provide a starting point if applied to the purpose of a just transition. One obvious measure is to implement new rounds of base closures through the Base Realignment and Closure process, especially including closures of overseas bases, and to shift the savings to just transition and civilian priorities. Closed bases and installations could serve as new sites for restored ecosystems and alternative energy generation (Crawford, 2019). Another measure would be to shift the focus of the Defense Production Act, which provides presidential authority to expedite and expand materials and services from U.S. industries, toward emergency preparedness. A specific focus would be to minimize hazards and impacts of climate change and protect and restore critical infrastructure—both built and natural—as needed for a just transition. In the context of climate emergency, a more direct approach to large-scale industrial planning could effectively reverse the function of the War Production Board, which was active during the wartime period. This mode of planning would instead convert factories from manufacture of weapons and military equipment to peacetime industries under public control, while conserving high-priority materials and limiting waste and luxury items.

It is important to emphasize in the context of emergency preparedness and climate resilience, the point is not to extend military control but rather to unequivocally transfer resources and responsibility of the existing military apparatus to civilian control (Barber and Bennis, 2020). For example, civilian organizations such as the U.S. Public Health Service or the Centers for Disease Control and Prevention would assume responsibility for medical staff, equipment, and facilities. All these methods and more will need to be integrated over time within a broad strategy of demilitarized industrial planning for conversion to a peace economy.

Assistance, Retraining, and Civilian Repurposing

The third general strategy for a just transition of the U.S. military centers on the contribution of the workers and their communities toward minimizing ecological catastrophe and providing for basic needs. This strategy therefore involves coordinating targeted assistance, retraining, and transitioning current military service members and contractors toward peace economy sectors, industries, and services. Ensuring social assistance and retraining provides the baseline for a demilitarized just transition. Here again, experience offers many examples, including the various GI Bills, Vocational Rehabilitation and Employment Assistance, the Veterans Retraining Assistance Program and Transition Assistance Program, and so on. Often under the charge of the Department of Veterans Affairs, these programs have demonstrated their ability to make a meaningful difference to people transitioning from military to civilian life by providing assistance and benefits for education, housing, business support, counseling, jobs training, and monthly subsistence payments. Funds could be reallocated from the DoD to the VA including the Office of Transition and Economic Development to dramatically scale up financial and logistical support for these transitions. This process must also involve redirecting economic activity and livelihoods. This would involve retraining public employees displaced by the phase out of military programs and offering them meaningful work in other federal, state, or local agencies, as engineers and mechanics, electricians, public health workers, accountants, and so on (D'Agostino and Rynn, 2019).

Many of these programs, historically and presently, emphasize private sector employment in anticipation that the market can supply the needed work. Here again, the present context requires a more targeted system of transition that aims to rectify existing social inequities and minimize ecological catastrophe. This means providing high-quality jobs to build and repair public infrastructure, construct public water and wastewater systems, clean up toxic and nuclear waste sites, restore ecological habitats, provide national and international disaster preparedness and relief, and build new and retrofit existing housing, libraries, and schools. Alongside these priorities, a just transition also calls for creating new opportunities for local manufacturing and production and worker-owned businesses and cooperatives, and providing technical and administrative assistance to communities facing these converging crises (D'Agostino and Rynn, 2019; Melman and Dumas, 1990).

An especially relevant opportunity follows from the growing calls for a re-envisioned and permanent Civilian Conservation Corps (CCC), or a Civilian Climate Corps as proposed in the first days of the Biden Administration. The original CCC, sometimes known as Roosevelt's Tree Army, operated from 1933 to 1942 as a New Deal public work relief program. The CCC was highly popular, employing millions, and completing a vast number of projects including planting trees, building parks and trails, fighting fires, and protecting soils. The CCC helped influence environmental programs and attitudes that continue to this day. Yet this program did not provide long-term, high-quality jobs, reinforced racial and gender inequities, and became militarized and subsumed under the war effort (Alexander, 2018; Heller, 2009; Maher, 2008). The just transition needed now would function differently. A reimagined CCC would create high-value work especially in ecological restoration and climate resilience, rectify inequities, and set a clear path for workers and communities to demilitarize in a way that is responsive to present and future needs (Aronoff et al., 2019). Boosted by funds reallocated from the military budget, a new Civilian Climate Corps could provide transitioning workers with well-paid union jobs as "conservation and resilience workers" who restore lands and waters, build green infrastructure, install solar panels and wind towers, sequester carbon in soils, protect biodiversity, clean up toxic waste, and much more (The White House, 2021). To align this initiative more closely with principles for a just transition, the program must center on the needs and priorities among Black, Indigenous, and People of Color (BIPOC), while providing skills needed for lifelong work in a demilitarized, post-fossil-fuel economy (Collier, 2021). Just as the military now serves as the nation's largest federal employer, this repurposed program could further serve as a key element of a federal public jobs guarantee program, thus ensuring a viable transition for military workers, families, and communities.

Military Repurposing could Significantly Advance a Just Transition

Implementing these steps of fiscal shift, economic conversion, and retraining and transition would have a profound benefit for workers and communities, while greatly reducing the additional risks and already occurring impacts of social and ecological catastrophe. A just transition approach brings this necessary focus to the process of demilitarization. The demographic profile of the U.S. military has dramatically diversified over the last half century, demonstrating the opportunities here for the many BIPOC people now doing this work. It is also recognized that veterans, BIPOC people, and people of Latin American descent often hold high levels of concern for climate change (Leiserowitz and Akerlof, 2010; Motta et al., 2021). Yet for many communities, military service provides a rare path out of inequitable conditions at home, while military recruitment is known to target low-to-middle income and poor communities (Steichen and Koshgarian, 2020). This repurposing would dramatically broaden the opportunities for high-quality, safe, and meaningful work among these communities and beyond.

Meanwhile, military personnel, bases, and their communities often suffer disproportionate levels of pollution, domestic violence, mental and physical health problems, alcohol and drug use, suicide, and other social and economic disparities. Contamination and pollution from military testing sites and weapons manufacturing hit local communities hardest, too often involving inadequate compensation. A repurposing would reduce these vulnerabilities by directly remedying and improving conditions for these communities. This work would start by cleaning up the many contaminated sites worldwide, and extend to a permanent peace economy that avoids the multi-generational and multi-dimensional harms and trauma of militarism.

As the U.S. military extends globally, so too would the effects of demilitarization through a just transition. Shifting from a role of securing fossil-fuel and energy dominance, a repurposed military under civilian control could activate a wave of global cooperation and solidarity, moving funding and personnel to support international aid and climate change adaptation and mitigation. Global agreements for nuclear weapon and fossil-fuel non-proliferation, poverty reduction, and ecological restoration may then receive the attention and support needed to make a real difference for people and planet. This approach is based on a radically different view of "national security" in recognizing the impossibility, immorality, and recklessness of the "fortress" response to climate change. Global climate change by definition is a global issue, not a *national* security one. Implementing a just transition urges a world view of collective solidarity and mutual interdependence. Demilitarization can enable support for displaced refugees and climate migrants, reduce tensions and military build-up in other nations, and ripple out to improving the lives of poor people worldwide (Siddique, 2021).

The obstacles and challenges of demilitarization are many. For one, we must stay critically engaged, lest this process be coopted as yet another opportunity to exert further military control over civilian life. The non-negotiable goal is rather to reclaim the assets of the military and reposition them under strict civilian control for non-military purposes. There are also the cultural biases and identities, and myths of the glory of war and militarism that require a committed, multi-faceted response to shift these narratives and develop real alternatives for achieving lasting security (Shifferd and Hiller, 2020). The vested interests, from fossil-fuel corporations, military contractors, Pentagon elites, and DoD officials, to federal, state, and local politicians will do all in their substantial power to resist cuts and downsizing. Yet no real transition to a peaceful and ecological future is possible without demilitarization. A significant leverage point is whether and how workers and their communities can be effectively and consistently engaged and help lead this just transition (Eisenscher, 2014).

Conclusion

The U.S. military is a major contributor to social and ecological catastrophe and an obstacle to achieving a just transition. It is therefore absolutely critical to recognize that (1) the military should not serve in the project of responding to ecological disasters, (2) demilitarization to civilian control must become a central feature of this response, and (3) a just transition offers a guiding frame for demilitarization to a peace economy. This chapter develops this approach by connecting key strategies for a coordinated, civilian- and public-sector planned demilitarization as a core element of a just transition. The primary strategies involve shifting funds and resources, advancing economic conversion and industrial planning, and providing retraining and civilian repurposing, each in a manner responsive to growing ecological and social crises. A just transition must therefore respond to the call of workers, unions, advocates, and supporters by implementing demilitarization as a fundamental element for realizing the vision for a just, peaceful, and ecological future.

Bibliography

Alexander, B.F. (2018) *The New Deal's Forest Army: How the Civilian Conservation Corps Worked*, Baltimore: Johns Hopkins University Press.

Alvarez, L. (2009) 'Army Data Show Rise in Number of Suicides', *The New York Times,* February 5.

Aronoff, K., Battistoni, A., Cohen, D.A. and Riofrancos, T. (2019) *A Planet to Win: Why We Need a Green New Deal*, New York: Verso Books.

Associated Press (2020) 'A look at recent attacks on U.S. military bases', *AP News*, May 21.

Bannerman, S. (2017) 'The fatal "new normal" for wives of veterans', *Women's Media Center*, March 8.

Barber, W. and Bennis, P. (2020) 'To fight this pandemic, we need to take over the military's resources—Not the other way around', *Newsweek|Opinion,* March 23.

Bechtel, C. (2017) 'Warriors, Scholars, Diplomats: The Role of Military Officers in Foreign Policymaking', *New Perspectives in Foreign Policy*, 14(8).

Belcher, O., Bigger, P., Neimark, B., and Kennelly, C. (2019) 'Hidden Carbon Costs of the "Everywhere War": Logistics, Geopolitical Ecology, and the Carbon Boot-print of the US Military', *Transactions of the Institute of British Geographers*, 45(1): 65–80.

Bischak, G.A. (1991) *Towards a Peace Economy in the United States: Essays on Military Industry, Disarmament, and Economic Conversion*, New York: St. Martin's Press.

Buxton, N. (2018) 'Don't turn to the military to solve the climate-change crisis', *The Guardian|Opinion|Climate Change*, June 3.

Buxton, N. (2017) 'Last line of defense: Are the US military the best hope for defending climate action under Trump?', *The Transnational Institute*, April 11. https://www.tni.org/en/article/last-line-of-defense-are-the-us-military-the-best-hope-for-defending-climate-action-under.

Buxton, N. (2015) 'The elephant in Paris—The military and greenhouse gas emissions', *The Transnational Institute*, November 25. https://www.tni.org/en/article/the-elephant-in-paris-the-military-and-greenhouse-gas-emissions.

Chappell, B. (2021) 'The Pentagon has never passed an audit. Some senators want to change that', *National Public Radio*, May 19. https://www.npr.org/2021/05/19/997961646/the-pentagon-has-never-passed-an-audit-some-senators-want-to-change-that.

Collier, C. (2021) 'Civilian Climate Corps: If you build it, will they come?', *Groundwork USA* February 4. https://groundworkusa.org/civilian-climate-corps-if-you-build-it-will-they-come/.

Conger, J., Femia, F. and Werrell, C. (2019) *A Climate Security Plan for America: A Presidential Plan for Combating the Security Risks of Climate Change*. Washington, DC: The Center for Climate and Security.

Crawford, N.C. (2019) *Pentagon Fuel Use, Climate Change, and the Costs of War*. Providence, RI: Brown University Watson Institute Costs of War Project.

D'Agostino, B. and Rynn, J. (2019) *Demilitarization and the Green New Deal: Reinventing the State for the Public Good*. Paper presented at the Philosophy, Politics and Economics Society Annual Meeting; 29 March, New Orleans, Louisiana, USA.

Deutch, J. (2020) 'Can the U.S. military reach zero carbon emissions?', *The National Interest*, December 21, https://nationalinterest.org/feature/can-us-military-reach-zero-carbon-emissions-174875, accessed 28 May 2021.

Distel, M. (2020) 'The U.S. military consumes more fossil fuels than entire countries', *FossilFuel.com* March 15, https://fossilfuel.com/the-u-s-military-consumes-more-fossil-fuels-than-entire-countries/, accessed 28 May 2021.

Department of Defense (2019) *Report on effects of a changing climate to the Department of Defense*. Washington, DC: Office of the Under Secretary of Defense for Acquisition and Sustainment, https://climateandsecurity.files.wordpress.com/2019/01/sec_335_ndaa-report_effects_of_a_changing_climate_to_dod.pdf, accessed 28 May 2021.

Dumas, L.J. (1995) *The Socio-economics of Conversion from War to Peace*. Armonk, N.Y.: M.E. Sharpe.

Eisenscher, M. (2014) *Just transition: From military addiction to economic sustainability*. US Labor Against the War (USLAW), https://worldbeyondwar.org/wp-content/uploads/2014/02/JustTransition.pdf.

Falk, R.A. (1973) 'Environmental Warfare and Ecocide—Facts, Appraisal, and Proposals', *Bulletin of Peace Proposals*, 4(1): 80–96.

Fernandez, M. (2019) 'On military bases, the dangers increasingly come from the inside', *The New York Times*, December 6.

Gourley, S.J. (2016) 'Possible causes of increased domestic violence among military veterans: PTSD or mefloquine toxicity?', *Montview Liberty University*

Journal of Undergraduate Research 2(1): Article 9, https://digitalcommons.liberty.edu/montview/vol2/iss1/9.

Green, F. and Gambhir, A. (2020) 'Transitional Assistance Policies for Just, Equitable and Smooth Low-carbon Transitions: Who, What and How?', *Climate Policy*, 20(8): 902–921, https://doi.org/10.1080/14693062.2019.1657379.

Heller, C.E. (2009) 'The US Army, the Civilian Conservation Corps, and Leadership for World War II, 1933-1942', *Armed Forces and Society*, 36(3): 439–453, https://doi.org/10.1177/0095327X09333944.

Hickel, J. (2018) *The case for reparations*. October 13, https://www.jasonhickel.org/blog/2018/10/13/the-case-for-reparations, accessed 3 June 2021.

Higgins, P. (2015) *Eradicating Ecocide: Laws and Governance to Prevent the Destruction of Our Planet* (2nd edition). London: Shepheard-Walwyn (Publishers) Ltd.

Hooper, C. and Hardey, S. (2020) *Suicide among veterans*. American Addiction Centers, 30 April, https://americanaddictioncenters.org/veterans/suicide-among-veterans, accessed 31 May 2021.

Hussain, M. (2019) 'War on the world: Industrialized militaries are a bigger part of the climate emergency than you know', *The Intercept*, September 15, https://theintercept.com/2019/09/15/climate-change-us-military-war/.

Institute of Medicine (2013) *Returning home from Iraq and Afghanistan: Assessment of readjustment needs of veterans, service members, and their families*. The National Academies Press, https://doi.org/10.17226/13499.

Jayaram, D. and Brisbois, M.C. (2021) 'Aiding or Undermining? The Military as an Emergent Actor in Global Climate Governance', *Earth System Governance*, 9: 100107. https://doi.org/10.1016/j.esg.2021.100107

Johnson, A. (2020) 'Framing climate change as a "national security priority" isn't a clever maneuver to get people to care—It's a centrist co-option strategy to bloat the budgets of ICE and the Pentagon', *Medium*, November 24, https://medium.com/@adamjohnsonNYC/framing-climate-change-as-a-national-security-priority-isnt-a-clever-maneuver-to-get-people-to-f13723840073.

Lee, B. (2020) *Reps. Lee, Pocan announce defense spending reduction caucus*. Congresswoman Barbara Lee website, 28 July, https://lee.house.gov/news/press-releases/reps-lee-pocan-announce-defense-spending-reduction-caucus, accessed 1 March 2021.

Leiserowitz, A. and Akerlof, K. (2010) *Race, Ethnicity and Public Responses to Climate Change*. Yale University and George Mason University. New Haven, CT: Yale Project on Climate Change.

Lieven, A. (2018) 'The only force that can beat climate change is the U.S. Army', *Foreign Policy*, January 9. https://foreignpolicy.com/2018/01/09/the-only-force-that-can-beat-climate-change-is-the-u-s-army/.

MacNeil, R. (2020) *The next US president should lean on the military to address climate change.* June 19, https://hdl.handle.net/2123/22598, accessed 31 May 2021.

Maher, N.M. (2008) *Nature's New Deal: The Civilian Conservation Corps and the Roots of the American Environmental Movement*, Oxford: Oxford University Press.

Melman, S. and Dumas, L.J. (1990) *Planning for economic conversion.* Peace Center Books, http://peacecenterbooks.com/planning-for-economic-conversion/, accessed 1 March 2021.

Motta, M., Ralston, R., and Spindel, J. 2021 'A Call to Arms for Climate Change? How Military Service Member Concern About Climate Change Can Inform Effective Climate Communication', *Environmental Communication*, 15(1): 85–98.

Movement Generation (2016) *From banks and tanks to cooperation and caring: A strategic framework for a just transition.* Movement Generation Justice and Ecology Project. https://movementgeneration.org/justtransition/, accessed 31 May 2021.

National Priorities Project (2021) *Trade-offs: Your money, your choices. Location: United States, Program: Military in 2021*, https://www.nationalpriorities.org/interactive-data/trade-offs/, accessed 1 March 2021.

Parenti, C. (2011) *Tropic of Chaos*, New York, NY: Nation Books.

Peattie, L.R. (1988) 'Economic Conversion as a Set of Organizing Ideas', *Bulletin of Peace Proposals*, 19(1): 11–20.

Peltier, H. (2019) *War Spending and Lost Opportunities.* Providence, RI: Brown University Watson Institute Costs of War Project.

Pemberton, M. (2018) 'The little agency that could have tamed the military-industrial complex', *The American Prospect* March 2. https://prospect.org/world/little-agency-tamed-military-industrial-complex/.

Pemberton, M. and Hartung, W. (2020) *From Swords to Ploughshares: Lessons Learned from Conversion Movements* [in Rethinking Unconstrained Military Spending]. United Nations Office for Disarmament.

Pollin, R. and Garrett-Peltier, H. (2011) *The US Employment effects of military and domestic spending priorities: 2011 update* (p. 12). UMASS: Political Economy Research Institute (PERI).

Reisch, N. and Kretzmann, S. (2008) *A Climate of War: The War in Iraq and Global Warming.* Oil Change International Report.

Ross-Brown, S. (2013) 'Armed lifeboat: Government's response to natural disaster', *Utne.Com* July 30. https://www.utne.com/politics/natural-disaster-zm0z13sozros/

Schneider, J. and Peeples, J. (2018). The energy covenant: Energy dominance and the rhetoric of the aggrieved. *Frontiers in Communication.* https://doi.org/10.3389/fcomm.2018.00005

Schwartzman, D. (2020) *The global solar commons, the future that is still possible: A guide for 21st century activists*. SolarUtopia.org. https://www.theearthisnotforsale.org/solarcommons.pdf.

Sharp, G. (1990) *Civilian-based Defense: A Post-military Weapons System*, Princeton, NJ: Princeton University Press.

Shifferd, K. and Hiller, P. (2020) *A Global Security System: An Alternative to War.* World Beyond War.

Siddique, A. (2021) 'Could Biden's climate policy invite more militarism?', *Foreign Policy In Focus,* February 22.

Steichen, L. and Koshgarian, L. (2020) *No Warming, No War: How Militarism Fuels the Climate Crisis—And Vice Versa*. Washington, DC: National Priorities Project / Institute for Policy Studies.

Swanson, D. (2010) 'Conversion is the way', *Let's Try Democracy* December 3. https://davidswanson.org/conversion-is-the-way/.

Swanson, D. (2021a) 'Where will Biden get the money?', *Roots Action* January 16. https://rootsaction.org/news-a-views/2405-where-will-biden-get-the-money.

Swanson, D. (2021b) 'US spends 11 times what China does on military per capita', *War is a Crime* February 24. http://warisacrime.org/2021b/02/24/u-s-spends-11-times-what-china-does-on-military-per-capita/.

Touchton, M. and Ashley, Amanda J. (2019) *Salvaging Community: How American Cities Rebuild Closed Military Bases.* Ithaca, NY: Cornell University Press.

Vine, D. (2019) *Lists of U.S. Military Bases Abroad, 1776–2019*, American University Digital Research Archive, https://doi.org/10.17606/vfyb-nc07, accessed 28 May 2021.

Warner, J. and Singer, P.W. (2009) *Fueling the 'Balance': A Defense Energy Strategy Primer*. Washington, DC: Brookings Institute.

Watson Institute (2021) *Pentagon fuel use, climate change, and the costs of war*. Watson Institute for International and Public Affairs, Brown University, https://watson.brown.edu/costsofwar/papers/ClimateChangeandCostofWar, accessed 28 May 2021.

Webb, W. (2017) 'U.S. military is world's biggest polluter', *EcoWatch* May 15. https://www.ecowatch.com/military-largest-polluter-2408760609.html#toggle-gdpr.

The White House (2021) 'Executive order on tackling the climate crisis at home and abroad', *Briefing Room*, January 27, https://www.whitehouse.gov/briefing-room/presidential-actions/2021/01/27/executive-order-on-tackling-the-climate-crisis-at-home-and-abroad/, accessed 1 March 2021.

World Beyond War (2020) 'We need $2 trillion/year for other things', *World Beyond War*, https://worldbeyondwar.org/2trillion/, accessed 1 March 2021.

Index

A. Dunlap and A. Brock (eds.), *Enforcing Ecocide*,
https://doi.org/10.1007/978-3-030-99646-8

Kolibri 360 GmbH
Ferdinand-Jühlke-Straße 7
99095 Erfurt, Deutschland
produktsicherheit@kolibri360.de